Springer Transactions in Civil and Environmental Engineering

Springer Transactions in Civil and Environmental Engineering (STICEE) publishes the latest developments in Civil and Environmental Engineering. The intent is to cover all the main branches of Civil and Environmental Engineering, both theoretical and applied, including, but not limited to: Structural Mechanics, Steel Structures, Concrete Structures, Reinforced Cement Concrete, Civil Engineering Materials, Soil Mechanics, Ground Improvement, Geotechnical Engineering, Foundation Engineering, Earthquake Engineering, Structural Health and Monitoring, Water Resources Engineering, Engineering Hydrology, Solid Waste Engineering, Environmental Engineering, Wastewater Management, Transportation Engineering, Sustainable Civil Infrastructure, Fluid Mechanics, Pavement Engineering, Soil Dynamics, Rock Mechanics, Timber Engineering, Hazardous Waste Disposal Instrumentation and Monitoring, Construction Management, Civil Engineering Construction, Surveying and GIS Strength of Materials (Mechanics of Materials), Environmental Geotechnics, Concrete Engineering, Timber Structures.

Within the scopes of the series are monographs, professional books, graduate and undergraduate textbooks, edited volumes and handbooks devoted to the above subject areas.

The proposal for each volume is reviewed by the main editor and/or the advisory board. The chapters in each volume are individually reviewed single blind by expert reviewers (at least four reviews per chapter) and the main editor.

Shyam Bahadur Singh · Devendra Narain Singh ·
Sanjay Kumar Ray · H. B. Nagaraj

Cricket Pitches

The Science Behind the Art of Pitch-Making—"An Integrated Pitch Management (I.P.M) Approach"

Springer

Shyam Bahadur Singh
Jharkhand State Cricket Association,
Board of Control for Cricket in India
(BCCI)
Ranchi, Jharkhand, India

Devendra Narain Singh
Department of Civil Engineering
and Geotechnical Engineering
Indian Institute of Technology Bombay
Mumbai, Maharashtra, India

Sanjay Kumar Ray
Regional Centre
ICAR-NBSS&LUP
Kolkata, West Bengal, India

H. B. Nagaraj
Department of Civil Engineering
B. M. S. College of Engineering
Bengaluru, Karnataka, India

ISSN 2363-7633 ISSN 2363-7641 (electronic)
Springer Transactions in Civil and Environmental Engineering
ISBN 978-981-99-2912-2 ISBN 978-981-99-2913-9 (eBook)
https://doi.org/10.1007/978-981-99-2913-9

Book reprinted from: PHD Dissertation *Critical Evaluation of Economic and Behavioural Analysis of Some Important Indian Cricket Pitches Based on Traditional or Conventional Methods Vs. Scientific Methods Based on Recent Research and Developments: A Case Study of the JSCA International Cricket Stadium Vs. The Keenan Stadium* – 30.10.2017
Date of signing publishing contract with the publisher – 22.05.2018

This Springer imprint is published by the registered company Springer Nature Singapore Pte Ltd.
The registered company address is: 152 Beach Road, #21-01/04 Gateway East, Singapore 189721, Singapore

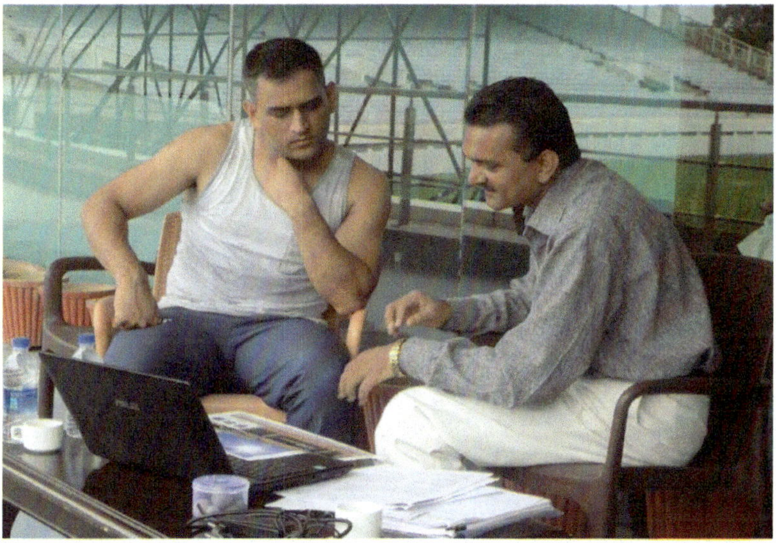

Photograph Showing the pitch interview with Shri Mahendra Singh Dhoni and the author and curator, Dr. S. B. Singh First session: 9.30 AM–11.15 AM, 15 July 2015. Second and third sessions: 9.45 AM–11.45 AM, 16 and 21 August 2015

It is an attempt to understand how the captains, players, coaches, curators, and groundsmen comprehend and analyse the cricket pitch behaviour (days before and during the matches) and whether the pitch behaviour can be standardised and quantified (through benchmarking or forming a data-based management system of a pitch profile, Pitch Quality Standards, a Pitch Behaviour Analysis Index, or Pitch Behaviour Forecasting) by examining or analysing its mineralogical, chemical, physical, or morphological compositions, weather variables, and different pitch preparation methods and techniques

Photograph Showing the Different Phases of the JSCA Stadium Complex "Place Where the Heart and Soul of This Book Lies"

Excavation of pitch soil—29.04.2010

Laying of a pitch square—June 2010

Ground inspection by Shri Daljit Singh, the ex-chairman, Pitches and Grounds Committee, BCCI—April 2010

Visionary behind the making of JSCA International Stadium Complex

Mahi discussing the general layout of JSCA Stadium Complex—06.11.2010

Author—in search of the Benchmark soils Gandhigram, Wardha, Nagpur—08.05.2016

Preface

Just by looking, feeling, analysing, and interpreting the finished match pitch surface, most of the predictions about the pitch behaviour during the matches are generally done by the coaches, managers, captains, commentators, or cricket enthusiasts. By feeling and observing the cracking, non-cracking, or crumbling patterns; the size, depth, and number of cracks; the colour of the finished surface; the pitch surface moisture levels; the colour of the grass cover on the pitch surface; etc., most of the speculation about the pitch surface behaviour during different sessions of the match is done.

Despite their extensive cricketing experience and knowledge, a wide range of variations in the forecast of pitch behaviour or match outcome is commonly observed, as pitch surface behaviour is not only dependent on a few superficial surface observations, but also, the synergistic and combined outcome of all the natural, controllable, and non-controllable major and minor factors available above or within the pitch profile that controls and regulates the pitch behaviour. Cricket pitch surface behaviour, analysis, interpretation, or forecasting is not only dependent on the face value of the finished pitch surface that generally one observes just before the toss. Observing the colour of the finished surface, surface hardness, moisture status, core sampling of the cricket pitch profile, cracking initiation, timing, and rate, cracking or crumbling patterns of the pitch profile, nature, size, and depth of cracks, status of the rolled-in grass and its colour, and ball rebound techniques are some of the general conventional techniques to access the pitch surface behaviour, which comes after a long cricketing experience or through testing.

Every major or minor factor or phenomenon (such as gravitation, capillary rise, surface tension, adhesion, moisture retention, and movement within the pitch profile) available just above and within the pitch profile plays an important role in shaping and defining the pitch surface behaviour at a point in time. To forecast pitch surface behaviour (other than judging the apparent values such as surface hardness, moisture levels, cracking patterns, grass density, and so on), analysis and interpretation of all major or minor soil properties in the form of morphological, physical, chemical, mineralogical, and engineering properties must be considered, as each has a distinct role to play. The invisible particulate matter available in the pitch soil aggregates (clay, silt, humus, air, water, predominant exchangeable ions, etc.) which is controlled and regulated by the natural forces (gravitation, adhesive force, capillary action, surface tension, evapotranspiration, solar radiation, dew, fog, mist, etc.) decides the pitch surface behaviour during various phases of the match.

The exchangeable ions on the predominant clay minerals (kaolinite, illite, or smectite) or the humus are present in minuscule amounts, but their effect on controlling the pitch surface is magnanimous. Al^{+3}/H^{+} ions, Ca^{+2}/Mg^{+2} or Na^{+}-dominant fine kaolinite or smectite-dominant clay pitch soil in a low, medium, or high pH range will behave entirely differently. Even the moisture suspended just above the pitch surface in vapour form (fog, mist) or as dew droplets on the turf grass or tightly adsorbed on the clay micelle as hygroscopic or capillary water plays an important role in providing variable pitch surface density during various phases of the match. Based on the morphological, physical, chemical, and mineralogical composition, every pitch soil has a different capacity to retain and release soil moisture by interacting with the available diurnal weather variations (temperature, RH, solar radiation, ET, etc.) and providing the desired pitch surface behaviour. The cricket pitch soils with the predominant colloidal minerals (smectite, illite, or kaolinite), organic humus, and dominant exchangeable cations (H^{+}, Al^{+3}, Ca^{+2}, Mg^{+2}, Na^{+}, or K^{+}) control the swelling and shrinking of the pitch profile, providing the desired variable densities by the loss or gain of moisture levels on the pitch surface or within the pitch profile. The weather variables (temperature, relative humidity, evapotranspiration, dew point, etc.) on and within the pitch profile control the retention and movement of pitch soil moisture and control the density, moisture levels, swelling, and shrinking levels of the pitch surface.

To understand, analyse, or interpret the complex nature of how pitch surfaces behave, every major or minor factor of pitch construction, pitch soil composition, pitch preparation methods, and weather variables must be taken into account, not just the finished match pitch surface.

Soil is made by natural processes called weathering, which take hundreds of years to happen. The physical and chemical properties depend on the type of rock they were formed from and other environmental factors. Hence, there is wide variation in its composition, and its properties are very complex. Because the soil used to prepare the cricket pitch is similar to that found in nature, the pitch surface will behave similarly (swelling, shrinking, cracking or crumbling patterns, bulk density, water holding capacity, etc.) as it did at the source from which it was excavated and transported. This is because environmental factors in the form of temperature, relative humidity, evapotranspiration rates, radiation, and moisture are more or less common in nature.

The more the groundsman or curator knows about the basic morphological, physical, chemical, mineralogical, and engineering properties of the soils that they are using on their pitches, along with the local weather and climatic variables, the more they will be able to comprehend and analyse the nature of the pitches precisely. Curators or groundsmen should not only concentrate on the above-mentioned soil reports but also on the geological origin and topography of the source soil, which is used to prepare the pitch. The natural weathering forces (temperature, water/ice, solar radiation, gravity, and time) that transformed the basaltic or granitic mother rocks into various types of soil will have a direct influence on the pitch's behaviour. The natural forces available and operating above and within the pitch profile also need to be understood by the curators so as to analyse and interpret the complex nature of pitch behaviour.

It is a difficult question as to how to access and predict the pitch behaviour before or during various phases of the match as the stakes of the players, coaches, managers, sponsors, broadcasters, etc., are greatly dependent upon them. Another question is whether pitch behaviour (based on a pitch profile's Pitch Behaviour Analysis Index (PBAI), Pitch Quality Index (PQI), and Database Management System (DBMS)) can be analysed, comprehended, and forecasted in the same way that weather or market (sen-sex) forecasting methods can. This book attempts to comprehend and analyse cricket pitch behaviour (days before and during matches) by examining and analysing its mineralogical, chemical, physical, and morphological characteristics, and whether pitch behaviour can be standardised and quantified (via benchmarking or forming a Database Management System for a pitch profile or Pitch Quality Standards or a Pitch Behaviour Analysis Index or Pitch Behaviour Forecasting).

The purpose of this book is to understand the science and logic behind the art of analysing the pitch surface behaviour and interpret the pitch behaviour through the Pitch Behaviour Analysis Index (PBAI), based on the Pitch Quality Index (PQI), and Database Management System (DBMS) of a cricket pitch profile. This book is a coordinated effort to understand the different techniques behind the different traditional or conventional arts of judging and analysing pitch behaviour, backed by recent research and developments in agronomy, engineering, and weather science.

The heart of a cricket pitch is the surface, which responds to different types of natural energy within and just above the pitch profile. Based on their morphological, physical, chemical, and mineralogical properties, weather variables, pitch preparation methods, and techniques adopted, these pitch surfaces provide the variations in pitch behaviour on which players show their mettle. It's a million-dollar question as to how and on what parameters these pitches behave (in a particular pattern, trend, or path), and whether this complex pitch behaviour (based on recent research and development in the area of pitch making and the experience of curators and players) can be mapped, comprehended, or forecasted (as a weather report or market forecast).

Through the journey of this book, let's try to understand and unravel the various hidden and invisible natural forces within and just above the pitch profile that directly or indirectly control and dictate the cricket pitch's behaviour during different sessions of the match.

Ranchi, India — Shyam Bahadur Singh
Mumbai, India — Devendra Narain Singh
Kolkata, India — Sanjay Kumar Ray
Bengaluru, India — Nagaraj H. B.

Acknowledgments

First, I would like to thank the late Amitabh Choudhary (IPS-VRS), who was the secretary of the BCCI, the longest serving president of the JSCA, and the Indian member representative on the ICC's chief executive committee. He was a great mentor for me.

I gratefully acknowledge Shri Mahendra Singh Dhoni (MSD) for providing 6 hours (2 hours in each of three sessions) of his valuable time and insight for assessing the cricket pitch behaviour through his vast pool of practical experience of playing on nearly all the continental and subcontinental pitches.

Special thanks to Shri Daljit Singh, who was the former chairman of the BCCI's Pitches and Grounds Committee, Shri P. R. V. Vishwanathan, who was a member of the BCCI's Pitches and Grounds Committee, Shri Ashish Bhowmick, the chief curator of the BCCI's Pitches and Grounds Group, and all the other curators and groundsmen from the BCCI and state associations who helped me learn more about making cricket pitches through their lifelong experience and learning.

I gratefully acknowledge Mr. Tony Hemming, the ICC accredited trainer and presenter, the chief curator, ICC Academy, Dubai (on 29 July 2016), for his marvellous presentation on the basic principles of science and logic behind the art of measuring and analysing the pitch surface compaction levels (with the help of the Clegg Hammer).

My sincere thanks also go to Dr. Surendra Kumar Singh, the Ex-Director, National Bureau of Soil Survey and Land Use Planning (ICAR-NBSS-LUP), Amravati Road, Nagpur; Dr. P. Chandran, Acting Director, Principal Scientist, and Head of the Soil Resource Studies (SRS) Division; Dr. D. K. Pal, Former Principal Scientist and Head of the SRS Division; Dr. Jagdish Prasad, Principal Scientist; Dr. R. P. Sharma, Scientist; and Shri V. T. Sahu, NBSS, who guided me and provided access to the laboratory and research facilities required for semi-quantitative and qualitative analyses .This was the first time in India that four cricket pitch soil samples belonging to Eden Gardens, Balangir, JSCA Ranchi, and Keenan Stadium were examined at the NBSS-LUP for their semi-quantitative and XRD (mineralogical) analyses. Without their precious support and guidance, it would not be possible to conduct this semi-quantitative and XRD analysis.

I gratefully acknowledge the funding I received for my Ph.D. dissertation from the BCCI and the JSCA. Shri Sanjay Patel, who used to be the honorary secretary of the BCCI; Shri Rajesh Verma, who used to be the honorary secretary of the JSCA; and Shri Ashok Kumar Singh, who is the CEO of the JSCA, deserve special thanks for sharing their vast knowledge of cricket with us.

Dr. (Prof.) Abdul Wadood, Chairman, Department of Soil Physics and Meteorology; Dr. D. K. Jha, Department of Horticulture; Dr. D. K. Sahi, Chairman, Department of Soil Science and Agricultural Chemistry; and Dr. B. K. Agarwal, Professor, Birsa Agricultural University, Ranchi, Jharkhand, have all been very helpful.

I also thank the Clay Mineral Society of India (CMSI), which gave me the chance to present my research papers during the National Conference on "Application of Clay Science in Agriculture, Environment, and Industry," organised by the Clay Minerals Society of India in collaboration with ICAR-NBSS and LUP and held in Kolkata on 7 and 8 August 2015 and ICAR-NBSS-LUP Land Use Planning (ISSLUP), for presenting my paper at the International Conference on "Integrated Land Use Planning for Smart Agriculture," held in Nagpur, India, on 10–13 November 2016, Nagpur, India.

No research is possible without the library, the centre of all learning processes. I take this moment to express my gratitude to the entire library's in-charge and officials (Birsa Agricultural University, Birla Institute of Technology, Ranchi, Jharkhand, ICAR-NBSS, Nagpur, IARI, New Delhi, etc.) for their services and support.

Lastly, I would like to thank my family for all the love, support, and encouragement they have shown me, especially my parents (Mother: late Sunaina Devi and Father: Hony/Capt Shri Krishna Singh), who raised me in spite of their financial hardship, and most of all, my loving, supportive, and encouraging wife, Mrs. Jyoti Singh, whose support during the entire stage of my journey of writing this book is so appreciated.

Above all, I owe it all to Almighty God for granting me the wisdom, health, and strength to undertake this book-writing task and enabling its completion. I deeply miss my mother, the late Sunaina Devi, who is not with me to share this joy.

Dr. Shyam Bahadur Singh
Curator; Board of Control for Cricket
in India (BCCI) and Jharkhand State
Cricket Association (JSCA)

About This Book

This book has several salient and unique features in terms of content and numerous newly developed concepts and phenomena that help the common cricket enthusiast understand the game of cricket and pitch behaviour based on basic scientific logic and understanding. To achieve the basic aims and objectives of this book, the chapters have been organised and structured as follows:

Chapter 1: The introductory chapter provides a definition and classification of cricket pitch behaviour (during the various phases of the game) as desired by the BCCI and ICC. This chapter provides an introduction and discusses problem definition, the first and most important step. It also tries to find out the gap between desirable and actual available pitch behaviour and how to bridge it. It also deals with the significance of this book and lays out the basic aims and objectives that will help the cricket fraternity in understanding the complex pitch behaviour and in unravelling the natural forces that indirectly or directly control the pitch behaviour.

Chapter 2 deals with the importance of clay minerals in defining, analysing, interpreting, and comprehending pitch behaviour. Further, it deals with various direct and indirect methods that will enable the pitch curators to analyse the predominant clay minerals and how to interpret and comprehend the pitch behaviour based on various index properties. In the year 2008, nearly 76 different cricket pitch soils being used in India for various domestic and international cricket matches were examined for their morphological, physical, chemical, and engineering properties by IIT, Bombay, in collaboration with the BCCI, but their analysis and scientific interpretation of these valuable data so as to understand the direct impact on cricket pitch behaviour were not done. Based on the scientific and statistical evaluation and interpretation of the above-mentioned data, a SWOT (strength, weakness, opportunity, and threat) analysis was done so as to find out the gap between what it is and what it ought to be.

Chapter 3: Based on the statistical and scientific analysis of 76 different cricket pitch soils being used in India for various domestic and international cricket matches based on their morphological, physical, chemical, and engineering properties by IIT, Bombay, in collaboration with the BCCI in the year 2008, a benchmark of cricket pitch soils has been developed for the first time in India.

Chapters 4–8 individually deal with the pitch soil chemical properties, pitch soil water, pitch turf grass, pitch soil organic matter, integrated rolling management, soil structure, and compressibility. This is an effort to decipher the impact of each and every major and minor component of pitch soil, which controls the pitch behaviour either in large or small magnitudes but acts as a critical factor in defining and shaping the pitch behaviour as a whole.

Chapter 9: Several real-life examples, pitch interviews, scenarios, and case studies as felt and observed by the curator during the preparation of various international (Test/ODI/T-20) and board matches or during the construction and renovations of new wickets have been included in each and every relevant chapter so as to analyse, interpret, comprehend, and justify the theoretical science with the existing practises involved in pitch construction, preparation, or judging the complex nature of pitch behaviour. Based on findings through the database management (DBM) of pitch profiles and the Pitch Behavioural Analysis Index (PBAI), various innovative and simple methods (Pitch Behaviour Forecasting methods) are devised that will enable to judge and comprehend the complex pitch behaviour in simple ways.

Data Collection Methods and Sources

Data collection to testify to the practical implications of various case studies and the newly developed concepts such as benchmarking of cricket pitch soils, integrated rolling management schedules, and various methods of analysing, comprehending, or predicting the cricket pitch behaviour through the Pitch Behavioural Analysis Index (PBAI), Pitch Behavioural Forecasting (PBF) methods, etc., was done through an integrated approach of primary and secondary data.

Primary data related with the pitch behaviour was collected through personal observations (while conducting and supervising several ODI, T-20, Ranji, and several board matches conducted by the BCCI/JSCA as a curator), questionnaires, and schedules (through the direct participatory methods at the curators certification programme, refresher courses, Level-1 curators certification courses, and the various yearly seminars and conferences conducted by the BCCI, ICAR, NCMSI, ICAR-NBSS, LUP, BIT, BAU, RAU, National Mineralogical Laboratory (NML) of CSIR Jamshedpur, etc.). The responses of the above-mentioned participants collected through surveys, questionnaires, and schedules (both physical and electronic responses through e-mails) were collected and tabulated in structured formats of questionnaires as the Pitch Quality Index (PQI) and Pitch Quality Standards (PQS). Direct response from the cricket players of national and international levels, curators, and experienced groundsmen representing the BCCI Pitches and Grounds Committee and the different states, coaches, match officials, and several learned professors and researchers directly connected and supervising this research (belonging to various related institutions as IIT-Bombay, IARI-New Delhi, ICAR-NBSS, LUP—groundsmen representing the BCCI Pitches and Grounds Committee and the different states, coaches, match officials, etc.) were taken as the primary data source to understand and unravel the truth, myth, and science behind the pitch behaviour.

As planned with Mahi, a structured performance is given. It includes a set of questions that Dr. S. B. Singh, the curator, BCCI, India, answered to explain the goals and aims of the investigation. The purpose was to understand and analyse how a captain of such international stature analyses and interprets the pitch behaviour before the match, at the time of the toss, or during the match.

Secondary data was collected with the help of the BCCI and IIT Powai (nearly 76 pitches from national and international matches were examined by IIT Powai for some physical, chemical, and engineering properties, but no analysis and interpretation were done before, and various books, research journals, research bulletins, seminar abstracts, etc., published by ICAR-NBSS and LUP Nagpur, etc., as mentioned in the reference and bibliography section, were interpreted and analysed for use in making good cricket pitches.

Case studies—for qualitative, intensive, in-depth investigative analysis, several case studies in the form of personal observations (analysis and interpretation while conducting and supervising several ODI/T-20, Ranji, and board matches organised by the BCCI/JSCA), visiting different sites (Purna Valley, Laheli-Rohtak, and several Indian international stadiums, etc.), and attending several conferences and seminars conducted by the BCCI/ICAR, the NBSS and LUP, the Clay Mineral Society of India (CMS), etc., are explained in this book.

The goal of these case studies was to find, confirm, measure, and explain all of the direct and indirect primary and secondary data and other factors that control cricket pitch behaviour.

Contents

About the Authors

Dr. Shyam Bahadur Singh is working as a faculty member and curator at Board of Cricket Control in India (BCCI)/Jharkhand State Cricket Association (JSCA). After graduating from Rajendra Agricultural University, Pusa, Samastipur, Bihar (1989–1993), he completed his master's degree in M.B.A. Marketing from SNSIBM, Ranchi (2007–2010), and his Ph.D. from Jharkhand Rai University (2014–2017). The title of his Ph.D. thesis was "Critical evaluation of economic and behavioural analysis of some important pitches laid on conventional or traditional methods Versus scientific methods based on recent research and developments: A case study of the JSCA International Cricket Stadium Versus the Keenan Stadium". This dissertation gave him a strong reason and objective to write a book and unravel the science behind the art of pitch making among cricket enthusiasts.

As a curator, Dr. Singh has conducted several Test Matches, One Day Internationals, T-20s, and many BCCI board matches since 2010. He has also cleared the curators basic certification course in 2013 and the Level 1 (2019) course conducted by the BCCI at the National Cricket Academy, Bengaluru.

Dr. Singh has published several papers related to cricket pitches in major national journals. His research papers have been presented and published at several national conferences organised by Clay Mineral Society of India in collaboration with Indian Council for Agriculture Research (ICAR), National Bureau of Soil Science (NBSS) and Land Use Planning (LUP), Nagpur, India and at BCCI seminars.

Dr. Singh has been actively involved in teaching, training, and active research on cricket pitch construction, preparation, and behavioural analysis while curating several international, national, and domestic cricket matches conducted by the BCCI and JSCA.

Dr. Devendra Narain Singh has been a faculty member of Civil Engineering at the Indian Institute of Technology, Bombay, since 1994. Before joining IIT Bombay, he served for short periods at the Central Road Research Institute, New Delhi, and the Indian Institute of Technology, Kharagpur. His early education was completed in Lucknow, UP, India. He obtained his bachelor's, master's, and doctoral degrees from IIT Kanpur in 1986, 1989, and 1993, respectively.

His teaching, research, and development activities are in diversified areas of **geotechnical engineering** (viz., soil mechanics, foundation engineering, environmental geotechnology, mechanics of unsaturated soils, soil characterization based on thermal and electrical properties, contaminant transport in porous media, mineralogical characterization, utilisation and recycling of industrial waste, geotechnical centrifuge modeling, etc.). He has published more than 300 technical articles. He has supervised several **doctoral** dissertations and **master's** theses. He has been successful in filing several **patents**, of which many have been granted.

Apart from teaching and research, Dr. Singh has been actively associated with some of the most prestigious business houses, as an in-house instructor and retainer consultant. He has worked on projects funded by the Bhabha Atomic Research Centre (BARC), the Indian Council of Agricultural Research (ICAR), the Department of Science and Technology (DST-TIFAC), the All India Council of Technical Education (AICTE), and the Atomic Energy Regulatory Board (AERB). With financial aid received from these and several other organizations, he has established a state-of-the-art **Environmental Geotechnology Laboratory** in the department.

Dr. Singh founded *Environmental Geotechnics* and has been its **Editor-in-chief**. He has been 'Editorial Board Member' of several journals of repute. He is the recipient of several awards and is a **Fellow** the **Indian National Academy of Engineering, American Society of Civil Engineers** (**ASCE**) and **Institution of Civil Engineers** (**ICE**).

Dr. Sanjay Kumar Ray joined the ICAR-National Bureau of Soil Survey and Land Use Planning (NBSS&LUP) as a scientist in 1991 and as a principal scientist in 2007. In 2015, he was appointed Head, ICAR-NBSS&LUP, Regional Centre, Jorhat. He obtained his bachelor's, master's, and doctoral degrees from the University of Calcutta.

He has about 29 years of professional experience working in various parts of India and abroad in natural resource management, soil genesis and mapping, soil classification, application of remote sensing and geographical information systems, soil chemistry and clay mineralogy, soil carbon sequestration, carbon modeling and climate change, soil and land quality, land evaluation, and land use planning.

He has published about 116 peer-reviewed research papers in national and international journals and several other types of technical publications during his career. He has also developed patents and copyrights from his work.

Dr. Ray has received about 14 awards, among which are the ICAR Team Award for Interdisciplinary Research, the Usha Zende Award for the best paper published in micro-nutrient research, the Fellow of the Maharashtra Academy of Sciences, the Fellow of the Clay Minerals Society of India, and the Fellow of the Indian Society of Soil Survey and Land Use Planning. He is a visiting scientist at the Soil and Water Science Department, Institute of Food and Agricultural Sciences, University of Florida, Gainesville, Florida, USA.

He is also a member of various committees of the Government of India's Department of Science and Technology, as well as the Boards of Management and Academic Councils of several universities, as well as Research Advisory Committees and Institute Management Committees of several institutes and organisations in India.

Dr. H. B. Nagaraj graduated in Civil Engineering from SJCE, Mysore, in 1986 and obtained his post-graduation in geotechnical engineering in 1991 from Karnataka Regional Engineering College, Surathkal (presently NITK, Surathkal). He completed his doctoral degree in soil mechanics from the Indian Institute of Science, Bengaluru, in 2000. Since 1997, he has been serving as faculty in Civil Engineering **at the B. M. S. College of Engineering, Bengaluru, India and currently he is heading the department**. He has keen research interests in "Engineering Behavior of Fine-Grained Soils" and "Alternate Building Materials".

Dr. H. B. Nagaraj has been active in research and has published more than 40 well-cited technical papers in international journals of repute and in a few important international conferences. He is an active reviewer for most of the leading international journals in the field of geotechnical engineering and building materials. He has been a recipient of many international awards for reviewing technical papers from PUBLONS in 2017, 2018, and 2019. He has also received the **Outstanding Reviewer Award** from two international journals of repute. Very recently, he has been successful in implementing his laboratory-scale research findings in the construction of a rammed earth structure in one of a residential building project in Bengaluru, a work that was taken up in collaboration with Made in Earth-Sustainable Architecture and Building Practices, Bengaluru, India. In 2016, he was invited to review **an international project proposal** (seeking funding) submitted by a leading materials manufacturing and research organisation to the **United States Department of Agriculture National Institute of Food and Agriculture under the grant programme Small Business Innovation Research (SBIR), USA**.

Dr. H. B. Nagaraj has successively supervised two doctoral dissertations, four Master of Science and forty master's degree theses. He has filed several patents, of which a few have been accepted. He has handled various externally funded projects, including those from DST and AICTE. He has also reviewed the BIS Code: Review of Geotechnical Provisions in Indian Seismic Code IS 1893 (Part 1): 2002 as a part of the GSDMA-IIT Kanpur project. He has delivered 20 lectures for VTU Edusat e-learning programmes on geotechnical engineering in

2009 (Program Nos. 10 and 11) and five lectures in 2018 (Program no. 24).

He has handled more than 50 major consultancy projects in the fields of geotechnical investigation and forensic geotechnics for major clients including STUP consultants, NCCCL, the Airport Authority of India, SEW Infrastructures Ltd., Reliance InfraTel Ltd., KRDCL, Bangalore Development Authority, Land Army Corporation, Southern Railways, and others. He recently managed two major industry-sponsored projects to use iron ore tailings from Rai Bahadur Seth Shreeram Narasingdas (RBSSN) Pvt. Ltd., Kariganur, Hospet, Karnataka, to produce building blocks for construction applications. He is presently supervising the practical implementation of the production unit, which is in progress. This will promote sustainability and sustenance of the construction industry in the area of implementation.

Acronyms and Abbreviations

BAU	Birsa Agricultural University
BCCI	Board of control for Cricket in India
BD	Bulk Density
BS	Base Saturation
BIT	Birla Institute of Technology
CBR	California Bearing Ratio
CEC	Cation Exchange Capacity
CH	Clegg Hammer
CIF	Cracking Intensity Factor
COLE	Coefficient of Linear Extensibility
COLC	Coefficient of Linear Contractility
CMSI	Clay Mineral Society of India
CSIR	Centre for Scientific and Industrial Research
DBMS	Data Base Management System
DBD	Dry Bulk Density
DDL	Double Diffused Layer
DI	Dispersion Index
DIP	Drop in Pitches
DNA	Deoxy ribose Nucleic Acid
DOSI	Depth of Significant Influence
DOT	Department of Transport
DP	Dew Point
EC	Equilibrium Coefficient/Electrolyte Content
EI	Expansion index
ESP	Exchangeable Sodium Percentage
ET	Evapo-Transpiration
FC	Field Capacity
FSI	Free Swell Index
HC	Hydraulic Conductivity/ Hygroscopic coefficient
HIS	Hydroxy interlayered smectites
HIV	Hydroxy interlayered vermiculite

IARI	Indian Agriculture Research Institute
ICAR	Indian Council of Agricultural Research
ICC	International Cricket Council
ICRISAT	International Crop Research Institute for Semi Arid Tropics
IIT	Indian Institute of Technology
IPM	Integrated Pitch Management
IOG	Institute of Groundsman-ship
IRM	Integrated Rolling Management
JSCA	Jharkhand State Cricket Association
HPAN	Hydrolyzed Poly Acrylo Nitrile
KIM	Kaolinite Illite Montmorillonite
LAI	Leaf Area Index
LL	Liquid Limit
MDD	Maximum Dry Density
MPA	Mass Per unit Area
NBSS	National Bureau of Soil Survey
NCMSI	National Clay Mineral Society of India
NCA	National Cricket Academy
NML	National Metallurgical Laboratory
ODI	One Day International
OMC	Optimum Moisture Content
PAWC	Pitch Available Water Content
PBAI	Pitch Behavioral Analysis Index
PGA	Poly Glactouronic Acid
PI	Plasticity Index
PL	Plastic Limit
PLV	Pitch Lifetime Value
PPM	Parts Per Million
PPT	Pitch Preparation Time
PQI	Pitch Quality Index
PQS	Pitch Quality Standards
PSD	Particle Size Distribution
PVC	Potential Volume Change
PSAWC	Pitch Soil Available Water Content
PWP	Permanent Wilting Point
RAU	Rajendra Agricultural University
RAI	Root Area Index
RAR	Root Area Ratio
RD	Root Density
RLD	Root Length Density
RH	Relative Humidity
RP	Rolling Potential
SAR	Sodium Adsorption Ratio
SCG	Sydney Cricket Ground
SEM	Scanning Electron Microscopy

SL	Shrinkage Limit
SOM	Soil organic matter
SOC	Soil Organic Content
SPAC	Soil Pitch Atmosphere Continuum
SSA	Specific Surface Area
STPP	sodium tri poly phosphate
SWOT	Strength, Weakness, Opportunity, Threat
TDS	Total Dissolved Solids
TRL	Transport Research Laboratory
USDA	United State Department of Agriculture
VAMA	Vinyl Acetate Maleic Acid
VSP	Volumetric Shrinkage Potential
WACA	West Australia Cricket Association
WHC	Water Holding Capacity
WDC	Water Dispersible Clay
XRD	X- ray diffraction
ZAVD	Zero Air Void Ratio

Chapter 1
Introduction

There are no published or written guidelines or quantifiable standards for the preparation or construction of a turf pitch. The majority of cricket pitches in the continental and subcontinental regions are typically prepared and set out based on the extensive experience that groundsmen and curators have amassed in the field of pitch making. By interacting with the weather factors present within and above the pitch surface, dominant clay minerals with exchangeable cations and a proportionate amount of clay, silt, and sand are viewed as the genes that control the pitch's behaviour by regulating its cracking or crumbling patterns, WHC, surface density, compression levels, etc. It is said that a uniform, high-quality pitch is necessary for playing excellent international cricket. A decent cricket field should allow for a competitive match between the bat and the ball. It ought to be suitable for stroke play and have a good pace, consistent bounce, and excellent ball carry. Above all, it should be goal-oriented, and the surface shouldn't ever become hazardous for the batsman or unplayable.[1]

The ICC operating manual (29.1)[2] states that preparing pitches requires considerable ability and is not simple. It is well known that the quality of the pitches that are made can differ greatly from country to country, venue to venue, and even game to game, depending on the weather. The ICC acknowledges that these variations are a component of what makes sports special. All pitches will only be evaluated based on how they perform. The goal is to create a balanced battle between bat and ball throughout the entire game, enabling the players to showcase all of their individual game skills at various stages of the match.[3]

[1] BCCI Curators Certification Course Manual, Second Edition, 2013, P. 15.

[2] ICC pitch and outfield monitoring process, October 2010 and ICC playing handbook, 2012–13.

[3] ICC playing handbook, 2012–13, standard test match playing conditions, p. 28.

S. B. Singh et al., *Cricket Pitches*, Springer Transactions in Civil and Environmental Engineering, https://doi.org/10.1007/978-981-99-2913-9_1

1.1 Definition of a Turf Pitch (a Turf Pitch Behaviour as Desired by the BCCI/ICC for Different Formats of the Match)

According to the ICC operating manual, an ideal pitch should have the following characteristics:

(i) For a five-day Test match (as desired by the ICC/BCCI) or four-day Ranji match:

Day 1: (1/5) Type: The pitch should be quite moist, with some green grass on the surface. A green surface will allow the ball to seam around a bit. The pitch should have a consistent pace and bounce (Fig. 1.1).

Days 2 and 3: (2/5 or 3/5 type)—The pitch will have dried out and should become more bouncier and quicker (provided it does not crack due to drying out too quickly). It will have lost its greenness, and so it should not seam around much. The ball should come nicely onto the bat, and the pitch should now be ideal for batting (Fig. 1.2).

Day 4 (Type 5): The surface should start to crumble and the cracks will start to open up. This will slow down the pitch. It will become less bouncy; the bounce will become inconsistent, and it will start to spin (Fig. 1.3).

Day 5 (5/5) Type: The above pattern will be accentuated as the pitch wears further (Figs. 1.4 and 1.5).

Note: 1/5 wicket type is the wicket conditions and behaviour as observed or desired on the first day of the Test match. The "type of wicket" is determined by the behaviour and conditions of the pitch on the first day of the Test match as opposed to what is generally observed or desired on the second day. Similarly, the 3/5, 4/5,

Fig. 1.1 Desirable pitch conditions on day one for a test/day matches (1/5)

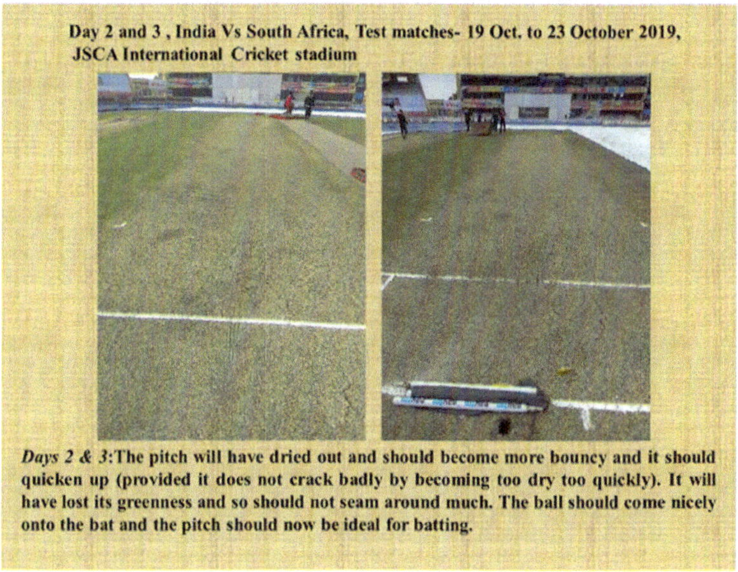

Fig. 1.2 Desirable pitch conditions on day two and three for a test/days match (2/5 and 3/5)

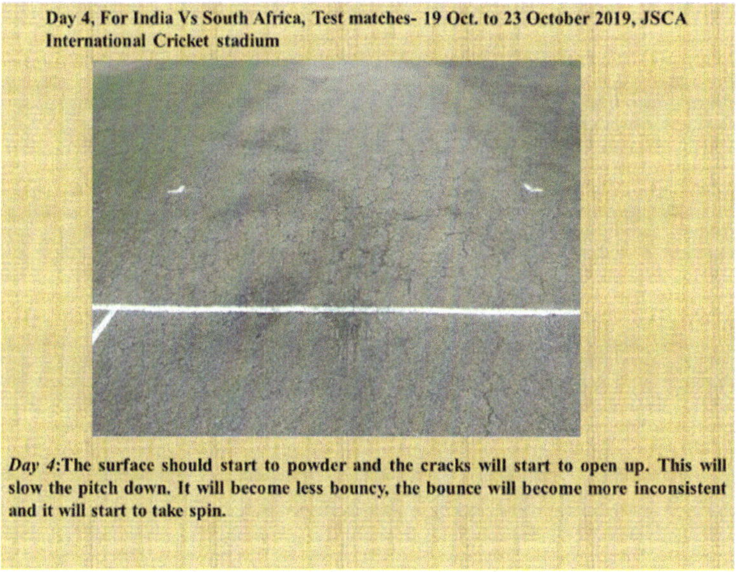

Fig. 1.3 Desirable pitch conditions on day four for a test/days match (4/5)

Fig. 1.4 Actual pitch conditions on day five (5/5), India versus South Africa, Test matches—19 October to 23 October 2019, JSCA International Cricket stadium, Ranchi, Jharkhand

Fig. 1.5 Actual pitch conditions on day five pitch surface having kaolinite-dominated fine red clay soil, for a Test match at M.A. Chidambaram stadium, Chepauk, Tamil Nadu

and 5/5 types of wicket are the behaviour and conditions of the pitch on the first day of the Test match as they are generally observed or desired on the third, fourth, and fifth day of the Test match.

(b) For a one-day limited-overs or T-20 game

The general requirement here is for a pitch that will favour the batsmen. The ideal would be to produce a pitch equivalent to that on the second or third day of a five-day game. This suggests that water should be withheld for one to two days longer than a five-day game to allow it to dry out for a day or two longer.[4]

1.2 Pitch Behaviour Controlling Factors

The above-mentioned desired variation in cricket pitch behaviour can be achieved through the integrated, cumulative, and synergistic effects of the following factors:

1. The format of the match, pitch preparation objectives, pitch construction, and pitch preparation techniques involved.
2. The cricket pitch soil type and quality (morphological, physical, chemical, or mineralogical properties), pitch turf grass density, type, and height, available pitch soil water within the profile, and amount of water applied to fully or partially saturate the profile (because these available and unavailable water within the profile, and roller type and rolling schedule to get the desired variable compaction levels.
3. Conditions of the microclimate and macroclimate (such as humidity, temperature, dew point, sunshine, wind speed, and direction), whether the pitch is new or has been used before, and the moisture level of the soil in the pitch profile before preparations begin.
4. Time in hand to prepare the pitch as the pitch preparation Watering, rolling, and drying through evapotranspiration through grass is a slow but dynamic natural process.

1.3 Various Controllable and Non-controllable Major and Minor Factors Controlling the Pitch Behaviour

There are hundreds of ways of preparing a wicket profile or a match wicket, but there are thousands of controllable and non-controllable factors through which the pitch surface expresses its behaviour. Their complex behaviour is the synergistic outcome of the type of pitch construction or pitch preparation process as defined by the number of pitch profile layers, depth, and quality of material used to construct a pitch profile; the morphological, physical, chemical, and mineralogical composition of pitch soil

[4] ICC playing handbook, 2012–13, standard ODI playing conditions, p.64 and 113.

aggregates; the turf grass type; density; pitch preparation methods and techniques; available soil moisture just above and below the pitch surface; and the effect of micro-climatic meteorological factors (such as temperature, relative humidity, weather, and climatic conditions available within and just above the pitch surface) on the behaviour of the pitch.

The curator always tries to manage the controllable factors (through pitch preparation methods and techniques) during the pitch preparation and minimise the effects of the non-controllable factors (climatic and weather variations) through different pitch-making schedules and techniques. By controlling and regulating various controllable and non-controllable factors available on and within the pitch profile based on skill and the scientific reasons behind the pitch preparation schedules and techniques, the desired behaviour of the pitch can be achieved. This thought can be enumerated and visualised as a concept called "integrated pitch management," which is a combination of controllable and uncontrollable factors involved in pitch preparation. Controllable factors of pitch preparations are in the hands of curators, but the impact of uncontrollable factors on pitch preparations can only be minimised. The important non-controllable factors that are not in the hands of the curators are the atmospheric and weather parameters in the form of solar radiation, temperature, relative humidity, radiation, wind speed, rainfall, evapotranspiration (E.T.) rates, dew, fog, mist, the type of soil and grass on the pitch, etc. Curators cannot eliminate the direct and indirect impacts of non-controllable factors in pitch preparation but can minimise their impact on pitch behaviour through various pitch preparation methods and techniques that they have mastered through their skills and experience.

1.3.1 The Important Controllable Factors in the Hands of the Curators that Affect Pitch Behaviour Are

1. Preseason pitch renovation schedules and techniques that rejuvenate and revive the overstressed pitches from the impact of rolling. This process decompacts and aerates the pitch profile required for the natural healing of the overstressed pitches. Most of the cricket pitch soil, which is high in smectitic clay with swelling and shrinking capabilities, recovers by decompacting and aerating itself during the preseason renovation stage.
2. Creating preseason block density of pitch profiles to make the curators' job easier during the cricket season and to achieve variable compaction levels of pitches with minimal effort.
3. Pitch preparation techniques based on his skills and experience and the type and format of game for which he is preparing the pitches. The total pitch preparation time available also affects the pitch's behaviour during the game.
4. Wetting and drying of the pitch profile, which directly affects and controls the pitch behaviour during different sessions of the match. The physical, chemical, mineralogical, and engineering properties of the pitch soil directly affect the pitch's behaviour, and the behaviour of this type of pitch soil is directly influenced by the type and quantity of water used during the pitch preparation.

5. Judicious and prudent use of rollers after knowing the different rolling techniques, rolling schedules, and compactive potential of his rollers and the pitch (soil texture, clay percentage, and mineral type) on which he is working. Always remember that rollers are the most important tool for a curator to control pitch behaviour.
6. Pitch grass, through which they control the pitch surface structure and the drying process of the pitch profile.

1.4 Pitch Surface Characteristics

At different levels of cricket, the pitch soil should have the following important qualities:

1. Swelling and shrinking-dominant clay mineral type with proportionate quantity of recommended silt and sand. The seat of charge in soils mostly lies in clays and fine silt fractions. Therefore, both the physical and chemical activities of soils are expected to be controlled by clays and silt. The bulk density (BD) of swelling and shrinking clays greatly varies because of their swelling and shrinking nature, which changes with moisture content. The soils have a high BD when dry and a low BD when they are in a swollen state, which varies between 1–2 gm/cm^3 depending on the moisture content (Jewitt et al. 1979). This is one of the main reasons that the cricket pitch soil should have swelling and shrinking-dominant clays, which induce the desired variations in the pitch's behaviour during the course of the match.
2. Have a plasticity that allows remoulding and compaction by rolling, thus giving a smooth surface.
3. Have a deeply dried hard surface that allows the elasticity of the ball to be seen, and preferably one that allows the ball to turn at some point (a hard clay surface is elastic in and of itself). In other words, an elastic surface will provide good bounce for the fast bowlers and turn for the spinners.
4. Have soil cohesion that provides vertical stability, preventing differential change in elevation of parts of the surface from crumbling or powdering. They should not break at depth, causing plating or the surface to flake.
5. Have the ability to recover from the compaction of preparation so that the soil structure and the Bermuda grass can regenerate. The soil must be cracked clay. When the soil cracks, it decompacts itself and allows air and water to enter the root zone of the grass, thus rejuvenating both the soil and the grass.
6. They should not crack or crumble excessively.
7. They should have a sufficiently high permeability when wet to ensure reasonable rates of water movement and leaching of salts. Different clays have different rates of hydraulic conductivity (the rate at which water moves through the soil), and a suitable soil has to have a reasonable hydraulic conductivity so that the water can move slowly right through the profile and move salts with it.

8. There should be a means of removing excess water so that aeration occurs, as well as the removal of salts through the drainage layer lying beneath. There needs to be a permeable layer directly under the clay soil to allow excess water and salts to drain out of the wicket soil into it.
9. Before the final preparation, there should be a completely uniform cover of grass, preferably couch grass, with roots that reach the same depth (at least 100 mm).This is essential to give uniform moisture levels and rates of drying of the surface and up to 100 mm.
10. It should be rich in nutrients, available essential minerals, and humus that is basically in a colloidal state.

The above-mentioned pitch surface characteristics for the different formats of the matches as desired by the ICC or the BCCI at the international or the domestic levels can only be achieved by having a scientific way of selecting good pitch soil by analysing its physical, chemical, microbial, and engineering properties over the long ways of the traditional style of making pitches in India, which is generally based on permutations and combinations and the skills transformed from one generation of groundsmen to another.

1.5 Desired Properties of Cricket Pitch Soils

Based on the scientific research and developments, the following morphological, physical, chemical, mechanical, and mineralogical properties are desired for the cricket pitch soils to induce and achieve the desired variable behaviour of the pitches.

1. Morphological Properties: Pitch soil colour, consistency, crack number and size, pitch profile depth, number of layers either 2, 3, or multi-layered.
2. Physical Properties: Some important physical properties generally considered for acceptance and rejection of cricket pitch soils are: particle size analysis for soil texture (sand, silt, and clay proportions), bulk density (B.D.), coefficient of linear extensibility (COLE), hydraulic conductivity (HC), water dispersible clay (WDC), crushing strength, cracking pattern, linear shrinkage, and emersion and dispersion indexes.
3. Chemical Properties: The pH and electrolyte content (EC), dominant exchangeable cations, soil organic content (SOC), extractable bases (base saturation), exchangeable sodium percentage (ESP) or sodium adsorption ratio (SAR), cation exchange capacity (CEC), EC total salts, extractable bases (base saturation).
4. Mineralogical Properties: X-ray diffraction analysis on clay samples to find out the structure and type of dominant clay minerals in the soil clay samples, such as kaolinite, illite, smectite, vermiculite.
5. Engineering Properties: Atterberg limits, soil strength, compatibility and compressibility, penetrability through the Cleg hammer test, the Procter or modified Procter test, shear strength (traffic ability), or the California Bearing Ratio (CBR) (Table 1.1).

Table 1.1 Summarises the desirable morphological, physical, chemical, mechanical, and mineralogical examinations to obtain the above-mentioned cricket pitch soil behaviour based on scientific research and developments

Properties	Desirable values	Relevance
Morphological properties		Pitch soil colour, consistency, cracks number and size, pitch profile depth, number of layers either 2, 3 or multi-layered
Physical properties		
Soil texture		
Clay (< 0.002 mm)	50–70%	Have high plasticity, elasticity, high water holding capacity, high compaction level, decompacts and grass growth and post-match recovery is fast
Silt (0.002–0.2 mm)	5–20%	
Sand-fine (0.02–2.5 mm)	10–20%	
Medium (0.25–1.00 mm)	0–10%	
Coarse (1–2.0 mm)	1%	
Crushing strength	0.8–1.6 MPA	To analyse the soil cohesion of strength to with stand the ball impact
Cracking pattern	2–5 per desirable	The width and number of cracks is essential
	6–10 pcs acceptable	For good grass growth and after match
	> 11 not acceptable	Recovery and reconsolidation
Linear shrinkage	8–15%	To analyse physically about the nature and type of clay having shrinking and swelling capacities
Emersion test	Value 6 desirable	To analyse physically about the dispersibility
Dispersion index	Value 6 is desirable	
FSI-free swell index	< 50	Are commonly used for identifying expansive clays and to predict the swelling potential
Expansion 0–20 21–50 51–90 91–130 130	Potential Very low Low Medium High Very high	
Chemical properties		

(continued)

Table 1.1 (continued)

Properties	Desirable values	Relevance
Sodium level	< 5 me%	High Na level reduces the hydraulic conductivity leading to more dispersion which causes pores to be blocked by migrating particles
EC		The soluble salts should be as low as possible, as excessive salts may possible cause crumbling of soil (Harris 1961) and effect the desired swelling and shrinking behaviour of the pitch profile
H^+, Al^{+3}, Na^+, Ca^{+2}, Mg^{+2}, K^+ (exchangeable cations), Ca^{+2}/Mg^{+2} Ratio		Predominant exchangeable cations at a particular pH levels control the swelling and shrinking behaviour of clay and the desired WHC, BD/compressibility of the pitch profile Calcium/magnesium carbonate help in binding soil particles, causing a greater cohesion between soil particles. This leads to greater compaction that increases the BD values
ESP	< 3–5	Deflocculates the pitch profile under saline conditions in which the pitch and grass recovery is very poor after the match
pH levels	6–7 in water and 5–6 in $CaCl_2$	Regulates the available nutrients and micronutrients in the soil also affects the swelling capacity of clay minerals
$CaCO_3$ or $MgCO_3$	< 0.15	Above causes the soil crumbling
	< 1%	Above causes the soil mulching
Organic matter/ humus	< 5%	Controls the clay through its colloidal behaviour and enmeshing the clay particles but the pitch profile loses the density if the organic matter content is higher than the prescribed levels. This phenomenon is generally observed in most of the North Eastern states and Kerala having high rainfall and high pH levels
Engineering properties		
Atterberg limit		For soil classification based on liquid, plastic, shrinkage limit and plasticity index
Clegg hammer test		To test the compaction level or surface hardness

(continued)

Table 1.1 (continued)

Properties	Desirable values	Relevance
	(207–424) g	Average readings
	< 200 g	Slow pitch
	(200–350)	Medium fast pitch
	> 350 g	Fast pitch
Mineralogical properties Both direct and indirect methods		To analyse the qualitative and quantitative nature cricket pitch soil-dominant clay minerals as kaolinitic, illitic or smectitic

1.6 Statement of Problems with the Cricket Pitches in India and Abroad

Some of the important Indian cricket stadiums listed above have been used for national and international tournaments since India's independence. These stadiums are built in a traditional way, with multiple layers (sometimes as many as five to seven) and no or very poor drainage systems. Recent research and developments show that 2- or 3-layer wickets are the best way to get the best results for the least amount of money and with the most sense.

Today, the BCCI and the state cricket associations have started a special campaign to find out what the difference is between the traditional way of making the pitch and the different scientific methods based on research and development. Nearly, all of the pitch soils used by the major cricket stadiums in India have been analysed on physical and chemical parameters by IIT Powai for the BCCI. The above-mentioned physical and chemical analysis report will help to understand the vast variability and its behaviour in important regions of India under different levels of climatic variability and zones. Synergistic efforts of the BCCI, along with the associated state cricket associations, and the Central and State Agricultural and Engineering Colleges may produce magical results regarding the nature and behaviour of great Indian pitches.

Rollers used for compaction have their maximum vertical density only up to a depth of 4 in. Nearly only 8 in. of sandy loam and river-washed sand (4 in. each) are required to act as a drainage layer. A thick layer of clayey soil covering more than 300 mm only creates pitch layering and acts as a source of perched water tables. Excess moisture is retained in this thick layer of pitch soil and affects the nature and behaviour of the playing surface. Most of the pitches will have to be renovated according to the latest prescribed scientific norms of pitch construction. Even the recently developed stadiums in the last five years have laid down their pitches without examining the mineralogical composition of pitch soils, without which the above-mentioned objectives of having a good pitch as found in the other playing continents cannot be achieved.

The total cost of laying a new multiple pitch (pitch square) and outfield (drainage system and irrigation facilities also) with scientifically approved layers comes out to be 5–10% of the total project cost. Newly laid pitches and outfields have nearly 10–15 years of pitch lifetime value (PLV) and economic life, during which they provide results that meet international standards. Its life is restricted due to the accumulation of dead organic matter and humus in the pitch grasses, which gradually reduce the clay content present in the pitch profile. Much air voids are created within the pitch soil due to the leaching of silt downwards within the drainage layer, thereby making the playing surface flat and slow. Further different layers occur within the pitch profile due to injudicious top dressing and the use of different pitch soil than that used in the initial stage of pitch preparation. Heavy salt accumulation is present within the pitch profile, thereby affecting the pH level and the nature and behaviour of the pitches.

Second, the clay soil used for pitch construction and the sandy loam and sand used to lay the drainage layer and the outfield layer are becoming scarce, expensive, and are seen as a mineral resource by most of the states in India. So, these materials should be used judiciously and according to the prescribed layers only. Undesired multiple layers of surface (not more than 300 mm) and subsurface (nearly 8–10 in. only) will only add to the project's cost.

Relaying or renovating the pitch and outfield due to the unscientific ways of its laying and maintenance will only provide different undesired behaviour and will have an additional recurring cost to the project.

1.7 Aim and Objective of This Book Based on Problem Identified

1. To understand the basic principles of science and logic behind the art of measuring and analysing the pitch surface behaviour and how to develop, analyse, and interpret the pitch behaviour through the Pitch Behaviour Analysis Index (PBAI) based on the Pitch Quality Index (PQI) and Database Management System (DBMS) of a cricket pitch profile.
2. To find out the economic feasibility of laying the pitches based on the selected soils after comparative analysis of their physical, chemical, mineral, and engineering properties. It is economically wise to spend a few thousand rupees on pitch and outfield soil tests and on their research and development before layering the pitches, which costs crores of rupees and is time-consuming.
3. Find the best pitch soil that can change the nature and characteristics of Indian pitches by comparing the morphological, physical, chemical, mineralogical, and engineering properties of the selected pitch soils.

4. To find out the gap between the old, traditional, uneconomical, unscientific, and illogical ways of pitch construction and preparation methods and what is required at the national or international level based on the latest research and developments.
5. To synchronise and coordinate the latest scientific knowledge and skills available in the agricultural, engineering, meteorological, and other parallel institutions, groundmen, and curators (at the local, national, or international arena) for pitch preparation and pitch construction methods. The comparative analysis of the morphological, physical, chemical, mineralogical, and engineering properties by the above-mentioned institutions may have a synergistic effect and can bring about drastic changes in the behaviour and character of Indian pitches.
6. To find out the best level of compaction, soil strength, and pitch surface density that can be achieved in situ at a given optimum moisture content (OMC) by analysing the engineering properties of pitch soils under observation (the Proctor or modified Proctor test, or CBR). It is also important to find out what the maximum level of compaction is for each type of pitch soil, either to find out which soil texture or mineralogy will give the best level of compaction for a fast, consistent bounce with the least amount of wear and tear.
7. To find out the best suitable pitch soil based on the physical, chemical, and mineralogical analysis and to find out whether two or more different soils with entirely different physical, chemical, and mineralogical behaviours can be mixed with each other to have a good synergetic effect on the pitch performance.
8. To analyse the impact of pitch grasses on the nature and behaviour of pitches.
9. To analyse the impact of various microclimate and macroclimatic conditions above the pitch soils and their impact on nature and behaviour.

1.8 The Book's Design to Accomplish the Aims and Objectives of This Book

1. Since 2009, 40 cricket fields with different types of pitch soil have been built at the Jharkhand State Cricket Association International Stadium (JSCA). Since then, several international (Test/ODI/T-20, etc.), board matches, and domestic matches have also been played on these pitches, which will serve as a reference sample for research based case studies. The existing pitch soil being used for the above-mentioned tournaments has given excellent results and is on par with the desirable pitch behaviour during the different formats of the matches.

Secondly, all the latest major machines, equipment, and manpower required for pitch construction and preparation are available to pursue the research-based case studies on which the practical implications of various skill, techniques, and process of pitch preparations can be scientifically testified and justified.

2. On a national level, the BCCI and some of the major affiliated cricket stadiums with different soils (black, brown, yellow, and red soil wickets) and agro-climatic conditions shall be taken as a reference sample so as to analyse, predict, and comprehend the pitch behaviour.
3. The collection of the required pitch soil sample will either be done by the different state associations under the BCCI or by the different affiliated districts under the JSCA.
4. All suggestions and advice will be taken from most of the skilled and technically qualified groundmen and curators regarding the various pitch-making skills and techniques that they have learned and acquired through their lifelong experience. This will be done either through a direct visit and interacting with them, through telephonic discussions, emails, or through standard questionnaires and schedules (annexed as Annexure—1, 2, and 2a). The purpose will be to blend the old traditional style and experience of pitch making with the more recent scientific ways based on research and development within India and outside the major cricket-playing countries.
5. Most of the prominent agricultural colleges and research institutes (such as the Indian Agriculture Research Institute (IARI) in New Delhi, the ICAR-National Bureau of Soil Survey and Land Use Planning (NBSS-LUP) in Nagpur) having their expertise in the fields of agronomy, soil science, soil physics, soil mineralogy, soil chemistry, meteorology, etc., shall be contacted so as to understand and justify the various scientific phenomena and processes involved in pitch preparation and construction.
6. Simultaneously, most of the prominent engineering colleges and research institutions that specialise in the fields of civil engineering, soil mechanics, soil geotechnology, etc. shall be contacted for their research on the soil compaction, consolidation, cracking, and crumbling patterns of soil, and the classification and identification of pitch soil based on various index properties and engineering properties of pitch soil.

1.9 Significance of This Book

1. In most of the continental and subcontinental pitches, the traditional style of pitch making with the least concentration on pitch soil sampling and its analysis on morphological, physical, chemical, mineralogical, and engineering properties is the general trend. It was done just by putting things together in different ways and using different pitching skills that had been passed down from generation to generation.

2. Even the type of soil used to prepare a pitch and the way it is layered (either in 3, 4, or 5 layers), the type of deep-rooted grass and how evenly it grows, the way it is rolled, and when it should be done to get the most compaction depend directly on the type and nature of the soil used.
3. Country teams like South Africa, England, Australia, and New Zealand have bouncy and fast pitches only as a result of a coordinated effort by the localised cricket associations and engineering and agricultural colleges. Their valuable research and development, based on the experience and skills of the groundskeeper and curator, either on pitch soil selection, layering, grass variety selection and maintenance, rolling techniques and patterns, maximum compaction level at the optimum moisture content, etc., can only produce fast and bouncy pitches in India at the international arena.
4. In India, scientific research and development with the involvement of centrally organised agricultural and engineering universities along with the BCCI and the local state units' associations working on individual localised and acclimatised soil and weather conditions can only bring magical results in the existing nature and behaviour of Indian pitches. A broad gap exists between what it is and what it ought to be. The old, traditional, unscientific style of pitch preparation and maintenance needs to be changed.

1.10 Summary

All three formats of cricket (Test matches, ODIs, and Twenty20s) are played on the same pitch, so why do the results differ? It is not only important to have a good-quality clay soil with a high percentage of clay contents, but the type of dominant clay minerals (either 1:1 or 2:1 kaolinite, illite, vermiculite, or smectite) is also very important. As a result of the process of human evolution, genes are the basic building blocks that build the nature and character of human beings. Soil type and its basic minerals (mica, kaolinite, or smectite) are the genes of any pitch soil, which define and govern the basic nature and behaviour of any pitch. This can only be achieved by having a detailed physical, chemical, mineralogical, and engineering property analysis of pitch soil before its final selection and use.

Of all the controllable and non-controllable factors interacting with each other depending on the pitch's morphological, structural, physical, chemical, or mineralogical properties, the pitch preparation process and techniques and the available weather variables within or above the pitch surface (such as temperature, relative humidity, evapotranspiration rates, solar radiations, altitude) act as a pivot around which all the other factors regulating the pitch's behaviour during the course of the game of cricket rotate. Clay, silt, humus, and predominant exchangeable cations as major and minor invisible and intangible nanoparticles in pitch soil control the nature and behaviour of the pitches by interacting with the non-controllable natural forces (gravity, solar radiation, wind, temperature, relative humidity, dew, fog, mist, etc.) and the natural phenomena (evapotranspiration, condensation, drying, desiccation, capillary action, etc.) available and operating within and just above the pitch profile. Through this book, let's try to unravel the impact of these major or minor invisible, intangible, and tangible components, factors, and the natural forces operating within and just above the pitch profile to control the cricket pitch's behaviour during the various phases of the match.

References/Bibliography

Adams WA, Gibbs RJ (1994) Natural Turf for sports and amenity, 1st edn. CAB International, Wallingford UK, 404 p

Adams WA, Tanavud C, Springsguth CT (1985) Factors influencing the stability of sports turf root zones. In: Lemaitre F (ed) Proceedings of the fifth international turf grass conference, Avignon, France, Institutional de la recherché agronomique, France, p 391

Adams WA, Baker SW, Carre MJ, Young RJ, James DM (2004) Pitch properties and performance. England and Wales Cricket Boards, London, UK

Adams WA, Baker SW, James DM, Young RJ (2005) Measuring and modeling the bounce and pace of county championship cricket pitches. Int Turf Grass Soc Res J 10:1021–1026

Aldous DE (2016) Handbook on international Turf management. CRC Press

Anderson RD, Sweeny JD, Williams AT (2012) An introduction to management science—quantitative approaches to decision making, Eleventh edition, CENGAGE Learning

Baker SW, Cook A, Adams WA (1998a) Soil characteristics of first class cricket pitches and their influence on playing performance. J Turf Grass Sci 74:63–79

Baker SW, Cook A, Binns DJ (1998b) The effect of soil type and profile construction on the performance of cricket pitches. I. Soil properties and grass cover during the first season of use. J Turf Grass Sci 74:80–92

Baker SW, Cook A, Binns DJ, Carre MJ, Haake SJ (1998c) The effect of soil type and profile construction on the performance of cricket pitches II. Playing quality during the first season of use. J Sports Turf Res Inst 74:93–107

Baker SW, Binns DJ, Cook A, Mooney SJ (2001a) The performance of cricket pitches in relation to soil type and moisture content. Int Turf Grass Soc Res J 9:469–478

Baker SW, Hammond LKF, Owen AG, Adams WA (2001b) Influence of Clegg hammer weight on the relationship between surface hardness and ball rebound for cricket. J Turf Grass Res Inst 77

Basma AA, Al-Homoud AS, Husein Malkawi AI, Al-Bashabsheh MA (1996) Swelling-shrinkage behavior of natural expansive clays. Appl Clay Sci 11(2–4):211–227

BCCI Curators Certification Course Manual (2013) BCCI, Second Edition

Brady NC, Weil Ray R (2008) Nature and properties of soil, Fourteenth Edition, Prentice Hall

Burdet L (1986) Turf cricket pitches, South Australian Cricket Association, Adelaide, 28 p

Bureau of Ureau of Indian Standards (1970) Indian Standard classification and identification of soils for general engineering purposes, BIS, New Delhi, (Reaffirmed 1987), IS 1498

ECB Staff (2007) Recommended guidelines for the construction, preparation and maintenance of cricket pitches and outfields at all levels of the game, England and Wales Cricket Board, Lord's Cricket Ground, London, UK

Ekwue EI, Lall DZ, Stone RJ (2006) Engineering properties of major soils used in cricket pitches in Trinidad. West Indies J Eng 28(2), Technical Paper

Gobat JM, Aragan M (2004) The living soil fundamental of soil science and soil biology. Wiley Mathey, Science Pub Inc.

Gumbs FA, Griffith SM, Shillingford G (1981) The design and physical features of the recently re-laid cricket pitch at the queens park oval, Trinidad. J Sports Turf Res Inst 58:16–20

Harris JR (1961) The crumbling of cricket pitches. The Australian Scientist, 01 April, 73–74

IOG/ECB (2006) Management and maintenance of the cricket pitch. The Institute of Groundsmanship, Milton Keynes, UK

James DM, Carre MJ, Haake SJ (2005) Predicting the playing character of cricket pitches. Sports Turf Eng 8:193–207

Kotler P, Keller LK, Koshy A, Jha M (2008) Marketing management, Twelfth Edition. Pearson

Lambe TW (1951) Soil testing for engineers. Wiley, New York

Lush WM, Cummings DJ, McIntyre DS (1985) Turf cricket wickets. Search 16:142–145

Malhotra KN, Dash S (2010) Marketing research—an applied orientation. Pearson Prentice Hall, Second Impression

McIntyre DK, Jackobsen B (1998) Drainage for sports turf and horticulture, Horticultural Engineering Consultancy, Canberra, Australia

McIntyre K, Don (2001) Cricket wickets—science vs fiction, Horticulture Engineering Consultancy, 5-Brimage Palace Kambah, ACT Australia

Mitchell K (2010) Cricket wickets. Copyright ©Queensland Cricket, pp 1–37

Nagraj TS, Srinivasamurty SR, Vastalya A (1994) Analysis and prediction of soil behavior. Taylor and Francis

Parsons S, The aeration of clay soils in cricket, engineering doctorate (Academic year 2008–2012), School of Applied Sciences, Cranfield University

Pearson J (2010) Basic guide to turf cricket pitch preparation. VCA

Salini U, Girish MS (2011) Suitability of soil for cricket pitch preparation. College of Engineering Trivandrum, Technical paper at NATCON-2011, Centre for Engineering Research and Development, Kerala, 28–29 Jan 2011

Sharma JK (2014) Fundamentals of business statistics. Vikas Publishing House PVT. LTD., Second Edition

Singh DN (2008) Physical, chemical, free swell index and Atterberg limit analysis reports of cricket pitch soils under the BCCI, IIT, Mumbai

Soil Resource Information of Different Agro- Eco Sub-Regions (ASER) of India for crop and soil carbon modeling (2016) Technical Bulletin, NBSS and LUP Publication 1035, ICAR, Nagpur

Stewart VI, Adams WA (1969) Soil factors affecting the control of pace on cricket pitches. In: Proceedings of the first international turf grass conference, pp 533–544

Taiton NM, Klug JR, Edmondson D, Campbell RK, Deventer PWV, Beer MJD (1993) Principles and practice of pitch preparation, United Cricket Board of South Africa

TCCB and NCA (1989) Recommendations on the preparation of cricket pitches, Test and county Cricket Board and National Cricket Association, Lords, London

Thornthwaite CW (1948) An approach toward a rational classification of climate. Geogr Rev 38:55–94

Wood C (2006) Recommended guidelines for the maintenance of cricket Pitches and outfields at all levels of game. England and Wales Cricket Board in association with The Institute of Groundsmanship

Chapter 2
Cricket Pitch Soil Mineralogy: The DNA of Cricket Pitch Profile

2.1 Introduction

Fine-grained soils used in cricket pitches are generally composed of clay minerals, non-clay minerals (silt and sand), and amorphous materials (oxides and hydroxides of mainly iron and aluminium; and humus). Because of isomorphous substitutions that occur during their formation (weathering process), clay minerals have a surface charge. This charged nature of clay minerals is responsible for complex soil behavior. It also determines the interactions with the liquid phases. These features together determine the plasticity, swelling and shrinking, compression, strength, and fluid conductivity of the fine-grained soils. As a result, understanding the complex nature of various continental and sub-continental pitches requires knowledge of soil clay mineralogy and the mechanisms that control the behaviour of clay minerals with variations in the pore medium environment.

All three formats of cricket (Test, ODI, and Twenty-20) are played on the same pitching surface, so why do the results differ? The type of predominant clay mineral present in the soil (kaolinite, illite, vermiculite, or smectite) present in the soil is very important in providing the desired pitch surface behaviour during the course of the match. Despite the lower percentage of clay, the presence of smectite type clay in the soil produces good bouncy and fast pitches when compared to pitches with clay minerals with a less expanding lattice structure, such as illite or kaolinite.

The seat of charge in soil mostly lies in clays, indicating mainly that physical, chemical, and mineralogical properties are controlled by the clay content. It can thus be envisaged that a high content of smectite-type clay minerals is the DNA of any cricket pitch, which defines its nature and behaviour as it controls the physical and chemical behaviour of the pitch's soil surface.

The desired physical characteristics (as plasticity, consistency, structure, cohesiveness, very large surface area, swelling and shrinking capacity, crushing strength, cracking pattern, linear shrinkage, dispersion Index, coefficient of linear extensibility etc.), chemical characteristics (soil pH, predominant exchangeable cations such as sodium, potassium, calcium and magnesium, cation exchange capacity, EC,

S. B. Singh et al., *Cricket Pitches*, Springer Transactions in Civil and Environmental Engineering, https://doi.org/10.1007/978-981-99-2913-9_2

total salts, exchangeable sodium percentage (ESP), organic carbon, and calcium carbonate) and engineering properties such as soil strength, pitch surface hardness/ density, compressibility, penetrability (through Proctor or modified Proctor test), shear strength (trafficability) or California Bearing Ratio (CBR), etc.; are directly governed by the quantity and the type of predominant clay minerals in cricket pitch soils.

Pitch soils with different clay mineralogy, either in the form of kaolinite, illite, or smectite, behave in a different manner irrespective of the same quantity of clay.

2.2 Classification of Clay Minerals Suitable for Cricket Pitch Soils

The three most predominant clay minerals in cricket pitch soils are kaolinite, micas (illite), and smectites, which are crystalline hydrous aluminosilicates, layer silicates, or phyllosilicates (Baruhisel and Bertsh, 1989). The basic structure of some clay minerals is enumerated in Fig. 2.1.

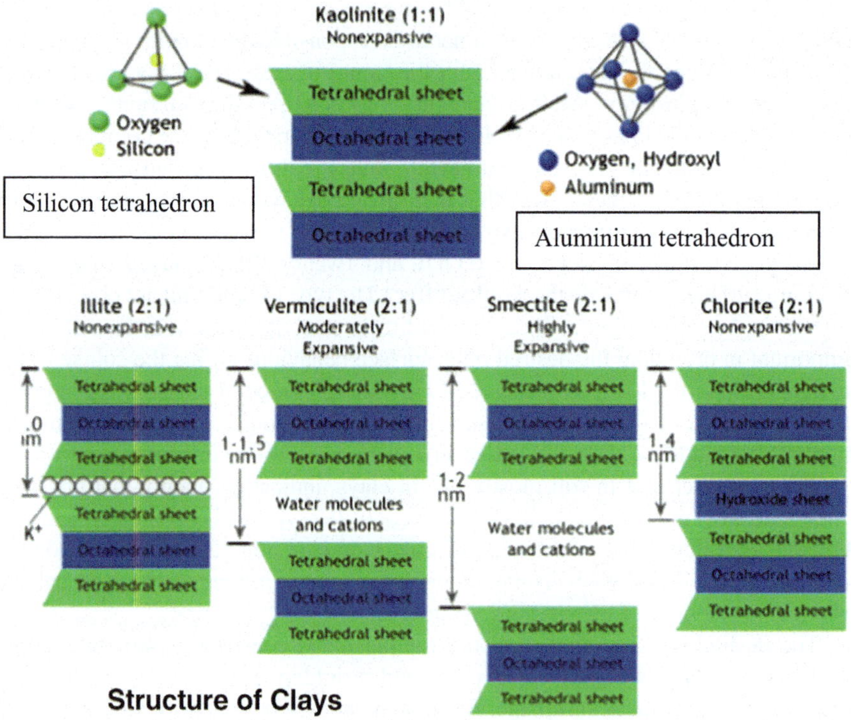

Fig. 2.1 Basic layer structure of predominant clay minerals in cricket pitch soils

The various factors of soil formation, such as climate, parent material, organisms, and topography, play a very important role in the formation of soils from minerals and organic matter (Jenny 1940). Kaolinite, a 1:1 phyllosilicate mineral, has very little tetrahedral and octahedral substitution. This bonding holds the 1:1 layer tightly together, leaving little to no interlayer space for the adsorption of cations or water (Dixon 1989). Because of the interlayer hydrogen bonding, the structure of kaolinite is fixed (as shown in Fig. 2.1), and no expansion can occur between the layers when the clay is wet. Cations and water do not enter between the structural layers of a 1:1 mineral particle. The effective surface area of kaolinite is restricted to its outer surface, or external surface area. This fact and the lack of isomorphous substitution in the mineral account for the relatively low negative charge and small capacity of kaolinite to adsorb exchangeable cations. Kaolinite crystals are usually hexagonal in shape and larger than most other clays. In contrast to some 2:1 clay minerals, 1:1 minerals like kaolinite exhibit less plasticity, stickiness, cohesion, shrinkage, and swelling and can also hold less water than other clays.

2:1 Layer Silicate Clays: The crystal units in 2:1 clay minerals are characterised by one octahedral sheet sandwiched between two tetrahedral sheets. Smectite and vermiculite are expanding minerals, while fine micas (illite) and chlorite are relatively non-expanding.

Smectites are three-layered minerals having a single octahedral sheet sandwiched between two silica tetrahedral sheets with a 2:1 lattice structure. The bond between adjacent unit structures of tetrahedral sheets is comparatively weak, and dipolar molecules of water (as shown in Fig. 2.2) can enter between the sheets, causing them to expand readily. Thus, soils containing substantial amounts of smectites will exhibit high swelling and shrinking characteristics. There will be no volume change if the moisture content of the clay element is unchanged.

The characteristically smectite group is known for the expansion of its interlayer, which occurs when water enters the interlayer spaces in relatively dry clay, forcing the layers apart. The flake-like crystals of smectite have a high amount of negative charge resulting mostly from isomorphous substitution of Al^{3+} ions with Mg^{2+} ions in the octahedral sheet. Some charge is also derived from the substitution of Al^{+3} ions

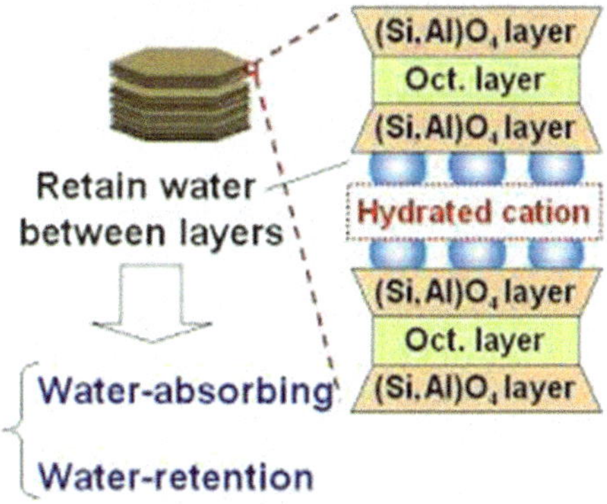

Fig. 2.2 Cross section of an interlayer clay micelle showing retention of water between layers

for Si^{+4} in the tetrahedral sheets. Because of these substitutions, the capacity to adsorb cations is very high (about 20–40 times that of kaolinite). The tetrahedral sheet of one smectite structural unit and the tetrahedral sheet of an adjacent unit are held together by bonds consisting of oxygen atoms. Consequently, adjacent layers are only loosely bound to each other by very weak oxygen-to-oxygen and cation-to-oxygen linkages. The hydration energy of the expansion of the interlayer is much higher than the weak Vander Waals forces between the structural cations and oxygen atoms, resulting in exchangeable cations and associated water molecules being strongly attracted to the spaces between the layers. The crystal expands as the water or organic solvents push apart the layer (as shown in Fig. 2.2). The internal surface area thus exposed by far exceeds the external surface area of these minerals and contributes to the very high total surface area. This expansion with interlayer water contributes to the very high degree of plasticity, stickiness, and cohesion for which smectites are well known. The adsorption of interlayer water also leads to severe swelling when soils high in smectite are wet, and the same soils shrink when dried.

Mica is a 2:1 clay mineral with a layer charge of about 2. Non-hydrated K^+ binds adjacent tetrahedral layers because it is of the correct size and coordination to occupy the holes in the tetrahedral layer, electro-statistically cementing the adjacent layer (Fanning et al. 1989). Although mica has a high layer charge, the tight bonding afforded by the K^+ causes the mica to behave similarly to kaolinite. The layers are bound together tightly (as shown in Fig. 2.1), allowing for little to no space for the adsorption of cations or water. Thus, mica is a non-expansive mineral like the kaolinites.

Vermiculite group: These are 2:1 minerals with an octahedral sheet dominated by Al^{3+}, though some are dominated by Mg^{2+} as well. The tetrahedral sheet of most vermiculites has considerable substitution of aluminum in the silicon positions, giving rise to a very large quantity of negative charge associated with these clays. The CEC of vermiculite usually exceeds that of all other silicate clays. The interlayer spaces of vermiculites usually contain strongly adsorbed water molecules, Al-hydroxy ions, and cations such as magnesium. However, rather than acting as a wedge to separate the units, this interlayer serves as a bridge to keep them together. Vermiculite does not have as much tetrahedral substitution of Al^{+3} for Si^{+4} as does mica, and thus has a layer charge of 0.5 to 1 (Douglas 1989). As a result, the interlayers of vermiculites are accessible to both cations and water. However, interlayer expansion is limited because vermiculite can absorb only two layers of water.

The degree of swelling and shrinkage is, therefore, considerably less for vermiculites than for smectites. As a result, vermiculites are classified as limited expansion clays because they expand more than kaolinite but less than smectites. Ions like K^+ fit perfectly into the interlayer space of vermiculite, fix K^+ very strongly, and get transformed from a 1.4 to 1.0 nm mineral. Whereas in smectites, most of the ionic substitution occurs in the octahedral layers, giving smectites a lower layer charge than the vermiculites, averaging about 0.2–0.5 (Table 2.1).

Table 2.1 Comparative analysis of some major properties of soil colloids

	Major properties of soil colloids			Surface area (m^2/g)		Inter layer
Colloid	Type	Shape	Size (um)	External	Internal	Spacing (nm)
Smectite	2:1, silicate	Flakes	0.01–1	80–150	550–650	01–2.0
Vermiculite	2:1, silicate	Plates, flakes	0.1-0.5	70–120	600–700	1–1.5
Kaolinite	1:1, silicate	Hexagonal crystal	0.1–5.0	5–30	0	0.72
Illites	2:1, silicate	Flakes	0.2–2.0	70–175	0	1
Humus	Organic	Amorphous	0.1–1.0	Variable		

Hydroxy inter-layered vermiculites (HIV) and hydroxy inter-layered smectites (HIS):

These are interstratified minerals with Al, Fe, and Mg hydroxyl polymers as the dominant chemical component in the interlayers of vermiculite and smectite minerals (Ray et al. 2008) and possess very little potential to shrink and swell (Barnhisel and Bertsh 1989). Smectites is the generic name given to all expanding 2:1 phyllosilicates with a layer charge of 0.2–0.6 (Bailey 1980; Ray et al. 2003). In smectites, cation substitution occurs either in the octahedral sheet or the tetrahedral sheet or in both. These isomorphous substitutions determine the properties and chemical composition of soils. Montmorillonites are a species of smectites and its formation is favored by are restricted leaching, so that Mg^{+2}, Na^{+}, and Fe^{+2} cations may be accumulated in the system. Thus, the formation of montmorillonite is aided by an alkaline environment, the presence of Mg^{+2} ions, and a lack of leaching. Such conditions are favourable in semi-arid regions with relatively low rainfall, particularly when evaporation is greater than precipitation. Under these conditions, enough water is available for the alteration process, but the accumulated cations will not be removed by flushing rain. Soil smectites are normally saturated with Ca^{+2} or Mg^{+2} in systems with pH > 7. Some other species of the smectite group of minerals are weigellite (Al^{3+} rich) and montmorillonite (Fe^{3+} rich).

2.3 Cricket Pitches: Why Clay Minerals?

1. Only proportionate clay, silt, sand, and organic matter composition with desirable predominant mineral types can produce desirable variations in cricket pitch behavior. The bulk density (BD) of deep shrink-swell soils varies greatly because of their swelling and shrinking nature. Since the shrink-swell phenomenon depends on soil moisture content, a change in moisture changes the BD values of the pitch profile. Further, moisture retention and its movement within the pitch profile are regulated and controlled by the available micro- and macro-climatic variables available within and above the pitch profile.

Cricket pitch soils with varying soil texture, particle-size distribution (PSD), clay percentage, and the predominant mineral and exchangeable ions present, as well as water holding capacity (WHC), generate differential pitch surface density due to the intermittent retention, movement, and loss of soil moisture (particularly capillary, hygroscopic, and film moisture) within and above the pitch surface, as well as changes in diurnal (day and night) weather conditions.

Because of the very large specific internal surface area and firmly held adsorbed water in the form of diffuse double layer present in the intra-layer lattice structure, as well as the capillary moisture present in the voids, a clay soil, which is potentially a swelling and shrinking type, serves as a reservoir of water at the start of a match. Gradual loss of soil moisture due to diurnal changes in weather variables (temperature, relative humidity, and radiation), E.T. thermo-osmosis, and vapour pressure generate variations in pitch surface density within the pitch soil profile, which regulate the nature and behaviour of the pitch as the game progresses day by day or session by session. This soil moisture, which exists as capillary water-adsorbed moisture and film moisture within the pitch soil rich in Smectite dominant clay mineral soil, provides equal opportunities for pacers, spinners, and batsmen by varying pitch surface density throughout the match. During the first and second days of the match, having high soil moisture within the pitch's soil profile along with uniform live grass favours the pacers and seamers as well as batsmen. As the match progresses and as the pitch profile gets dried (desiccated) due to the evapo-transpirational loss of moisture from the pitch surface, it assists the spinners. The adsorbed hygroscopic moisture (which has a higher density, surface tension, and viscosity than ordinary water) within the smectite clay lattice acts as an unavailable water for the grass roots present in the pitch soils and acts as a permanent source of moisture content desired for obtaining the maximum initial density as well as variable density of the pitch profile during the course of the match (especially in Test matches). Fine clay soils retain at least 15–20% of the hygroscopic moisture as compaired to the coarse-grained sandy soils which retains only 1–2% as as hygroscopic moisture content. Retention and movement of this capillary, hygroscopic and film moisture within the pitch profile in vapour form generates the desirable variations in pitch behaviour during different phases of the match.

2.3.1 *Mechanics of Swelling of Clay Soils Suitable for Cricket Pitches*

Although all clays shrink and swell with changes in moisture content, none approach the magnitude of swelling and shrinking exhibited by smectites. Small size, large internal surface area, and a double diffused layer allow smectites to absorb several times their weight in water (Borchardt 1989).

The fundamental concept of clay swelling is the balancing of forces of interaction among the clay surface, ions, and water. Interlayer expansion of smectites has been commonly attributed to an electrostatic repulsive force between double layers of adjacent platelets. Layer silicates have a maximum hydration state beyond which they do not expand, which can be explained by hydrogen bonding (Farmer) and Russell (1971). Inter layer expansion is initially driven by the hydration energy of exchangeable cations. Other swelling mechanics are crystal strain relaxation and entrapped air pressure.

Smectitic and vermiculitic clay minerals and their group of minerals present in soil are able to expand and contract in one dimension as they absorb water and organic compounds or dry like a stack of papers that can become taller. The structural layer of these types of clays has a resultant negative charge, which attracts water molecules or other polar entities into the inter layer area, causing expansion. These water molecules are generally aligned around positively charged ions or cations such as Na^+ or Ca^{+2} (inter layer cations) and next to the layer. The amount of swelling or shrinking is related to the type of interlayer cations present in the lattice structures. Sodium-rich smectite clays expand more than those containing calcium.

Terzaghi (1927) proposed that the elastic rebound of crystals distorted during compression would result in expansion upon loading. Swelling may also result from the buildup of pressure in trapped air of an advancing wetting front (Emersion 1964; Parker et al. 1980, 1982). In general, a Montmorillonite saturated with divalent cations such as Ca^{+2} will swell by 100%, or from 1 to 2 nm. Conversely, a montmorillonite saturated with a monovalent cation such as Na^+ swells to infinity (Norrish 1954).

However, natural soil smectites do not exhibit as much expansion as Na saturated smectites because of

1. Inter-stratification with non-expansive clay minerals and
2. Soil systems are rarely monoionic and are typically saturated with Ca^{+2} and Mg^{+2}.

   (Borchardt 1989)

According to Mielenz and King (1955) two mechanics are involved in the swelling of clays a relaxation of effective compressive strength related to enlargement of capillary films, and osmotic ambitions of water by clay minerals with expanding lattice.

Complete saturation is not necessary to initiate swelling. A slight increase in moisture content in the magnitude of 1–2% is sufficient to cause detrimental swelling (Fig. 2.2). Heaving of expansive soil may occur in the absence of free water. Vapor transfer plays an important role in providing the means for the volume increase of expansive soils. Water vapour at a higher temperature than its surroundings migrates towards the cooler area to equalise the thermal energy of the two areas. When water reaches the cooler area, generally beneath the surface structures, condensation can take place and provide sufficient moisture to initiate swelling. Complete saturation is not required to result in a large volume change as most of the soil water is lost as gravitational and capillary water through the forces of gravity and evapotranspiration. Smectites can swell to several times their own size in the presence of ions like Na^+, which makes soils rich in smectites difficult to manage. However, it has been determined that such sodium-rich fine smectitic soils with high pitch surface density may be suitable for laying cricket pitches to some extent.

The pitch soil composed of fine-textured, medium-textured, or coarse-textured soil with different proportions of predominant minerals will behave differently during the course of the game due to their differences in their water retention and holding capacities, thereby having different bulk densities and pitch surface densities during the course of the game.

2. Cricket pitch soils having high clay content with preferably smectite type clay mineral have greater swelling and shrinking capacities helps the pitch grass to regenerate quickly (after the completion of match), and thus, such kind of pitches are always available for the next match in a short span of time. On the other hand pitch and grass recovery is very poor in cricket pitches with soils dominant in non-swelling and less shrinking type of clay mineral. Most of the cricket pitch soils dominant in kaolinitic or illitic clay mineral suffer from these type of perennial problems in which replanting of turf grass after every cricketing season becomes mandatory. In addition to this high intensive labour and mechanical cost are involved to regenerate the pitch profile and the grass in such type of pitches.
3. The pitch soil with high swelling and shrinking clay content generates deep wide cracks within the pitch profile which helps in exchanging of certain poisonous gases (carbon mono-oxide, carbon di-oxide and sulphur di-oxide, etc.) which generally accumulates within the grass root zones during the heavy compaction by rollers. More densely compacted cricket pitch surface have reduced pore space, smaller average pore radii and reduced connectivity between pores which directly affects the soils ability to transport water and active diffusion and exchange of gases potentially leading to anoxic or hypoxic environment and the possible toxic build up of waste products (Lipiec and Hatano 2003; Horton et al.; Stepnewski et al. 1994). Less or no mechanical measures are required to de compact and aerate the pitch profile if they are prepared with clay soils having dominant swelling and shrinking clay type (as they have the natural ability to aerate themselves due to their natural swelling and shrinking behaviour).
4. Accumulation of undesired heavy salts and minerals present in the irrigation water used for pitches gets easily leached down the pitch profile through the wide cracks developed on the pitch surface having fine smectite dominant cracking clay. Undesired accumulation of salts within the pitch profile adversely affects the availability of essential elements for good grass growth. It also affects the desired pitch surface behavior of the pitch profile by reducing the shrinking and swelling capacity of pitch soils.
5. Irrigation water use efficiency is much higher in pitches with high clayey soils of the shrink-swell type because it retains much more water for a longer period of time due to its high water holding capacity and very low water permeability (due to fine clay soils' extremely low hydraulic conductivity). The amount of water required to irrigate smectitic type clayey pitches is less than that required for kaolinite dominant fine clay or loamy soils with low water holding capacity and high hydraulic conductivity.

6. Pitch soils high in clay along with humus are essential as clay-humus interaction (both in colloidal form) helps in retaining a good clay structure and in controlling many of the desirable physical and chemical behaviours of pitch surfaces.

2.4 How Does the Pitch Soil Clay Mineralogy Determine the Physical and Chemical Behavior?

Pitch soils with different clay mineralogy, either in the form of kaolinite, illite, or smectite, behave in a different manner irrespective of the same quantity of clay. Soil physical properties such as plasticity, cohesion, swelling, shrinkage, dispersion, and flocculation are greatly influenced by the quantitative and qualitative nature of their mineralogical composition, as the nature of clay minerals has a profound effect on the nature, size, and stability of the aggregates (Russel 1973). Because of the clay mineral composition of fine-grained soils, their behaviour under a given pore-medium environment is more physio-chemical than purely physical. Changes in pore medium chemistry have an opposite effect on the liquid limit, sediment volume, shear strength, and compressibility behaviour of kaolinite and montmorillonitic clayey soils.

The pore size distribution is the result of primary soil particle aggregation. The development of inter-particle binding contributes to the stability of soil aggregates and may take place through cations, organic molecules, water, and also between clay-coarse grains (Biswas and Jagan Nath 1982). Clay minerals have a high specific surface area, electrical charges, the ability to disperse and become flocculated, to swell and shrink, and a variety of particle shapes. Because of their high specific surface areas and charged nature, these particles form strong associations with other small and/or charged soil particles. Hagin and Bodman (1954) reported that the stable aggregates are formed with smectite but not with kaolinite. The shape and structure of the soil aggregates are greatly influenced by the type of clay minerals. Smectite induces the formation of sphere-like aggregates, whereas kaolinite induces platy structure (Baver 1956). High CEC of the smectite mineral is responsible for high soil aggregation, as evidenced by the highly aggregated black soils as compared to other soils of non-smectitic mineralogy (Biswas et al. 1961; Biswas and Karale 1974). Hydrogen bonding appears to be operative with kaolinite, but the attraction of surfaces due to other intermediate cations is weak because of its low cation exchange capacity.

Pitch soils containing more than 15% clay exhibit plasticity (that is, pliability and the capacity to be molded). This property is due to the plate-like nature of the clay particles and the combined lubricating and binding influence of the adsorbed water. Thus, the particles of plastic easily slide over each other, much like panes of glass with films of water between them.

Cohesion indicates the tendency of clay particles to stick together. This tendency is primarily due to the attraction of the clay particles to the water molecules held between them. Hydrogen bonding between the clay surface and water, and also among water molecules is the attractive force responsible for cohesion. Smectites and fine-grained micas exhibit a much higher degree of cohesion than kaolinite or hydrous oxides. Humus, by contrast, tends to reduce the attraction of individual clay particles to each other.

Swelling and shrinking are the third and fourth most important physical characteristics, which solely depend upon the quantity and quality of the mineralogical composition of pitch soils. Some clay rich in smectites swells when wet and shrinks when dry. Later, because of swelling, such soil is likely to close up and become more impervious than the one dominated by kaolinite or fine-grained micas. Vermiculite is intermediate in its swelling and shrinking capacities. Some swelling is owing to the penetration of water between the crystal layers, resulting in intra-layer expansion. Most of the swelling, however, results from water attracted to the colloids and to the ions adsorbed by them, which form tiny air bubbles entrapped as the water moves into the extremely small pores of these soils, forming the double diffused layer (DDL). Apparently, swelling, shrinkage, cohesion, and plasticity are closely related. They depend not only on the clay mixture present in the soil and the dominant adsorbed cations, but also on the nature and amount of humus that accompany the inorganic colloids.

Dispersion and flocculation: Clay particle dispersion is caused by the repulsion of negatively charged particles towards one another; the particles move away from one another. Dispersion is encouraged by the large number of water molecules associated with each micelle and by the adsorbed cations. Highly hydrated monovalent cations, such as sodium, that are not tightly held by the micelles enhance clay dispersion; more tightly held cations, such as Ca^{++} and Al^{+3}, inhibit it.

The ability of common cations to flocculate is in the general order of:

$$Al^{+3} > H^{+} > Ca^{+2} > Mg^{+2} > K^{+} > Na^{+}$$

The colloidal complexes of humid and sub-humid regions' soils are dominated by aluminum, hydrogen, and calcium, while those of semi-arid regions are high in calcium ions. All of these ions enhance flocculation.

Despite the fact that the exchangeable cations and anions present in the swarm of adsorbed DDL are in minute amounts, the effect they have on pitch behaviour via flocculation and de flocculation is enormous. A deflocculated soil structure with smectitic clay dominated by Na^{+} or Mg^{+2} as exchangeable cations, resulting in high ESP, EMP, COLE, WDC, compressibility, low hydraulic conductivity (HC), and air voids, is desirable for a cricket pitch profile because it generates a high pitch surface dry density during the course of the match.

2.5 Classification and Categorization of Cricket Pitches Based on the Predominant Clay Minerals and the Exchangeable Cations

1a. Very high swelling and shrinking Na^{+}/Mg^{+2} and moderate Ca^{+2} and Mg^{+2} dominant fine clay smectites at strongly acidic pH ranges (pH > 8.5, EC 4, high ESP (> 5), SAR, BS, CEC, WHC, dispersion index, and good flocculated structure) with non-proportionate silt and fine sand proportions. Example: the soil of the Perth and Harvey cricket pitches. very wide surface cracks with a very high water holding capacity (due to the hydration of clay) and dry density of the pitch surface express their pitch surface behaviour through large cracks and a very high pitch surface dry density with very low air voids within the pitch profile. Pacers are generally favored, while batsmen and spinners have limited options on this type of pitch surface. After matches, recovery of the pitch surface and the turf grass is moderate.

1b. Ca^{+2} and Mg^{+2} dominant fine clay smectites with proportionate silt and fine sand proportions at an alkaline pH range (pH neutral to moderately alkaline, EC > 4, moderate ESP, SAR, high BS, CEC, WHC, good flocculated structure); examples: Brisbane, MCG, and Merri Creek cricket pitch soils. wide surface cracks on the pitch surface with a very high dry density Following a match, the pitch surface and turf grass recover quickly.

1c. Moderate swelling and shrinking Ca^{+2} and Mg^{+2} dominate fine silty clay, illite, and smectites (mixed minerals) at an alkaline pH range (pH neutral to moderately alkaline, EC > 4, moderate ESP, SAR, high BS, CEC, WHC, flocculated structure) with non-proportionate silt and fine sand. express their pitch behaviour through multiple medium-sized (plate-sized) cracks during the initial stages of the match, but crumbles and assists spinners during the later stages of the match due to the very high proportion of silt and moderate WHC of the pitch surface. provides a diverse range of pitch surface behaviour in which pacers, batsmen, and spinners all have an equal opportunity to perform in a limited number of match days.

2. Non-swelling and less shrinking kaolinite, an Al^{+3} or H^{+} ion dominant fine clay with a low pH range, CEC, BS, WHC, and compressibility, with proportionate or non-proportionate silt and high to medium sand proportions In general, such pitch surfaces exhibit their behaviour due to low pace and bounce with crumbling (powdery) pitch surfaces, and pitch surface and turf grass recovery after a match is extremely poor.

2.6 Methods of Identification of Swell-Shrink Clays Suitable as Cricket Pitch Soils

It can be analysed using the following methods:

Field identification methods: This is done by visual inspection. Some of the important field identification criteria indicating the presence of expansive types of minerals suitable for cricket pitch soil are as follows (Figs. 2.3, 2.4 and 2.5):

1. Polygonal type of surface cracks in the dry season.

Fig. 2.3 Polygonal cracks in clay soil from an open field identified to be used to construct cricket pitches

2. Soil peds when freshly broken may have smooth shiny surface. These can also be observed when the partially dry soil is rubbed over smooth objects.

Fig. 2.4 Shinny peds and black colour of clay dominant pitch soils excavated from Mokama Tal area, Bihar

3. The wet samples of the soil are sticky and very hard when dry. They have a color which is slightly dark black or greyish tinge.

Fig. 2.5 Black to grey coloured soil being excavated from Mokama Tal area, Bihar and from Kasigiri Farm House, I.C.A.R. Regional Research centre for the North East Region, Lamphelpat, Imphal

2.6.1 A Method of Distinguishing Clay, Silt, and Sand by the Hand-Feel Method

Prepare a walnut-sized sample of moist soil into a uniform, putty-like consistency by kneading. While squeezing and kneading the sample, one can note its malleability, stickiness, and stiffness, all properties that are associated with the clay content. If the soil contains a high silt content, one can feel the soil sample to be smooth and silky, with little stickiness or resistance to deformation. A soil with a significant content of sand gives a rough and gritty feeling and makes grinding noises when rubbed near one's ear. A significant amount of clay can be identified by being able to make a long ribbon of soil when an attempt is made to squeeze a ball of properly moistened soil between the thumb and the side of the forefinger. One can make the ribbon as long as possible until it breaks from its own weight (Fig. 2.6; Table 2.2).

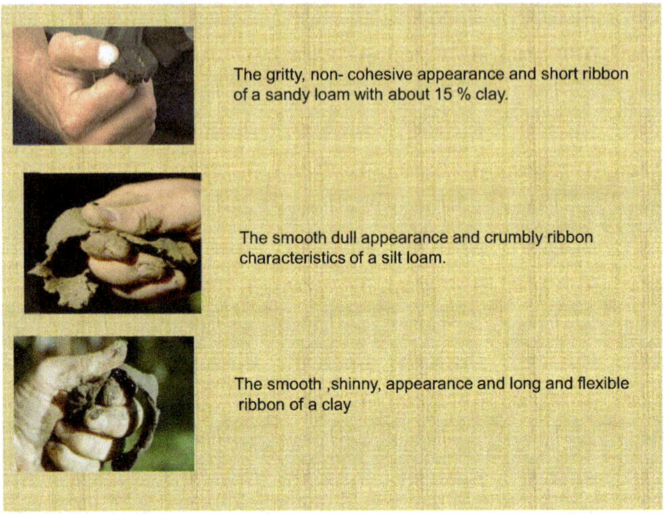

Fig. 2.6 Identification of soil texture by hand feel method

Table 2.2 Observations by feel method can be interpreted as follows

1. Soil will not cohere into a ball and fall apart	Sand
2. Soil forms a ball, but will not form a ribbon	Loamy sand
3. Soil ribbon is dull and breaks off when less than 2.5 cm long and	Loamy soil
(a) Grinding noise is audible; grittiness is prominent feel	Silt loam
(b) Smooth, floury feel prominent; no grinding audible	Sandy loam
(c) Only slight grittiness and smoothness; grinding not clearly audible	Loam
4. Soil exhibits moderate stickiness and firmness, forms ribbons 2.5–5 cm long and	
(a) Grinding noise audible; grittiness is prominent feel	Sandy clay loam
(b) Smooth, floury feel prominent; no grinding audible	Silty clay loam
(c) Only slight grittiness and smoothness; grinding not clearly audible	Clay loam
5. Soil exhibits dominant stickiness and firmness, forms shiny ribbons longer than 5 cm, and	
(a) Grinding noise is audible; grittiness is dominant feel	Sandy clay
(b) Smooth, floury feel prominent; no grinding audible	Silty clay
(c) Only slight grittiness and smoothness; grinding not clearly audible	Clay

Source Brady, Nyle C. and Weil Ray R., Nature and Properties of Soil, Fourteenth Edition, Prentice Hall, 2008

2.6.2 How to Identify Silt and Clay in the Field?

The desirable cricket pitch soil should have predominantly fine smectite clay, moderate fine sand, and little silt. The distinction between silt and clay cannot be based on particle size because the particle size is too small and difficult to be distinguished by the naked eye. As a result, their presence in soil is detected indirectly through physical tests.

There are four field identification tests that can easily conducted in the field to identify the presence of silt and clay. These are—DDDT.

1. Dry strength test
2. Dilatancy or Shaking test
3. Toughness to plasticity test
4. Dispersion test

1. Dry Strength Test

The dry strength provides one basis for distinction. A small briquette of soil is moulded and allowed to dry in the air. It is then broken, and a small fragment about 1 cm in size is pressed between the thumb and forefinger. The effort required to break the fragment provides a basis for describing the strength as "very low," "low," "medium," "high," or "very high." Clay fragments can be broken only with great effort, whereas silt fragments crush easily.

2. Dilatancy/Shaking Test

Since silts are considerably more permeable than clays, the dilatancy or shaking test may also be used to distinguish between the two materials. In this test, a small amount of soil is mixed with water to a very soft consistency in the palm of the hand. The back of the hand is then lightly tapped. If the soil is silty, water rises quickly to its surface and gives it a shiny or glistening appearance. The water then flows back into the soil pat if it is deformed, in some cases by squeezing and in others by stretching, leaving the surface dull. Usually, the greater the proportion of clay in the sample, the slower the reaction to the test. The reaction can be described as fast, slow, or none at all.

3. Toughness to Plasticity Test/Consistency Test

The property of plasticity is characteristic of clays and may be used as the basis for a simple field test. At certain moisture contents, a soil that contains appreciable quantities of clay can be deformed and remoulded in the hand without disintegrating. Thus, if a sample of moist soil can be manipulated between the palms of the hands and fingers and rolled out into a long thread, it unquestionably contains a significant amount of clay. As moisture is lost during continued manipulation, the soil approaches a non-plastic condition and becomes crumbly. Just before the crumbly state is reached, highly plastic clay can be rolled into a long thread with a diameter of approximately 3 mm, which has sufficient strength to support its own weight. Silt, on

Fig. 2.7 Consistency/plasticity test done by Dr. Basantha Singh, Soil Scientist, I.C.A.R. Regional Research centre for the North East Region, Lamphelpat, Imphal, Curator and Mr. Singam Priyananda Singha, Hony. Secretary, Manipur Cricket Association (MNCA)

the other hand, can seldom be rolled into a thread with a diameter as small as 3 mm without severe cracking and is completely lacking in tensile strength unless small amounts of clay are present. The record of a simple plasticity test should indicate not only whether a plastic thread can be formed, but also the toughness of the thread as it nears the crumbling stage. This condition is described as weak and friable, medium, or tough (Fig. 2.7).

4. Dispersion Test

The fourth procedure, known as the dispersion test, is also useful for distinguishing between silt and clay, and for making a rough estimate of the relative amounts of sand, silt and clay in a material. A small quantity of the soil is dispersed with water in a glass cylinder or test tube and then allowed to settle. The coarser particles fall out first and the finest particles remain in suspension the longest. Ordinarily sand settles in 30–60 s. Materials of silt size settles in 15–60 min, whereas that of clay size remains in suspension for at least several hours and usually for several days unless the particles of clay combines in groups or floccules. A small lump of soil is taken in a beaker with distilled or de-ionized water and kept for observations for few hours. The rate and amount of dispersion is rated in the test. The soil is categorized based on dispersion patterns. This test does not give quantitative measures of the soil properties, but they indicate water holding capacities and hydraulic conductivity.

Dispersion index vary from 0–16.

DI = 16 indicates complete dispersion,

DI = 8 indicates dispersion only after remolding,

DI = 0 indicates extremely stable soil which is unaffected by remolding.

A. On a scale of 1–10 is rated as 1—The soil has rapidly lost all structure and water become cloudy as it reverts back to primary particles.

B. Soil is rated as 2—The soil has broken down into particles less quickly and water has quickly becomes cloudy.
C. Is rated as 6—A halo of cloudiness is formed around outside the soil periphery, which holds its shape for long period of time.
D. Is rated as 8—Soil is completely stable, there is no dispersion and the water remains clear.

   08 and above is considered too hard and not very suitable for promoting healthy grass cover.

Generally, field identification of pitch soils is initially done only for acceptance or rejection of the pitch soil samples based on the prescribed morphological and physical parameters. The final acceptance of pitch soil samples based on basic field identification techniques must be followed by the prescribed morphological, chemical, mineralogical, and engineering test parameters.

2.7 Laboratory Identification Methods for Selection of Cricket Pitch Soils

There are number of ways of laboratory tests that are useful in identifying expansive soils suitable for constructing cricket pitch surface. Generally, these can be categorized as mineralogical identification, direct and indirect methods.

1. Mineralogical methods
2. Direct methods
3. Indirect methods

2.7.1 Mineralogical methods

To identify the predominant minerals present in the pitch soils quantitatively and qualitatively X-ray Diffraction (XRD), Infrared Spectrograph, Scanning Electron Microscopy (SEM), Differential Thermal Analysis (DTA), Dye Adsorption, silica sesqui-oxide ratio and Cation Exchange Capacity (CEC) are generally used.

X-Ray Diffraction Techniques for Predominant Clay Mineral Identification and Composition—X-ray diffraction (XRD) is a technique to investigate the organisation of solids, usually at the atomic scale. Although many newer analytic techniques have been applied to the study of layer silicates in recent years, XRD remains the single most useful approach for the identification of minerals. Soil minerals consist of three-dimensional arrays of atoms that are arranged in regularly spaced planes. X-rays (short-wavelength electromagnetic radiation) are well suited to probe these planes. X-ray diffraction in minerals occurs in accordance with Bragg's law:

$$n\lambda = 2d \sin\theta$$

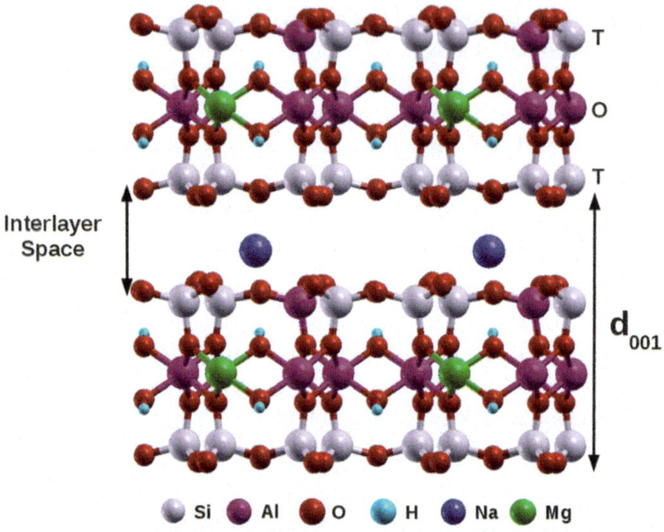

Fig. 2.8 Cross section of an interlayer clay micelle showing the *d*-spacing

where λ is the wavelength of monochromic R-rays, *d* is interplanar spacing, θ is the critical angle for the constructive interface of scattered rays, and n is the integer. The accepted unit for lattice spacing is the angstrom unit (1 Å = a@ 10^{-10} m), which corresponds to the unit of X-ray wavelength. The value *d*(001)/*n*, called lattice spacing, can be measured from the results of X-ray diffraction analysis. If $n = 1$, then *d*(001) represents the first-order diffraction spacing. If $n = 2$, the second-order diffraction spacing is represented by *d*(001)/2. The series of *d*/*n* values obtained, together with the intensity of the X-ray diffraction peaks, are diagnostic for the identification of mineral species (Fig. 2.8).

A popular method of mounting samples for X-ray analysis is the preparation of oriented samples on microscopic glass slides or on porous ceramic plates. A clay suspension is made properly and pipette onto the slide so that approximately 15–25 mg of clay are transferred per 10 cm^2. After the sample has been allowed to dry at room temperature, it is ready for analysis with a direct recording x-ray spectrometer, in which X-ray patterns are printed on charts. Generally, clay minerals exhibit *d*-spacings in the range 3–30 Å, which corresponds to 2–30° θ. *d*(001) planes have the highest intensity of diffraction maxima. The first-order diffraction peak (001) is typically used in conjunction with the second-order diffraction peak as diagnostics for mineral species identification.

Common cementing agents in soils include alkaline earth carbonates, organic matter, oxides and hydroxides of iron, oxides of magnesium, and amorphous silica and alumina. The removal of these materials is advantageous not only from the standpoint of obtaining efficient dispersion of soil particles prior to size fractionation but also from the standpoint of improving X-ray diffraction quality.

Since different cations may retain different amounts of water during hydration (Barshad 1950; Norrish 1954; Mielenz 1955), it is imperative that a clay sample prepared for diffraction analysis be homoionic to ensure that expansion as a result of hydration will be uniform within all crystals of a species. Also, since clay samples are commonly analysed after drying in the air, it is advisable to exchange-saturate the clay with a cation, which will minimise changes in inter-layer water adsorption due to fluctuations in relative humidity. Magnesium (Mg), which allows relatively uniform interlayer adsorption of water by expandable layer silicates, and potassium (K), which specially restricts interlayer adsorption of water by vermiculite, are most commonly used for exchange saturation (Hellman et al. 1943; Mac Ewan 1944; Walker 1957).

2.7.1.1 Criteria for Differentiating the Dominant Minerals in Cricket Pitch Soils Using X-Ray Diffraction (XRD)

Soil always contains a number of mineral species. Many of the species can be identified easily by their distinctive diffraction maxima from a single X-ray pattern. Layer silicate species, however, have many similar or identical structural features, which make their differentiation and identification more difficult. In general, no two minerals have exactly the same distances between planes of atoms; diffraction angles are distinctive for individual minerals. On this basis, XRD is a powerful tool for the identification of soil minerals in all size fractions. Because the interplaner spacings of some layer silicate minerals are variable and depend upon the saturating cation and the presence of water or other water molecules, thoroughly tested protocols have been developed to assist in the XRD characterization of clay minerals (Whittig and Allardice 1986).

Kaolinite produces a *d*(001) XRD reflection at 0.715 nm that is destroyed after de-hydroxylation by heating the mineral at 550 °C for 2 h (as shown in Figs. 2.13, 2.14, 2.15 and 2.16 and Table 2.3).

As can be seen from the *d* spacings listed in Table 2.3, several of the minerals have similar, or even overlapping, diffraction peaks. In such cases, pretreatments of samples are required to distinguish these minerals. The following are the most common methods for pre-treating samples prior to analysis: (1) K-saturation; (2) Mg-saturation; (3) solvation of Mg-saturated samples; and (4) heating at 550 °C. The first three methods stated have the purpose of distinguishing between expanding and non-expanding minerals (Table 2.3). Potassium saturation will normally collapse the inter-micelle spacings and the *d* spacings to 1.

7–2.0 nm, as exhibited by expanding smectites, may collapse to 10 Å. Reconstitution of the *d*-spacing to 1. 7 nm can be achieved by solvation of Mg-saturated samples. None of these treatments will have any effect on the *d*-spacing of non-expanding minerals, e.g., kaolinite.

Table 2.3 *d*-spacing (001) of XRD reflection ranges for different predominant minerals present in different cricket pitch soils

Pre treatments	Kaolinite	Illite/mica	Vermiculite	Smectite	Chlorite	HIM
Mg-saturated, air dried	0.715	1	1.4–1.5	1.4–1.5	1.4–1.5	
Mg-saturated, 54% RH	0.715	1	1.4–1.45	1.5	1.42	1.4–1.5
Mg-Ethylene glycol solvation	0.715	1	1.4–1.45	1.65–1.7	1.42	1.4–1.7
Glycerol solvated	0.715	1	1.4–1.45	1.75–1.8	1.42	1.4–1.7
K-saturation, air dried	0.715	1	1	1.24–1.28	1.4	1.1–1.4
K-saturation heated at 105 °C	0.715	1	1	1–1.05	1.4	1.1–1.4
350 °C	0.715	1	1	0.98–1.04	1.4	1.1–1.4
550 °C	Collapses	1	1	0.98–1.02	1.4	1.1–1.4

Note All *d*(001) spacing's are in nano-meters (nm), *HIM* hydro-oxy-inter-layered minerals

Heating at 550 °C is usually done to distinguish among vermiculite, chlorite, and kaolinite. Vermiculite may have interlayer hydroxyl alumina complexes in its structure. These inter-layers will be destroyed by heating at 550 °C, with the consequent collapse of the d-spacing of vermiculite from 1.4 to 1.0 nm. On the other hand, heating at 550 °C will have no effect on chlorite, though it may produce a second-order peak at 0.7.2 nm. This peak can be confused for kaolinite. However, heating at 550 °C will make kaolinite amorphous for X-ray diffraction analysis, and both the 0.7.13 and 0.3.56 nm peaks of kaolinite will disappear.

XRD is commonly used to identify smectites such as montmorilonite due to their characteristic swelling in polar organic compounds. Magnesium-saturated and air-dried smectites display XRD peaks at about 14 Å (Table 2.3; Figs. 2.13, 2.14, 2.15 and 2.16). When they are solvated with glycerol or ethylene glycol, on the other hand, the organic molecules are adsorbed and form two sheets in inter-layer region. Formation of a stable, two layer water complex between plates of air dried, Mg saturated members of the montmorillonite and vermiculite series results in an inter-atomic spacing of approximately 1.4 nm, between (001) planes. This spacing, caused by layer expansion, distinguishes these species from non-expanding 2:1 layer silicates, the interatomic spacing of which is approximately 1 nm.

Detection of non-expandable 1.4 nm chlorites (2:2 layer silicates) that may be present requires further diagnostic tests for positive differentiation from 1.4 nm vermiculite. Vermiculite will collapse to a non-expanded structure (1.0 nm) when saturated with K (Barshad 1948), whereas chlorites will be unaffected. The difference in behaviour of these species when saturated with K is utilised as a means for their differentiation.

Rich and Obenshain, Tamura, Klages, and White have found that some soil vermiculites contain non-exchangeable interlayer aluminium (as a hydroxy complex), which inhibits lattice contraction on saturation with K. Heating a K-saturated sample at 550 °C for 2 h will destroy the hydroxy complex and permit contraction of the vermiculite layers. The heat treatment will not affect the chlorite.

A diffraction spacing of approximately 1.4 nm obtained from a Mg-saturated, air-dried sample may be contributed by smectite, vermiculite, chlorite, or a mixture of these species. Solvation with glycerol allows separation and positive identification of smectite. Saturation with K similarly allows separation of vermiculite from chlorite, which does not collapse.

The heating of a sample to 550 °C serves two important functions. It causes the collapse of vermiculites, which contain non-exchangeable interlayer aluminium hydroxyl complexes, and destroys the kaolin minerals. When chlorite is present in a sample, it normally yields a second-order maximum of kaolinite (0.715 nm). If the spacing, obtained from an unheated sample, disappears or decreases in intensity after heating at 550 °C, the presence of kaolinite is confirmed.

This treatment expands the $d(001)$ spacing to 16–18, depending on the layer charge of the smectite layer (Figs. 2.13, 2.14, 2.15 and 2.16): layers with a lesser charge tend to expand more. Potassium saturation of smectite layers collapses them (reversibly) to approximately 1.0-nm (001) spacings after excess water molecules have been driven out by heating the sample at 110 °C for 2 h.

Vermiculite: Like smectites, vermiculite displays $d(001)$ spacing of about 1.4–1.45 nm upon Mg saturation. Because the layer charge of vermiculite is greater than that of smectite, the layer does not expand in the presence of polar organic solvents. The repeat distance of 1.4 nm represents one sheet of glycerol in interlayer position, not two as in smectite. On the other hand, discrete vermiculite crystals act like smectite layers upon potassium saturation, collapsing irreversibly to $d(001)$ spacings of 1 nm.

If illite exhibits D (001) XRD reflections at about 1 nm, so do Mica, Magnesium-saturated, and Glycerol-solvated samples. This d-spacing does not change upon saturation by different cations, by treatments, or by solvation with polar solvents.

Chlorite/Hydroxy-inter-layered vermiculite and smectite-

In magnesium-saturated samples, chlorite is identified by an XRD peak at 1.4 nm that does not change in response to glycerol or ethylene glycol salvation or heat treatment. Vermiculite or smectite crystallites that contain hydroxyl-Al compounds in their interlayer spaces may produce XRD reflections that range from 1.4 to 1.7 nm, depending on whether glycerol is present or not. Upon heating, water is generally driven out of hydroxyl-interlayer minerals. In weakly inter-layered minerals, the d (001) spacing of a K-saturated sample collapses to about 1 nm after heating at 550 °C for more than 2 h (as observed in Figs. 2.13, 2.14, 2.15 and 2.16). When the Al interlayers are extensive, however, heating does not result in complete collapse. In fact, the degree to which the layers are propped open by a hydroxyl aluminium precipitate and resist collapse on increasingly severe heat treatments is a diagnostic indicator of the abundance of hydroxyl-inter-layered minerals.

2.7.1.2 Case Study 1

The objective is to identify, analyze, and interpret the quantitative and qualitative nature of the predominant clay minerals based on the above-mentioned X-ray diffraction techniques.

On the joint efforts by the Board of Control for Cricket in India (BCCI), the Jharkhand State Cricket Association (JSCA), and the Indian Council of Agriculture Research (ICAR) National Bureau of Soil Survey (NBSS) and Land Use Planning (LUP), Nagpur, Maharashtra (dated May 2–11, 2016), this was the first ever initiative undertaken in India to understand, identify, and interpret the predominant clay minerals through XRD for cricket pitch soils. Four representative cricket pitch soil samples from the Eden Gardens (Kolkata), Balangir (Odisha), Keenan (Jamshedpur), and JSCA Ranchi (as in Figs. 2.13, 2.14, 2.15 and 2.16; Table 2.3) were selected to identify the quantitative and qualitative nature of the predominant clay minerals at the SRS Division, Indian Council of Agriculture Research (ICAR) National Bureau of Soil Survey (NBSS), and Land Use Planning (LUP), Nagpur, Maharashtra (Figs. 2.9, 2.10, 2.11 and 2.12).

Fig. 2.9 Extraction of fine clay by hydrometer method by Dr. P. Chandran, Head and Principal Scientist, SRS Division, ICAR NBSS and LUP, Nagpur along with Laboratory Technician Shri V. T. Sahu in mineralogical laboratory

Fig. 2.10 Cricket pitch soil samples of 1—Eden Gardens (Kolkata), 2—Balangir (Odisha), 3—Keenan, Jamshedpur—to be analyzed for mineralogical composition through X-ray diffraction method

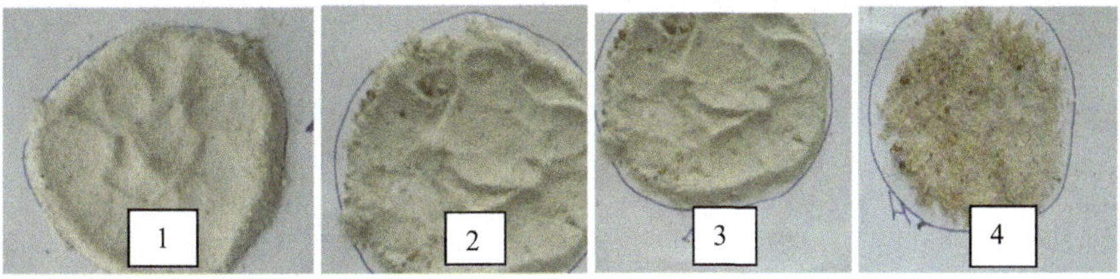

Fig. 2.11 Extracted fine/medium sand form samples of 1—Eden Gardens (Kolkata), 2—Balangir (Odisha), 3—Keenan, Jamshedpur and 4—JSCA Ranchi, (after removal of common cementing agents-soluble salts, alkaline earth, carbonates, organic matter, oxides and hydro-oxides of iron, amorphous silica, alumina, silt and coarse sand particles)

Methodology: The clay fractions (2.0–0.2 μm) were separated from soil samples after dispersion according to size segregation procedure of Jackson (1956). The clay samples were flocculated with KCL and $CaCl_2$. Calcium saturated samples were further glycolated using ethylene glycol (EG).

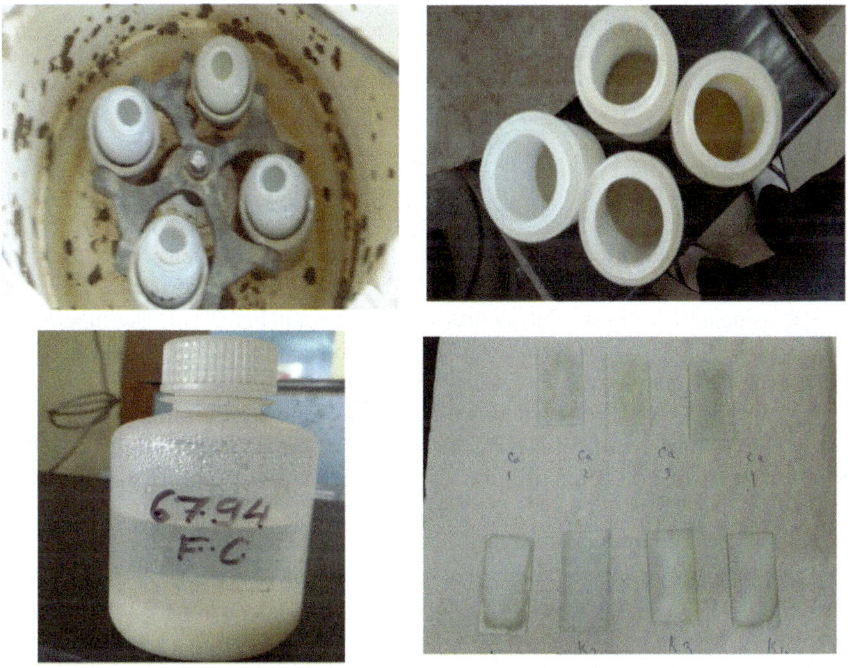

Fig. 2.12 Different steps involved during the process of X-Ray Diffraction analysis and semi-quantitative estimates of clay minerals for some of the important cricket pitch soils in India at the Indian Council of Agriculture Research (ICAR) national Bureau of Soil Survey (NBSS) and Land Use Planning (LUP), Nagpur, Maharashtra, dated 2–11 May 2016

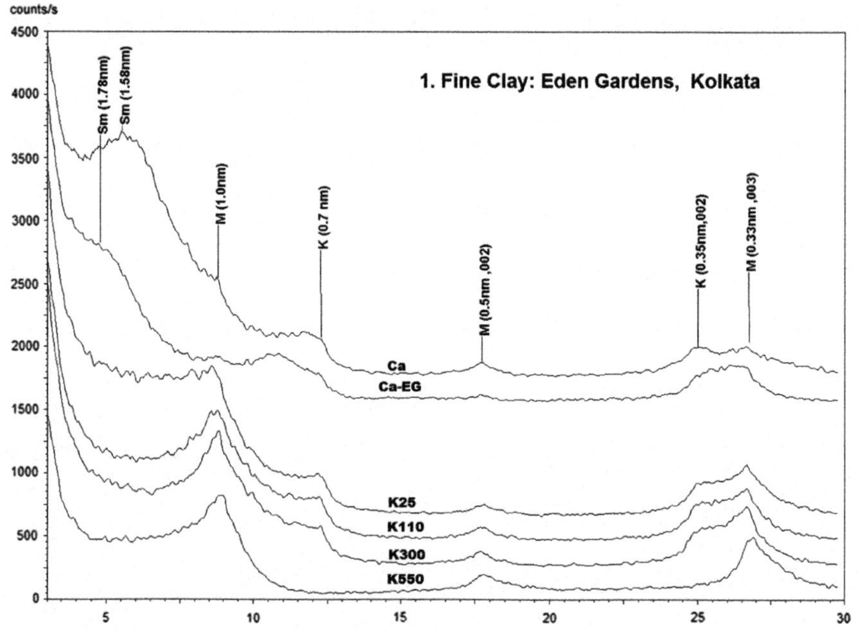

Fig. 2.13 Representative XRD diagrams of Ca-saturated and glycolated Fine Clay (CaEg) of representative cricket pitch soil of Eden Gardens, CAB, Kolkata. Sm = Smectite, M = Mica, K = Kaolinite, K25/110/300/550 = K-saturated and heated at 25, 110, 300 and 550 °C. *Source* ICAR-NBSS-LUP, Amravati Road, Nagpur, 2016

Potassium-saturated clay samples were heated at 300 °C for one hour and again up to 550 °C for one hour. Fine clay fractions (0.2 μ) were separated from the 2 μ clay fractions by agitating a volume of clay fraction water suspension up to 10 cm depth in a Robinson pipette at 3000 RPM for 23 min at a recorded temperature of 380 °C using a REMI heavy duty centrifuge (as shown in Fig. 2.12). The supernatant was separated as fine clay, which was dried and placed in slides. The powered and Ca/K saturated samples (as shown in Fig. 2.12) were put to X-ray diffractometry using a Philips X-ray diffractometer with Cu/Ka radiation at a scanning speed of 1–2 θ/min. Clays were identified according to the procedure outlined by Jackson (1956). Semi-quantitative estimation of minerals was made as per the procedure of Brown (1961).

Analysis and interpretation of the XRD graph patterns and the semi-quantitative estimation of predominant minerals and exchangeable cations (1) Eden Gardens (Kolkata), (2) Balangir (Odisha), (3) Keenan, (4) Jamshedpur, and (5) JSCA Ranchi, are as follows:

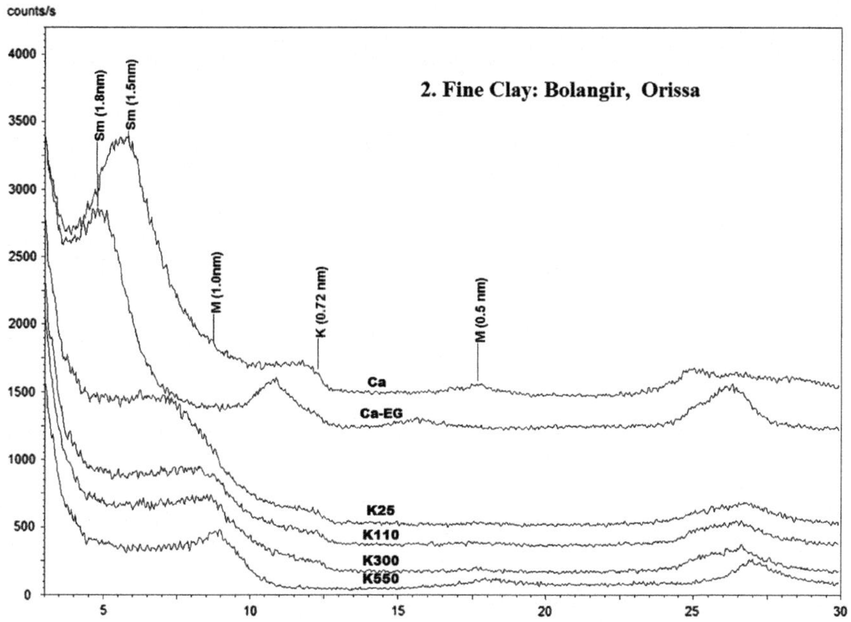

Fig. 2.14 Representative XRD diagrams of Ca-saturated and glycolated Fine Clay (CaEg) of representative cricket pitch soil of Belangir, Orissa. Sm = Smectite, M = Mica, K = Kaolinite, K25/110/300/550 = K-saturated and heated at 25, 110, 300 and 550 °C. *Source* ICAR-NBSS-LUP, Amravati Road, Nagpur, 2016

X-ray diffractograms of the total clay (TC) (2000 nm) and fine clay (FC) (< 200 nm) fractions indicate the presence of kaolinite, mica/illite, vermiculite, smectite and chlorite from peaks at 0.7 nm (Ca saturated clays), 1 nm (Ca saturated clays), 1.4 nm (glycolated Ca-clays), 1.75–1.8 nm (glycolated Ca-clays) and 1.4 nm (K saturated clays heated at 550 °C respectively in diffractograms. Semi-quantitative estimates of X.R.D. data for TC and FC fractions (Table 2.4) indicate that smectite is the dominant mineral in all the fractions except in Keenan soils, where the FC fraction is dominated by smectites but the TC fraction is dominated by kaolin. This was established by the basal reflection around 1.4 nm of the Ca-saturated sample, which expanded to around 1.7 nm on glycolation (Figs. 2.13, 2.14, 2.15 and 2.16). The presence of kaolin was detected by the peak at 0.7 nm, which persisted on K treatment and heating at 300 °C but disappeared on heating at 550 °C.

X-ray diffractograms of Ca-saturated and Ca-EG treated fine clay samples of Ranchi, Eden, and Balangir black soil showed strong reflection at 1.5 and 1.7 nm, respectively, indicating clear evidence of the presence of smectite minerals (Figs. 2.13, 2.14 and 2.15).

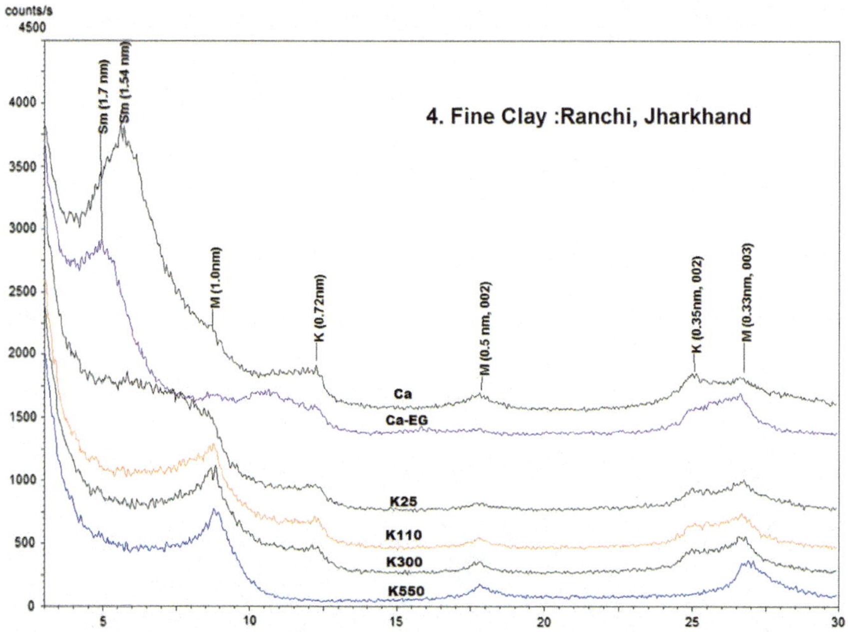

Fig. 2.15 Representative XRD diagrams of Ca-saturated and glycolated Fine Clay (CaEg) of representative Cricket pitch soil of JSCA Int. Stadium Complex, Ranchi, Jharkhand. Sm = Smectite, M = Mica, K = Kaolinite, K25/110/300/550 = K-saturated and heated at 25, 110, 300 and 550 °C. *Source* ICAR-NBSS-LUP, Amravati Road, Nagpur, 2016

The smectites are fairly crystalline and show higher order reflections, but are mostly hydroxyl-interlayered, as evidenced by the high peak sizes of the Ca-saturated samples compared to the Ca-EG-saturated samples and the tailing in the low angle side of the 1 nm peak when K-saturated samples are heated at 550 °C (Figs. 2.13, 2.14 and 2.15). Such hydroxy-inter-layered smectites are formed at pH levels well below 7, which are maintained during their formation in humid climate conditions (Pal et al. 2021b).

The reflection at 1.0 nm, which resisted collapse even after heating up to 550 °C, indicates the presence of hydroxy-interlayered vermiculite and hydroxy-interlayered mica, as well as mica/illite which consists of both muscovite and biotite micas, as confirmed by the more uniform 001/002 peak height of the basal reflection of mica/illite at 1.00 nm.

On K-saturation at 250 °C, the fine clay smectite shifted to a peak area greater than 1.1–1.2 nm, indicating the presence of both low and high charge density smectites (Ray et al. 2003; Pal et al. 2021a) and its dioctahedral nature, as confirmed by Ray et al. (2006) for similar samples. The presence of a small amount of vermiculite is not perceptible in the glycolated fine clay saturated with Ca ions at a 1.4 nm peak.

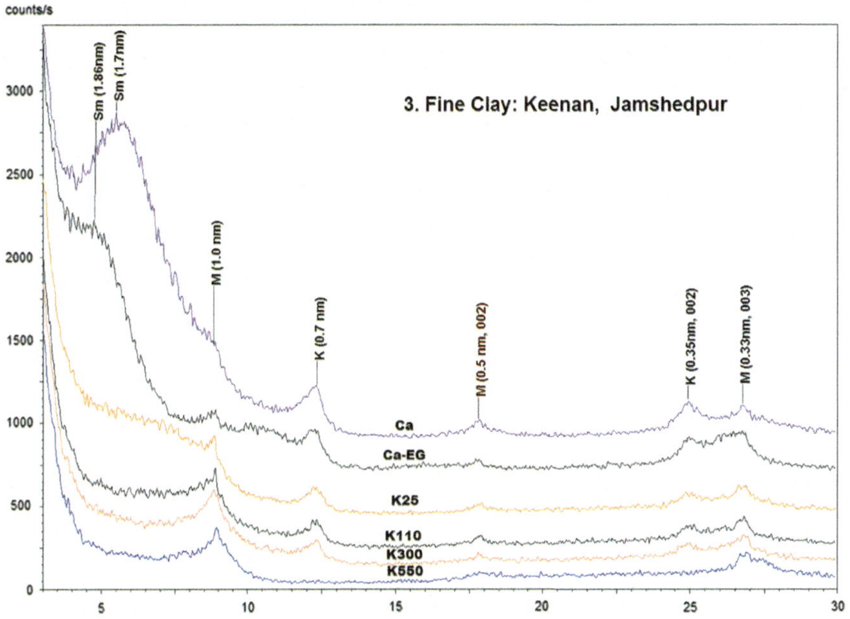

Fig. 2.16 Representative XRD diagrams of Ca-saturated and glycolated Fine Clay (CaEg) of representative cricket pitch soil of Keenan Stadium, Jamshedpur Jharkhand. Sm = Smectite, M = Mica, K = Kaolinite, K25/110/300/550 = K-saturated and heated at 25, 110, 300 and 550 °C

But it is detected by the reinforcement of the 1.0 nm peak of mica when K-saturated samples are heated at 110 and 300 °C. The observed 0.7-nm peak is not of pure kaolinite but is of kaolin, as evidenced by its broad peak, which does not collapse upon heating to 550 °C but reinforces the 1.00-nm mica peak. This suggests that the 0.7 nm peak is interstratified with 0.7–0.14 nm minerals. The 1.4 nm mineral is smectite, which is hydroxyl-interlayered, because the 1.00 nm peak after heating to 550 °C shows tailing down the low-angle side of the 1.00 nm peak (Figs. 2.13, 2.14 and 2.15).

The presence of HIS, mica, vermiculite, and kaolinite in the cracking clay soils of Jharkhand is almost similar to the vertisols developed in the alluvium of the Deccan basalt areas of the peninsula (Bhattacharya et al. 1993, 1997; Pal et al. 1989, 2012a) and extra-peninsular region (Ray et al. 2006).

Semi-quantitative estimates (Tables 2.3 and 2.4) of clay minerals in the fine clay of the Eden Gardens, Balangir (Odisha), JSCA (Ranchi) shows calcium dominant smectite as the predominant mineral followed by moderate to high levels of and mica and kaolin mineral levels in Keenan, Jamshedpur pitch soil sample (Table 2.5).

Table 2.4 Semi-quantitative estimates of clay minerals for cricket pitch soils of Eden Garden, Balangir (Odisha), JSCA (Ranchi), and Keenan stadium (Jamshedpur)

	Relative	Percentage			
		Smectite	Mica	Kaolinite	Remarks
Eden Garden, Kolkata	FC	60.3	22.5	17.2	The dominant mineralogy is smectite with moderate proportion of mica and kaolin
	TC	40.8	28.2	31	
Bolangir, Odisha	FC	76.8	11.6	11.6	The dominant mineralogy is smectite with less proportion of mica and kaolin
	TC	89.4	0	10.6	
JSCA, Ranchi	FC	72.8	12	15.2	The dominant mineralogy is smectite with moderate proportion of mica and kaolin
	TC	44.7	30.8	21.2	
Keenan, Jamshedpur	FC	61.6	16	22.4	The dominant mineralogy is smectite with high proportion of mica and kaolin
	TC	27	26.9	45.6	

Source ICAR-NBSS-LUP, Amravati Road, Nagpur, 2016
FC fine clay, *TC* TOTAL CLAY

Table 2.5 Exchangeable bases of the pitch samples

S.No.	Sample name	Exchangeable bases (meq/100 g of soil)				Sum
		Ca	Mg	Na	K	
1	Eden Garden	22.68	8.13	2.30	0.57	33.69
2	Bolangir	37.58	7.10	2.99	0.38	48.05
3	Keenan	13.66	8.70	0.82	0.34	23.52
4	JSCA	22.92	4.80	1.09	0.57	29.38

Source ICAR-NBSS-LUP, Amravati Road, Nagpur, 2016

2.7.1.3 Cation or the Base Exchange Capacity

Cation or the Base Exchange capacity method has been commonly used to estimate shrink swell potential of soils. The capacity of soils to adsorb and exchange ions (cations and anions) varies greatly with the type, amount of clay and the organic matter. CEC is defined as the amount of cations species at pH 7 and is expressed as millieqivalents/100 g or as $cmol_c/kg$ (Figs. 2.17, 2.18 and 2.19; Table 2.6).

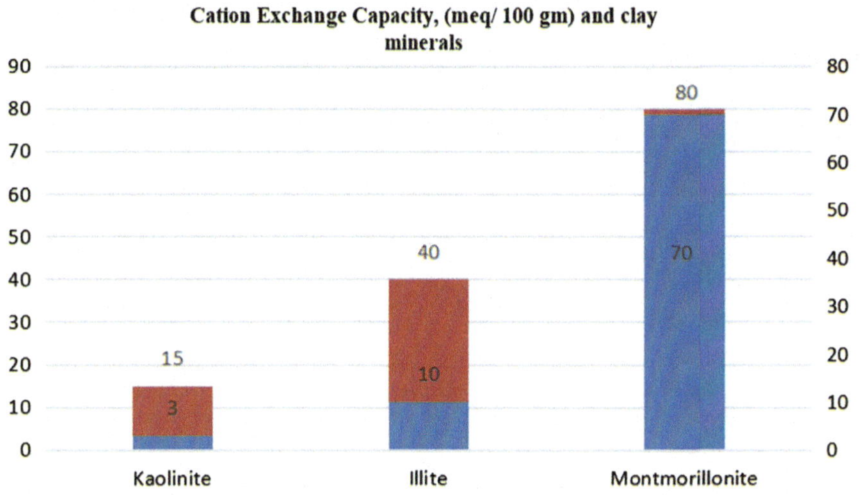

Fig. 2.17 Identification of clay mineral type based on the cation exchange capacity values—properties of important clay minerals, Woodward-Cyde and Associates, 1967

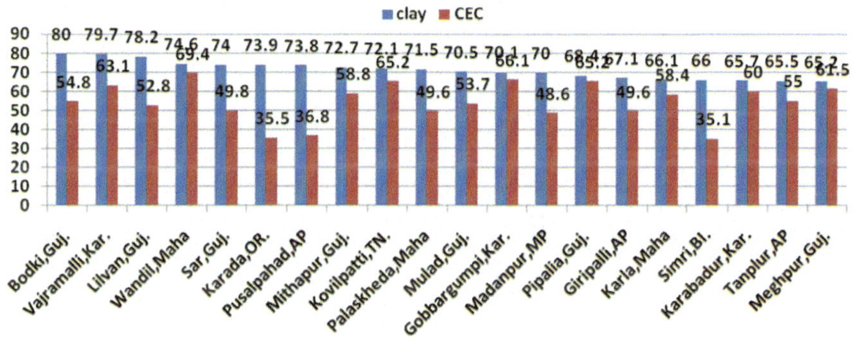

Fig. 2.18 Selected bench marked pitch soil samples with high clay contents (in descending orders) and having moderate to high CEC (35–69.5) values showing the presence of moderate expanding illites to high swelling and shrinking smectite clay minerals

2.7.1.4 The CEC/Clay Content Ratio

The CEC/Clay content ratio has been used as an auxillary parameter to assess clay mineralogy and is used for clay activity classes in Soil Taxonomy (Soil Survey Staff, 1996) (Table 2.7).

The current practice of the compositional analysis of clay phase of fine-grained soils includes XRD, DTA, SEM, Infrared Spectroscopy and the like, which require highly sophisticated instrumentation (Mitchell and Soga 2005). In addition, these methods involve tedious and time-consuming processes like soil sample preparation, conducting tests under specified conditions, and judicious interpretation of the test

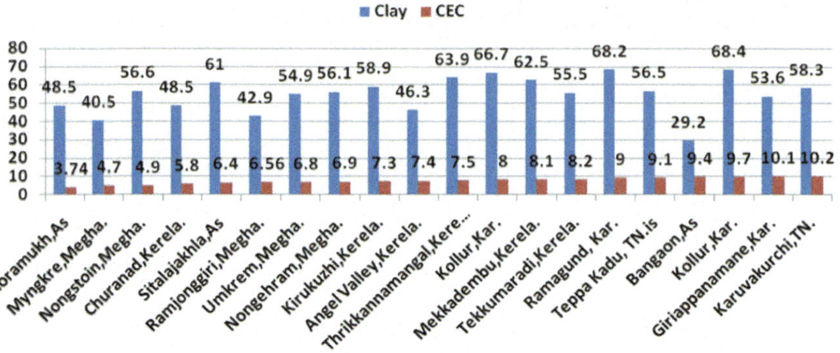

Fig. 2.19 Selected bench marked pitch soil samples with high clay contents but having low CEC (less than 15) values showing the presence of non swelling and shrinking kaolinite clay minerals

Table 2.6 Soil texture and their corresponding CEC values

Soil texture	CEC ($cmol_c kg^{-1}$ of soil)
Sand—light color	3–5
Sand—dark color	10–20
Loams	10–15
Silt loams	15–25
Clay and clay loams	20–50
Organic soils	50–100

Table 2.7 The relationship between the CEC, clay content and clay activity classes

CEC/clay content ratio	Clay activity classes
> 0.7	Montmorilonite or vermiculite
0.3–0.7	Mixed or illite or chlorite
< 0.3	Kaolinite or hallosite

Source Summer E. Malcolm, Olson, C. G., Thompson, M. L., Wilson, M. A., Handbook of Soil Science, CRC Press, Baca Raton, London, pp. 112–113

results. Most of the laboratories cannot be expected to have such a high level of expensive, sophisticated instrumentation and interpretation technology. There are numerous simple and accurate direct or indirect criteria available to characterise soil expansibility via index properties, which can be used as an indicator of soil mineralogy (Sridharan and Prakash 2000).

2.7.2 *Direct Methods*

Direct methods are the most accurate way to determine the predominant minerals present in a given pitch soil sample and are preferably carried out on representative undisturbed samples. These are typically carried out by measuring the swelling pressure and volume change of the soil, as well as performing a free swell test on disturbed samples. Several laboratory methods have been developed to directly determine the swelling of a soil as its moisture content changes. Pitch soils shrink and swell because the moisture content changes from dry to moist and vice versa. Direct measurement includes the free swell index (FSI), free swell ratio (FSR), expansion index, California bearing ratio (CBR), potential volume change (PVC), and coefficient of linear extensibility (COLE).

A. **Free Swell Index (FSI)** are the most commonly used test used to identify the expansive clay and predict the swelling potential (Holtz and Gibbs 1956). 10 cm^3 dry soil passing the No. 40 sieve (< 425 μm) is transferred to a graduated cylinder and filled with water. The free swell is determined by comparing the initial volume with the final volume. IS: 2720 (Part 40) 1977 gives a criterion to predict the expansivity of soils, based on their Free swell Index (FSI)

$$\mathrm{FSI} = V_d - V_k / V_k @ 100$$

where V_d is the sediment volume of 10 g of oven dried soil passing through a 425 μm Indian Standrad (IS) sieve placed in a 100 ml graduated measuring jar containing distilled water, and V_k is the sediment volume of a 10 g of oven dried soil passing through a 425 μm IS sieve placed in a 100 ml graduated measuring jar containing kerosene (Figs. 2.20, 2.21 and 2.22).

Fig. 2.20 The classification of swell-shrink soils based on free swell index

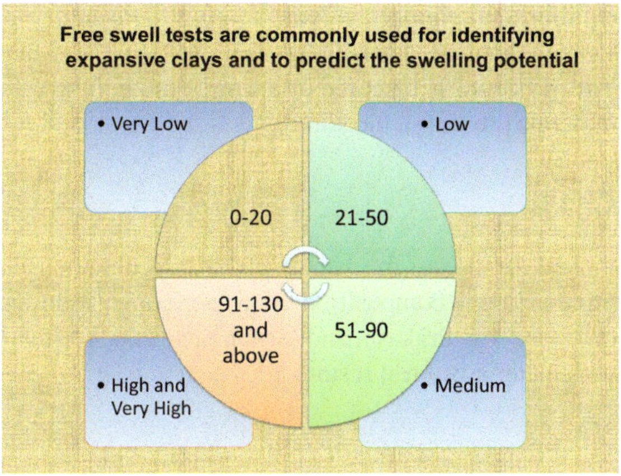

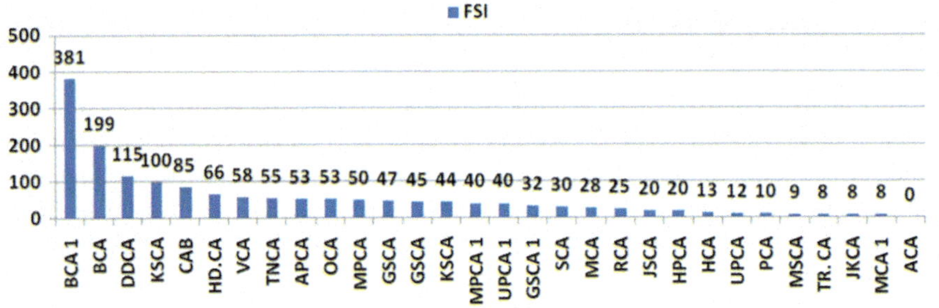

Fig. 2.21 The free swell index (FSI) of some of the important cricket pitches of India. *Data source* Singh (2007)

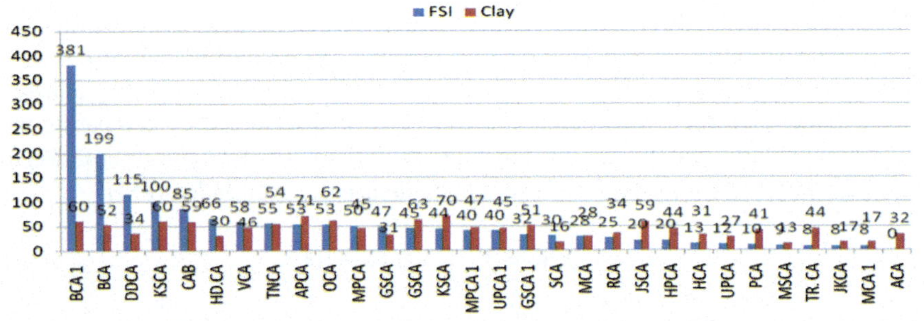

Fig. 2.22 FSI (in descending order) to Clay for some of the important cricket pitch soils of India. *Data source* Singh (2007)

The sediment behavior of montmorillonitic and kaolinitic soils in distilled water and kerosene or carbon tetra chloride (a non-polar fluid) are of opposite nature. This can be effectively used for classifying the degree of expansibility of natural fine-grained soils and also the mineralogical composition of such soils. Prakash and Sridharan (2004) have defined a term Free Swell Ratio (FSR) as the ratio of equilibrium sediment volume of 10 g of oven dry soil passing through a 425 μm sieve in distilled water (V_d) to that in CCL_4 (V_k) after an equilibrium period of 24 h. This process has emerged as a simple methodology to predict the predominant clay minerals present in the soil.

$$\mathrm{FSR} = V_d / V_k$$

Making use of the values of FSR so obtained, the degree of expansibility and mineralogical composition of fine-grained soils can obtained by the guidelines proposed by Prakash and Sridharan (2004) after validating the same with the oedometer swell test results.

Table 2.8 Classification and identification of soils based on FSR (Prakash and Sridharan 2004)

FSR	Clay type	Soil expansivity	Dominant clay
			Mineral type
< 1	Non-swelling	Negligible	Kaolinitic
1–1.5	Mixture of swelling and non-swelling	Low	Mixture of kaolinitic and montmorillonitic
1.5–2	Swelling	Moderate	Montmorillonitic
2.0–4.0	Swelling	High	Montmorillonitic
> 4	Swelling	Very high	Montmorillonitic

B. **Expansion index (EI)**—Soil passing through No-4 sieve is compacted into a consolidometer ring and a 6.9 kPa surcharge is applied. Volume change is measured over a 24 h period. EI = 100^*h* @ F, where *h* is percent swell and F is the fraction passing the No-4 sieve. EI values less than 50 are considered to have low expansion potential, and EI > 91 Have the high expansion potential.

Table 2.9 Expansion Index range and expansion potential of soils

Expansion index	Expansion potential
0–20	Very low
21–50	Low
51–90	Medium
91–130	High
> 130	Very high

Source American Society for Testing and Materials (ASTM)

C. California Bearing Ratio (CBR)—vertical swell is determined by measuring the Penetration resistance. Soils are compacted into CBR cylinders and soaked in water for several days. The samples are drained and the penetration test is performed.

D. The coefficient of linear extensibility (COLE) is a method to characterise the swelling and shrinking potential of expansive soils. It is a measure of the change in sample dimensions from the moist to the dry state and is estimated from the bulk density of the clod at a suction of 5 psi and oven-dry moisture conditions. The COLE is an important engineering property of soils that relates to the content of undisturbed samples and consistency to characterise soil swell-shrink behaviour (Franzmeir and Ros 1968; Grossman et al. 1968). Cole is an indicator of clay mineralogy. According to Greene-Kelly (1974), the increase in smectite and clay content increases the COLE values and shrink-swell potential of the soil.

Table 2.10 Range of COLE values to determine soil swell-shrink potential

COLE value range	Swell-shrink potential
< 0.03	Low
0.03–0.06	Moderate
0.06–0.09	High
> 0.09	Very high

Thomas et al. (2000), Vanght et al. (2006)

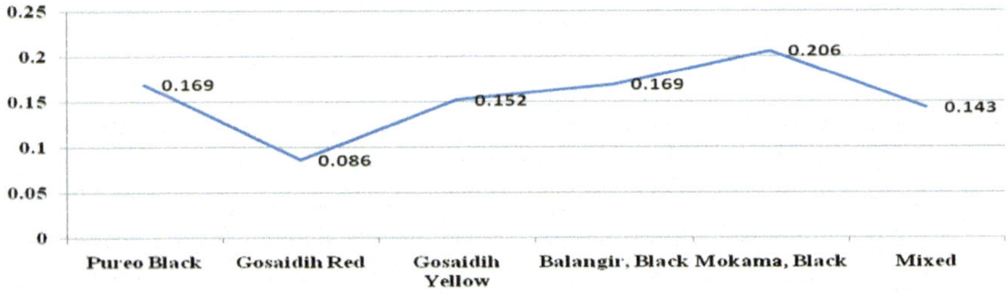

Fig. 2.23 COLE vale of different pitch soils used at the JSCA Cricket Stadium, Complex, Ranchi, Jharkhand. *Data source* ICAR National Bureau of Soil Science (N.B.S.S.), Nagpur

From Fig. 2.23 pitch soil samples from Pureo black, Ranchi, Balangir black from Odisha and Mokama black soil sample from Bihar having COLE value of 0.169, 0.152 and 0.206 shows the smectite as predominant minerals (as inferred from Table 2.8 also) and illite/kaolinite as a predominant clay minerals in Gosaidih red and yellow soil samples and mixed soil sample having COLE value as 0.86, 0.152 and 0.143.

2.7.3 *Indirect Methods*

Indirect Methods are the simplest method for identifying expansive clay soils. Indirect methods comprise determination and analysis of particle size, Atterberg limits and clay Activity (Fig. 2.24).

Fig. 2.24 Soil clay minerals based and their specific surface area

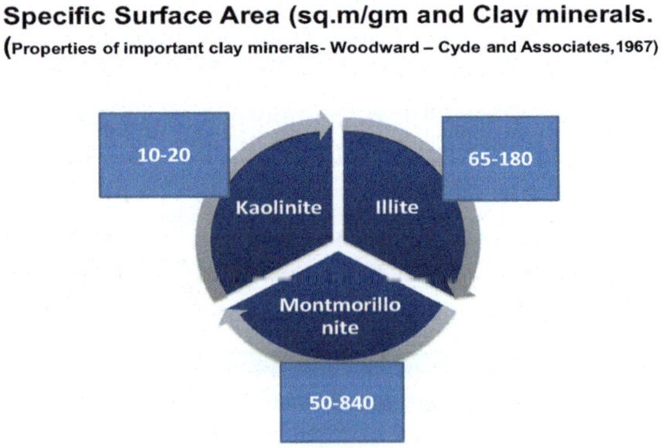

2.7.3.1 Classification and Analysis of Clay Mineral Soils Based on Atterberg Limits

Borchardt (1989), Davidson and Page; Ross and Franzmeir (1968), Haiek and Karathanasis (1985), Singer and Schafer (1976) identified smectite content as the soil property controlling shrinking and swelling behavior. The key to all expansive soil classification and analysis is the method of measuring swell potential, which may be measured directly in a swell test or indirectly determined by correlation with other tests. Swedish scientist Albert Atterberg (1911) originally defined six limits of consistency, but Casagrande borrowed only three limits, namely the liquid limit, the plastic limit, and the shrinkage limit for geotechnical applications. Among them, only liquid limits and plastic limits are used for the classification of fine-grained soils. At least two parameters, namely liquid and plastic limits, are required to define the plasticity of clays. Fine-grained soils can exist in any of several states. Each state depends on the amount of water in the soil system.

Liquid limit (LL) is the percent of water content at which the soil has such small shear strength that it flows to a close grove of standard width when jarred in a specified manner. At the moisture content of LL, the distance between the particles is such that the force of interaction between particles is sufficiently weak to allow easy movement of particles relative to each other. It is dependent on the amount and type of clay.

It can be noticed from Table 2.11 that soils with the highest level of LL contain high levels of smectite clays and montmorillonite minerals, especially those specially saturated with sodium. On the other hand, kaolinite and illites generally have low LL (Figs. 2.25 and 2.26).

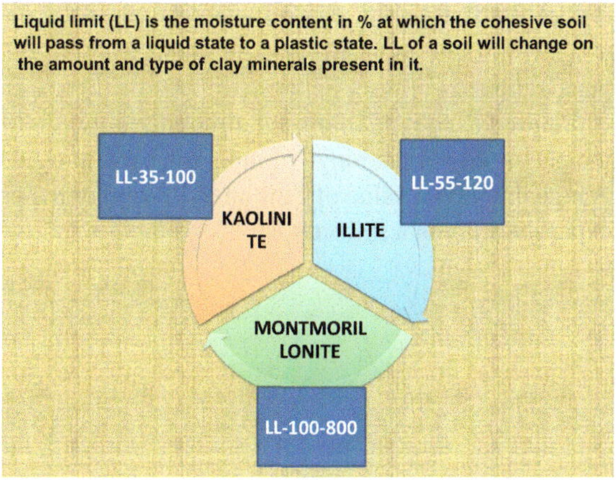

Fig. 2.25 The classification of type of clay minerals based on the liquid limit range

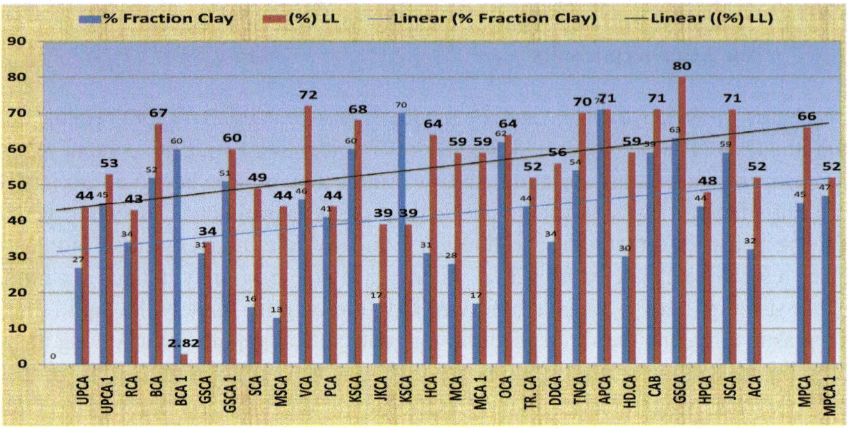

Fig. 2.26 The clay versus liquid limit range of some of the important cricket pitch soils of India. *Data source* Singh (2007)

It is interesting to mention that the mechanisms controlling the liquid limit of kaolinite and montmorillonitic soils are entirely different from each other. While the liquid limit of a kaolinite soil is controlled by the particle arrangement and inter-particle attractive forces, that of a montmorillonitic soil is controlled by the double layer thickness. Hence, higher liquid limits do not necessarily mean expansive montmoriollonitic soils. This is the reason why the liquid limit and the related index properties, without consideration of clay mineralogy, cannot satisfactorily indicate soil expansibility. Instead, they may present an entirely different and incorrect picture.

The "plastic limit" (PL) is the moisture content at which the soil stiffens from a plastic state to a semi-arid and friable state. It is the amount of moisture in a soil that allows it to be rolled in a 3 mm-diameter thread without breaking. It is thus at the lower end of the range of water content at which a clayey soil is in a plastic state (Hillel 1980). Further, soils with a high specific surface area (SSA) can retain water at a much lower moisture content than kaolinite; illite can be attributed to a lower plastic limit.

Plasticity Index (PI)—is the difference between the LL and the PL and indicates the magnitude of the range of moisture content over which the soil remains plastic. A high numerical value of PI is an indication of the presence of a high percentage of expansive clay in a soil sample.

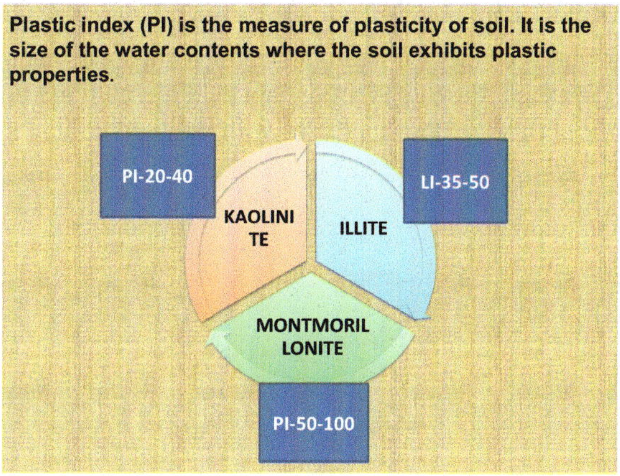

Fig. 2.27 The classification of type of clay minerals based on the PI—Plasticity Index range

It is also observed from Table 2.9 and Fig. 2.27 that high numerical value of PI can be attributed to presence of high percentage of expansive clay in the pitch soil. In this context, range of PI for different minerals present in the soil is pictured in Fig. 2.28.

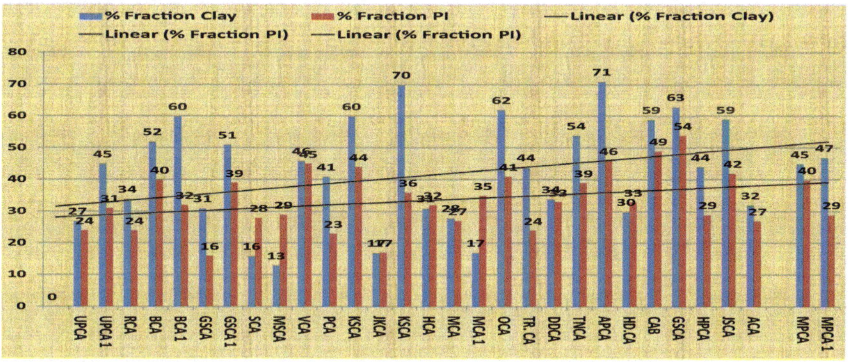

Fig. 2.28 Relationship between clay and plasticity index range of some of the important cricket pitch soils of India. *Data source* Singh (2007)

It can be observed in this figure that more PI values may be attributed to the presence of montmorillonite in such pitch soil as compared to other minerals, namely kaolinite and illite.

Shrinkage Limit (SL)—the maximum water content below which a decrease in water will not result in a decrease in soil mass volume. If a mass of soft, saturated soil gradually dries, its volume is reduced by an amount equivalent to the volume of water lost by evapotranspiration. This volume change is caused by capillary forces acting on the surface of the soil mass. At a certain water content, the sum of the capillary forces reaches its greatest value and the volume change ceases. On further drying, the water begins to draw into the interior of the soil, whose colour then changes from dark to light. The surface of the desiccating soil shows a characteristic pattern of shrinkage cracks. Finer the size of particles of pitch soil, greater is the amount of shrinkage. Based on the SL data in Table 2.11 and Fig. 2.29, it can be observed that the shrinkage limit increases with the type of clay mineral present in a soil (Table 2.11).

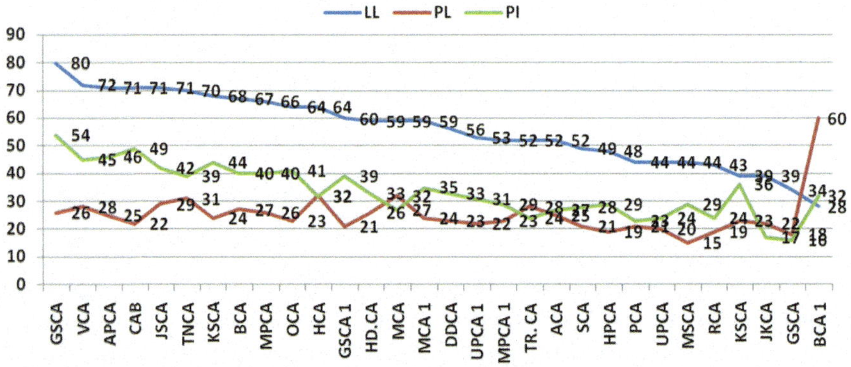

Fig. 2.29 Atterberg limits for some of the important cricket pitch soils of India

Table 2.11 Atterberg limits of some clay minerals

Mineral	Exchangeable	LL	PL	PI	SL
	Cations	%	%	%	%
Montmorillonite	Na	710	54	656	9.9
	K	660	98	562	9.3
	Ca	510	81	429	10.5
	Mg	410	60	350	14.7
	Fe	290	75	215	10.3
Illite	Na	120	53	67	15.4
	K	120	60	60	17.5
	Ca	100	45	55	16.8
	Mg	95	46	49	14.7
	Fe	110	48	61	15.3
Kaolinite	Na	53	32	21	26.8
	K	49	29	20	
	Ca	38	27	11	24.5
	Mg	54	31	23	28.7
	Fe	56	35	21	

Data Source Cornell (1951), *LL* liquid limit, *PL* plastic limit, *PI* plasticity index, *SL* shrinkage limit

2.7.3.2 Clay Activity

Skempton (1953) was the first to observe that the plasticity of soil is related to the amount of clay fraction present in the soil. Some major activities (such as plasticity and volume change) and the resulting behaviour of soils are thought to be directly related to the amount of colloidal particles in soils (Nelson and Miller 1992). The amount of water adsorbed in a soil is greatly influenced by the amount of clay that is present in the soil. The plasticity index of a soil is a function of both the type of clay mineral and the amount of clay mineral present in the soil (Skempton 1953). For this reason, Atterberg limits (Fig. 2.30), clay content, and pH have been combined into a single parameter called the Activity Ratio (A), developed by Skempton (1953). The activity of clay is defined as (Fig. 2.30):

$$\text{Activity}\ (A) = \text{Plasticity Index}/\text{Clay content}\ (\%)$$

where, PI = Plasticity Index, Skempton suggested three classes of clays according to clay Activity as

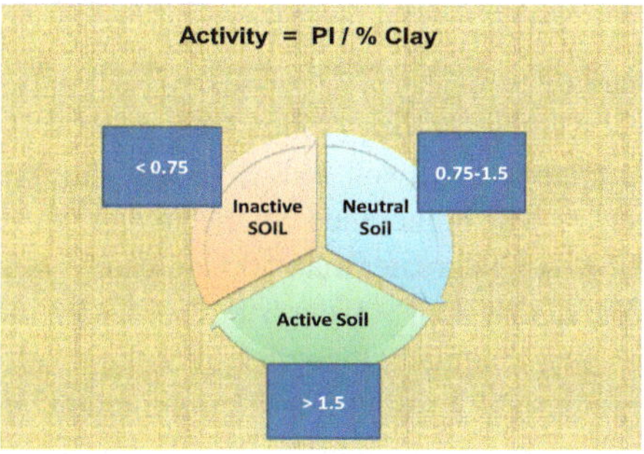

Fig. 2.30 Activity range of some clay minerals

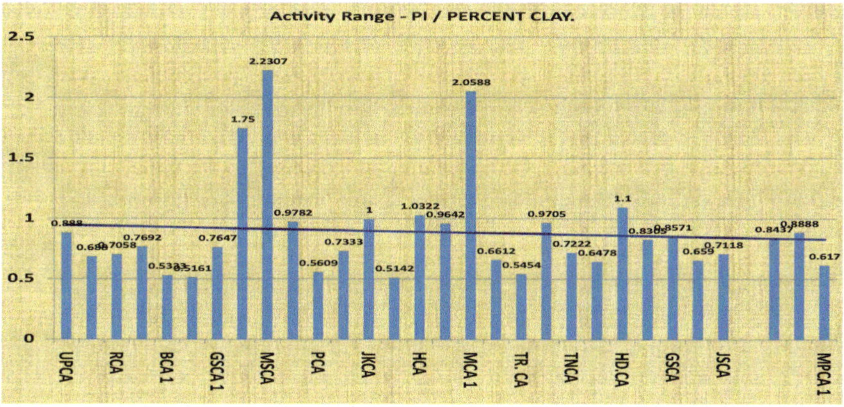

Fig. 2.31 Activity range of some of the important cricket pitch soils of India. *Data source* Singh (2007)

Based on variations in clay activity, pronounced clay-like behaviours (i.e., higher compressibility, swelling, and shrinking characteristics) have been observed in soils with higher activity values. A calcium-saturated montmorillonite has an estimated activity value of 1.5, whereas a sodium-saturated montmorillonite has an activity value greater than 7. Furthermore, Kaolinite has been estimated to exhibit a value of about 0.3–0.46 as compared to Illite, which has an activity value of about 0.9–1.

With reference to the above variation in clay activity, several physical, chemical, and mineralogical soil properties have been reported by researchers in the past to influence the shrink-swell behaviour of soils, with no one accurately predicting shrink-swell potential for all soil types. Index properties such as liquid limit, plasticity index, and related parameters cannot satisfactorily predict the soil swelling and shrinking behaviour as they do not consider the effect of clay mineralogy. The free soil ratio (FSR) method, in addition to predicting the soil expansivity more realistically and satisfactorily, gives additional information about the nature of the predominant clay mineralogy of soils.

Based on the direct and indirect methods of identification of predominant clay minerals, the cricket pitch soils should have a high cation exchange capacity (CEC), a high specific surface area (SSA), a high free swell ratio, a high coefficient of linear extensibilty, a high liquid limit (LL) and plasticity index (PL), and a high clay activity range.

Another property either Expansive Soil Index (ESI) which is the sum of swelling 2:1 minerals, Swell Index, Liquid Limits and CEC gives ESI ratings for each soil series. The higher the ESI, the greater the swelling and shrinking potential of the soil. ESI ratings greater than 500 indicate very high swelling and shrinkage potential. An ESI of less than 500 describes soils with moderate to high swell and shrink potential.

Soil scientists recognise that the swelling and shrinking behaviour can be best predicted by examining a combination of morphological, physical, chemical, and mineralogical properties. Field identification of pitch soil should only be used as a preliminary test so as to accept or reject the pitch soil samples. Further, selected samples should be analysed for their morphological, physical, chemical, mineralogical, and engineering properties so as to attain the desired pitch surface behavior. Before final pitch soil selection, the entire interrelated mineralogical, direct and indirect tests should be combined, analyzed, and interpreted.

Ca^{+2}/Mg^{+2} saturated fine smectite dominant fine clays with proportionate silt and sand ratios, high FSI or FSR, COLE (via direct methods), LL, PI, CEC, clay activity, and low SL (via indirect methods) are the desired index properties for cricket pitch soils.

Table 2.12 Comparative analysis of mineralogical (XRD, CEC), direct (FSI, FSR, COLE) and indirect methods (Atterberg limits, clay Activity, ESI ratings) to distinguish between the available predominant swelling or non-swelling clay minerals

Mineral type	A. Mineralogical	Methods		B. Direct	Methods	C. Indirect	Methods				
Dominant clay mineral	XRD	CEC	CEC/Clay	FSR	COLE	Atterberg	limits			Activity	ESI ratings
Type and expansibility			% ratio	< 1	< 0.05	LL	PL	PI	SL		
Kaolinite, non-swelling	7.5	3–15	< 0.3			35–100		20–40		< 0.75	
Illite (Mica), mixed				1–1.5	0.05–0.15	55–120		35–50		0.75–1.5	< 500
limited swelling	9.9–10.1	10–40	0.3–0.7								
Montmorillonite/ Smectite	14–18	70–80	> 0.70		> 0.15	100–800		50–100		> 1.5	> 500
Swelling-Moderate				1.5–2							
High				2–4							
Very high				> 4							

XRD X-ray diffraction measured in d-spacing, Amstrong (Å), CEC—Cation exchange capacity measured in mill equivalents/100 g soil, Woodward Cyde and Associates, 1967, FSR—Free swell ratio is equilibrium sediment volume of 10 g of oven dried soil passing a 425 μ sieve in distilled water to that in carbon tetra chloride or kerosene, Atterberg limits in percentage, Clay Activity is ratio of plasticity index/clay content % developed by Skempton in 1953, ESI ratings—sum of swelling 2:1 minerals, FSI, LL and CEC

2.8 Case Study 2

The primary goal of this case study is to determine the quantitative and qualitative nature of the dominant clay minerals using direct (FSI/COLE values) and indirect (Atterberg limits) methods, as well as a critical analysis of the physical, chemical, mineralogical, and engineering properties of nearly 76 pitch soil samples used to host National and International cricket matches in India.

In 2007, the Board of Control for Cricket in India (BCCI), all state cricket associations, and IIT, Bombay collaborated to examine the physical, chemical, and engineering properties of nearly 76 cricket pitch soils hosting domestic and international matches in India. These laboratory data have been considered a primary data source to analyse and interpret the predominant clay minerals present in cricket pitch soils (through the various direct and indirect methods discussed above). Data is annexed to this book for a better understanding of the full physical, chemical, and engineering properties (Figs. 2.32, 2.33, 2.34, 2.35, 2.36, 2.37, and 2.38).

Table 2.13 Physical, chemical and Atterberg limits for some of the important cricket pitch soils of India

S.no.	Asson	Location	Clay	Silt	Sand	OM	Color	USCS	LL	PL	PI	FSI	pH
1	UPCA	Lucknow	27	64	9	0	Grey	CL	44	20	24	12	7.13
2	UPCA 1	Kamla Club Ground, Kanpur	45	53	2	4	Light brown	CH	53	22	31	40	6.82
3	RCA	SMS Stadium, Rajasthan, SMS	34	49	17	2	Grey-light brown	CL	43	19	24	25	7.09
4	BCA	Motibaug, Baroda	52	38	10	5	Red	CH	67	27	40	199	7.14
5	BCA 1	Alembic, Baroda	60	32	8	2	Red	CH	28	60	32	381	7.86
6	GSCA	Sardar Patel Gujarat Stadium, Motera	31	40	29	2	Red	CL	34	18	16	47	7.3
7	GSCA 1	Sardar Patel Gujarat Stadium, Motera	51	48	1	1	Brown	CH	60	21	39	32	6.94
8	SCA	SCA Ground, Khanderi, Saurashtra	16	49	35	4	Dark Brown	CH	49	21	28	30	7.28
9	MSCA	Subrata Roy Sahara Stadium, Gahunje, Pune	13	45	42	0	Dark brown	CL	44	15	29	9	7.28
10	VCA	Nagpur, Maharashtra	46	34	20	2	Red-dark brown	CH	72	28	45	58	6.44
11	PCA	Mohali	41	49	10	5	Grey	CL	44	21	23	10	7.56
12	KSCA	M. Chinnaswamy Stadium, Bangalore	60	37	3	6	Brown	CH	68	24	44	100	7.91
13	JKCA	Sonwar, Srinagar	17	79	4	4	Light brown	CL	39	22	17	8	7.45
14	KSCA	Jawaharlal Nehru International Stadium, Kochi	70	25	5	1	Brown	CH	39	23	36	44	6.16
15	HCA	Bhiwani, Haryana	31	68	1	0	Yellow-brown	CH	64	32	32	13	7.12
16	MCA	Wankhade Stadium, Mumbai	28	52	20	2	Red	MH	59	32	27	28	5.76
17	MCA 1	Bandra Kurla Complex Stadium Mumbai	17	67	16	1	Red	CH	59	24	35	8	6.53

(continued)

Table 2.13 (continued)

S.no.	Asson	Location	Clay	Silt	Sand	OM	Color	USCS	LL	PL	PI	FSI	pH
18	OCA	(Bhubhaneshwar)	62	37	1	0	Light brown	CH	64	23	41	53	6.62
19	TR.CA	PTG Cricket Ground, Agartala	44	49	7	1	Grey-Brown	CH	52	28	24	8	5.24
20	DDCA	Ferozshah Kotla Ground, Delhi	34	61	5	1	Grey	CH	56	23	33	115	7.75
21	TNCA	MA Chidambaram Stadium, Chennai (Black Clay)	54	38	8	4	Dark Brown	CH	70	31	39	55	7.07
22	APCA	YSR-ACA-VDCA Stadium, Vishakapattnam, Andhra Pradesh	71	14	15	9	Dark grey	CH	71	25	46	53	6.99
23	HDCA	Rajiv Gandhi International Stadium Hyderabad	30	40	30	7	Brown	CH	59	26	33	66	6.8
24	CAB	CAB	59	41	0	3	Grey	CH	71	22	49	85	6.49
25	GSCA	Porvorim Ground, Goa	63	26	11	6	Dark grey	CH	80	26	54	45	6.86
26	HPCA	HPCA Cricket Stadium, Dharmshala	44	50	6	4	Grey-light brown	CL	48	19	29	20	6.94
27	JSCA	JSCA International Cricket Stadium, Ranchi	59	34	7	5	Brown	CH	71	29	42	20	6.09
28	ACA	Nehru stadium (Soil-2)	32	64	4	5	Light brown	CH	52	25	27	0	6.49
29	MPCA	Holkar Stadium, Indore	45	50	5	1	Brown	CH	66	26	40	50	7.63
30	MPCA 1	Capt. Roop Singh Stadium, Gwalior	47	44	9	1	Brown	CH	52	23	29	40	7.43

Data source Singh (2007)
Clay less than 2 μ; Silt 2–75 μ; Sand 0.075–4.75 mm; OM organic matter; Cl clay with low plasticity; CH clay with high plasticity; LL liquid limit; PL plastic limit; PIplasticity index; FSI free swell index

Table 2.14 The particle size distribution for some of the important cricket pitch soils of India: in descending order of clay percentage

S.No.	Association	Clay	Silt	Sand	S.No.	Association	Clay	Silt	Sand
1	APCA	71	14	15	16	TR.CA	44	49	7
2	KSCA	70	25	5	17	HPCA	44	50	6
3	GSCA	63	26	11	18	PCA	41	49	10
4	OCA	62	37	1	19	RCA	34	49	17
5	BCA 1	60	32	8	20	DDCA	34	61	5
6	KSCA	60	37	3	21	ACA	32	64	4
7	CAB	59	41	0	22	GSCA	31	40	29
8	JSCA	59	34	7	23	HCA	31	68	1
9	TNCA	54	38	8	24	HD.CA	30	40	30
10	BCA	52	38	10	25	MCA	28	52	20
11	GSCA 1	51	48	1	26	UPCA	27	64	9
12	MPCA 1	47	44	9	27	JKCA	17	79	4
13	VCA	46	34	20	28	MCA 1	17	67	16
14	UPCA 1	45	53	2	29	SCA	16	49	35
15	MPCA	45	50	5	30	MSCA	13	45	42

Data source Singh (2007)

Table 2.15 Fhysical and chemical composition of some of the test playing cricket centers in India

Association	% fraction			OM	Soil color	USCS	(%)			(%)	pH	EC
	Clay	Silt	Sand	(%)			LL	PL	PI			
	< 2 μm	2–75 μm	0.075–4.75 mm							FSI		us/cm
UPCA 1	45	53	2	4	Light brown	CH	53	22	31	40	6.82	240.3
GSCA	31	40	29	2	Red	CL	34	18	16	47	7.3	203.1
GSCA 1	51	48	1	1	Brown	CH	60	21	39	32	6.94	159.7
VCA	46	34	20	2	Red-dark brown	CH	72	28	45	58	6.44	112.1
PCA	41	49	10	5	Grey	CL	44	21	23	10	7.56	136.6
KSCA	60	37	3	6	Brown	CH	68	24	44	100	7.91	303.9
MCA	28	52	20	2	Red	MH	59	32	27	28	5.76	76.31
OCA	62	37	1	0	Light brown	CH	64	23	41	53	6.62	90.72
DDCA	34	61	5	1	Grey	CH	56	23	33	115	7.75	449.4
TNCA	54	38	8	4	Dark Brown	CH	70	31	39	55	7.07	303.4
HD.CA	30	40	30	7	Brown	CH	59	26	33	66	6.8	284.4
CAB	59	41	0	3	Grey	CH	71	22	49	85	6.49	422.2

Data source Singh (2007)

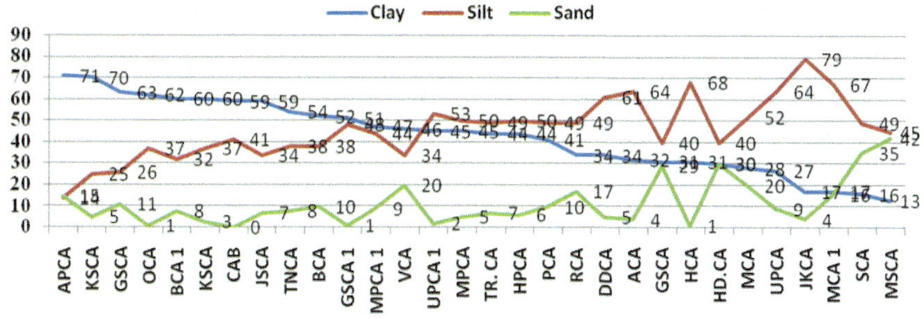

Fig. 2.32 Particle size distribution for some of the important cricket pitch soils of India

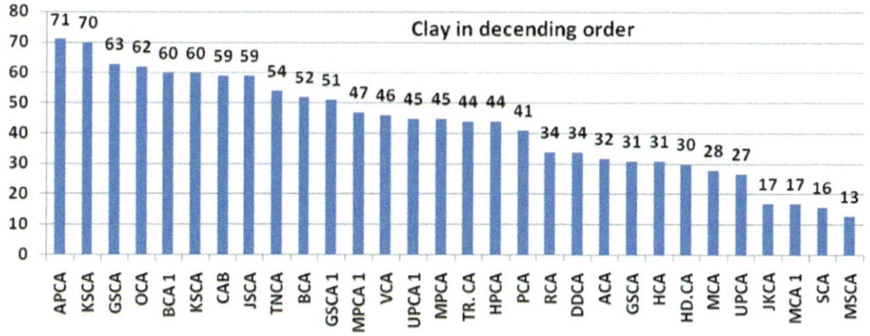

Fig. 2.33 Clay percentage for some of the important cricket pitch soils of India: in descending order of clay percentage

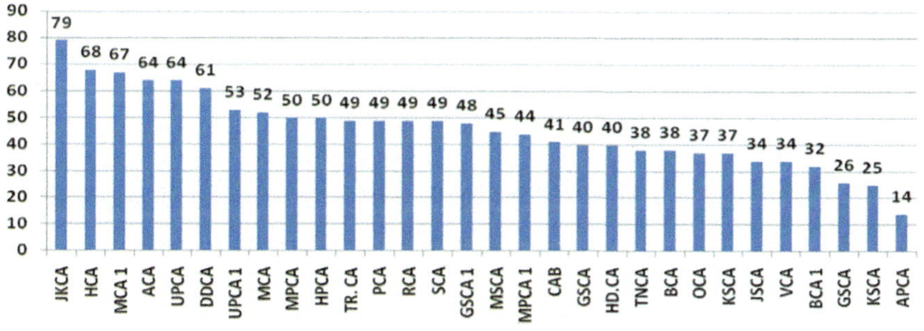

Fig. 2.34 Silt percentage for some of the important cricket pitch soils of India: in descending order of silt percentage

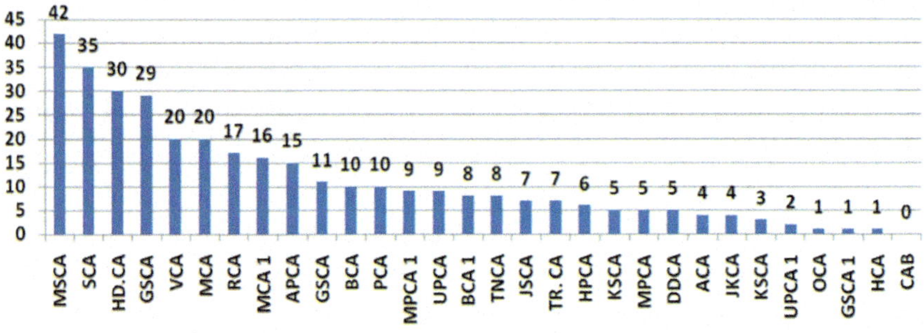

Fig. 2.35 Sand percentage for some of the important cricket pitch soils of India: in descending order of sand percentage

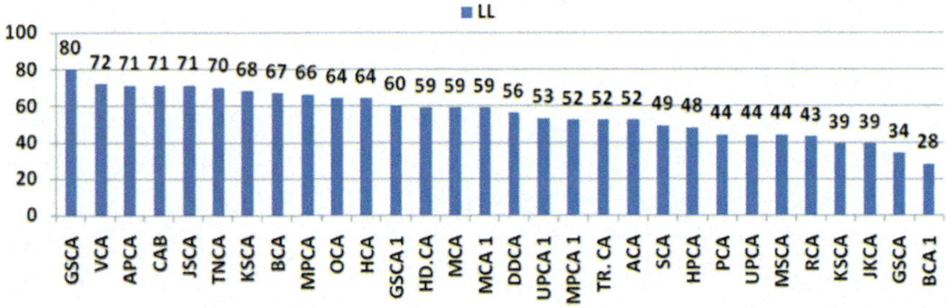

Fig. 2.36 Liquid limits (in descending orders) for some of the important cricket pitch soils of India

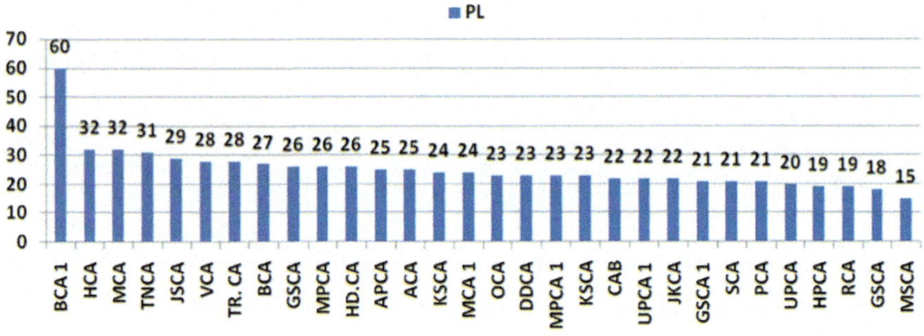

Fig. 2.37 Plastic limits (in descending orders) for some of the important cricket pitch soils of India

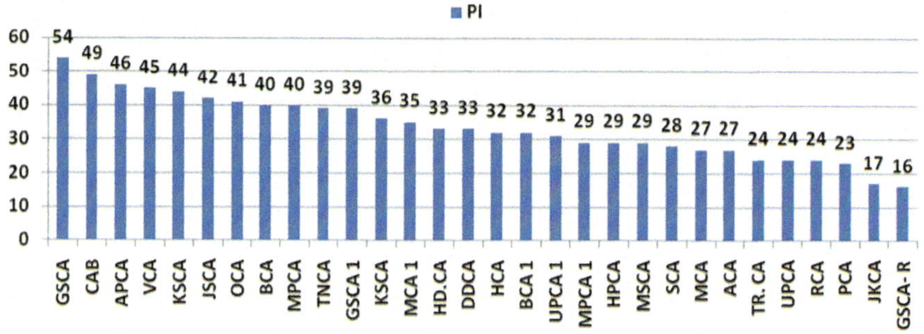

Fig. 2.38 Plasticity index (in descending orders) for some of the important cricket pitch soils of India

Analysis of 30 selected pitch soil samples examined by IIT-Bombay. *Data source* Singh (2007) (Figs. 2.39, 2.40, 2.41 and 2.47).

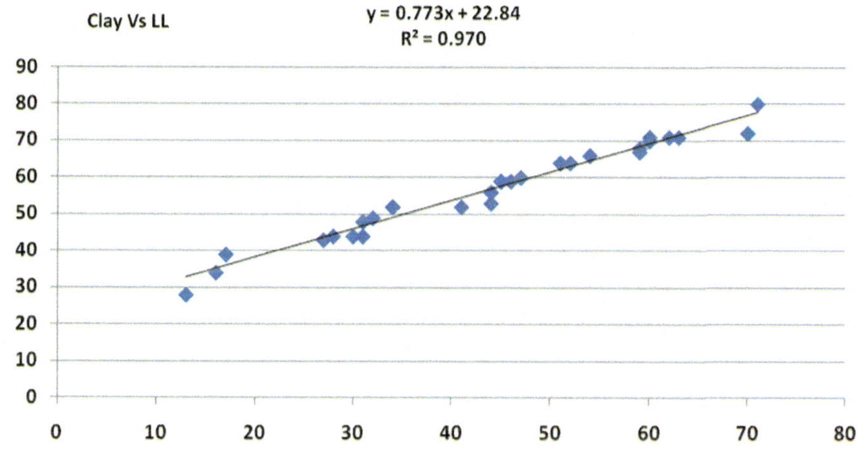

Fig. 2.39 Correlation coefficient and regression analysis of clay versus liquid limit for selected 30 benchmarked pitch soil samples

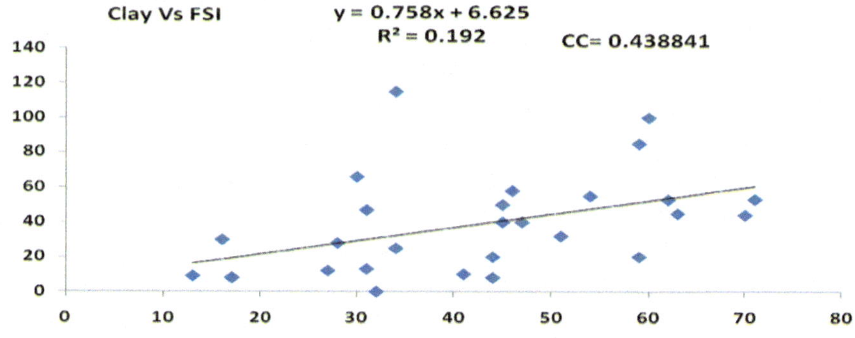

Fig. 2.40 Correlation coefficient and regression analysis of clay versus FSI for selected 30 benchmarked pitch soil samples

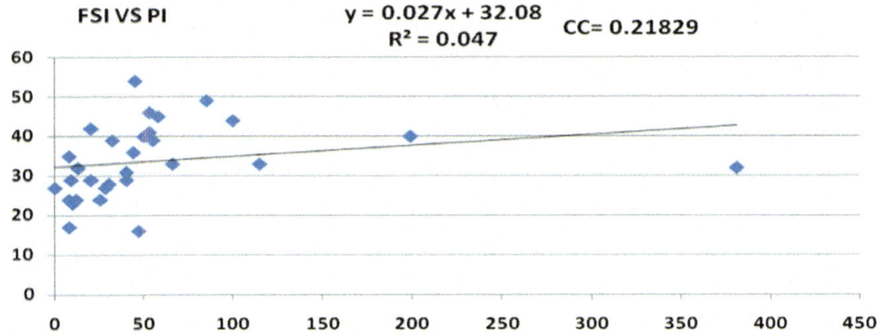

Fig. 2.41 Correlation coefficient and regression analysis of FSI versus PI for selected 30 benchmarked pitch soil samples

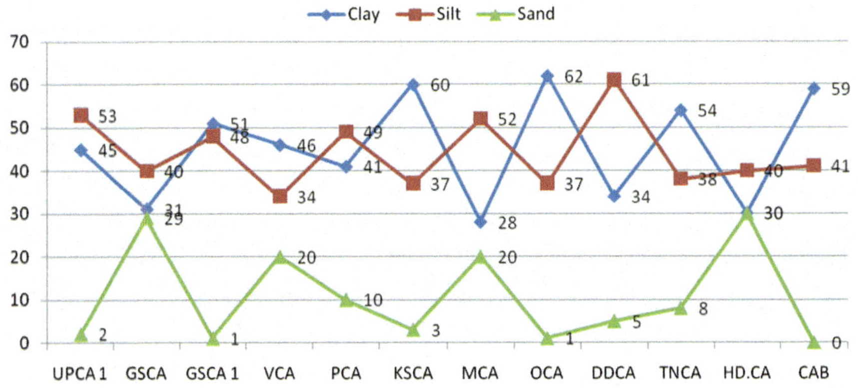

Fig. 2.42 The particle size distribution of some of the Test playing cricket centers in India. *Data source* Singh (2007)

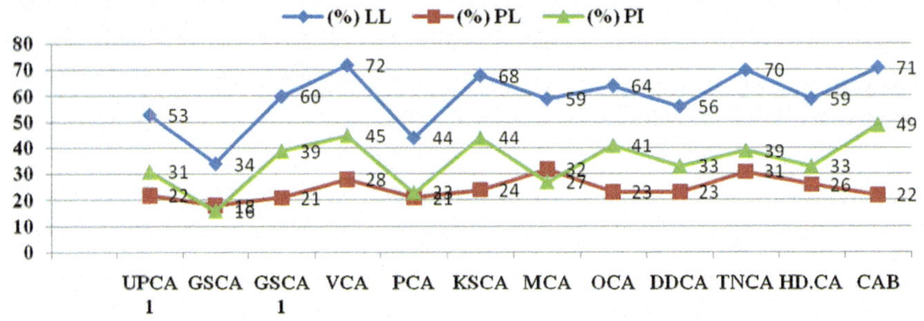

Fig. 2.43 The Atterberg limits (liquid limit, plastic limit and plasticity Index) of some of the Test playing cricket centers in India. *Data source* Singh (2007)

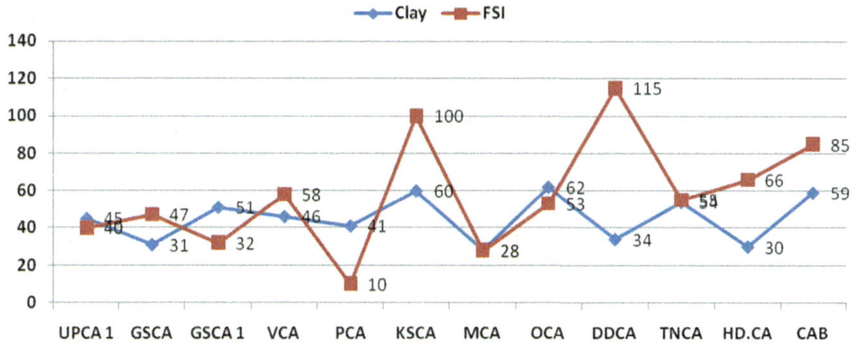

Fig. 2.44 The clay versus FSI of some of the test playing cricket centers in India. *Data source* Singh (2007)

Figure 2.45, 2.46 and 2.47 showing correlation coefficient and regression analysis of Clay versus the Atterberg limits for some of the important Test pitches of India. *Data source* Singh (2007).

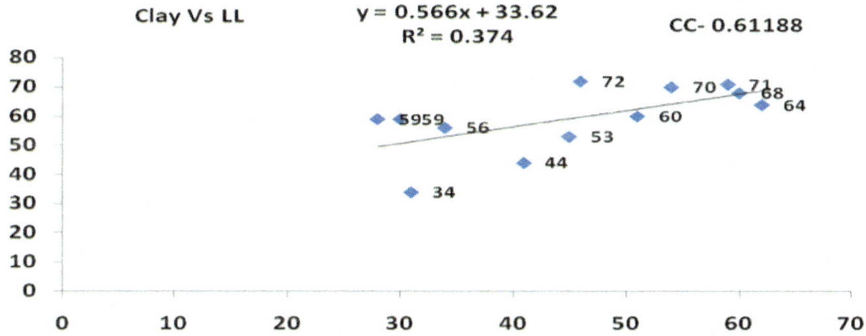

Fig. 2.45 Correlation coefficient and regression analysis of clay versus the liquid limit for some of the important test pitches of India

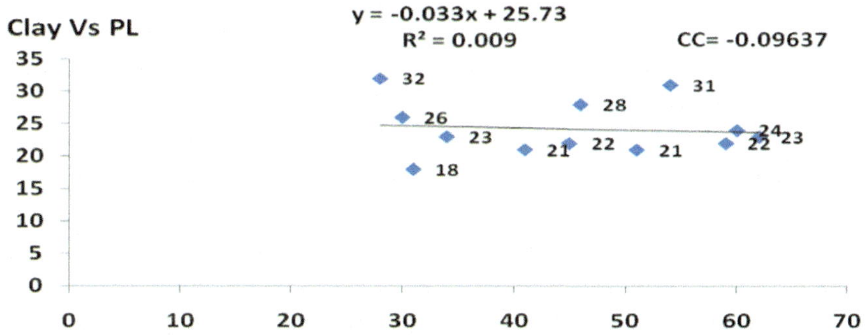

Fig. 2.46 Correlation coefficient and regression analysis of clay versus the Plastic limit for some of the important test pitches of India

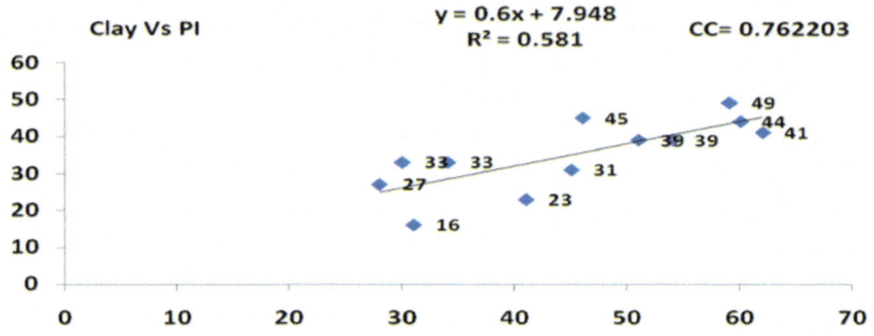

Fig. 2.47 Correlation coefficient and regression analysis of clay versus the plasticity index for some of the important test pitches of India

Figures 2.48, 2.49 and 2.50 showing correlation coefficient and regression analysis of FSI versus the Atterberg limits for some of the important Test pitches of India. *Data source* Singh (2007).

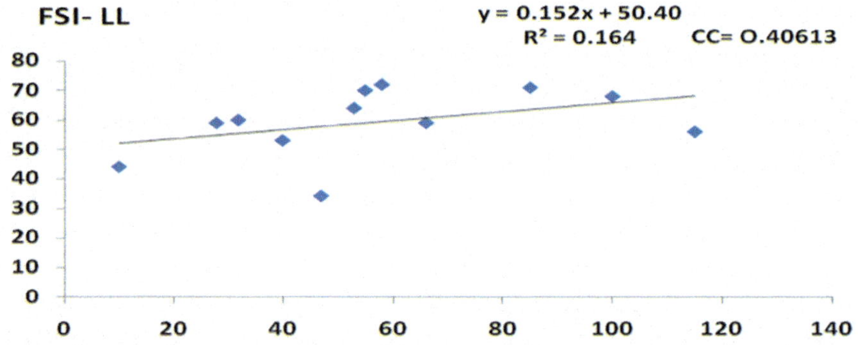

Fig. 2.48 Correlation coefficient and regression analysis of FSI versus the liquid limit for some of the important test pitches of India

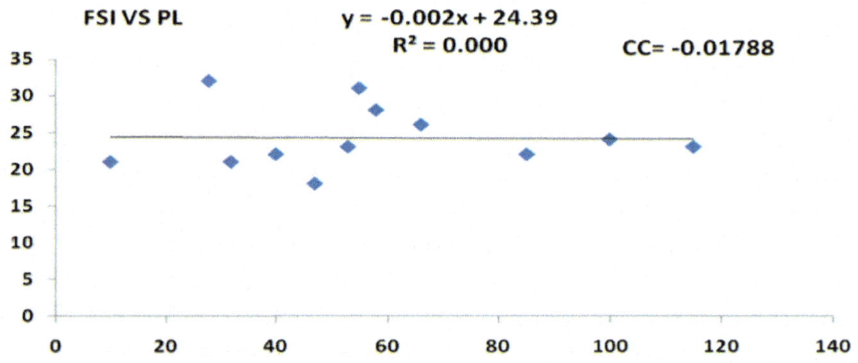

Fig. 2.49 Correlation coefficient and regression analysis of FSI versus the plastic limit for some of the important test pitches of India

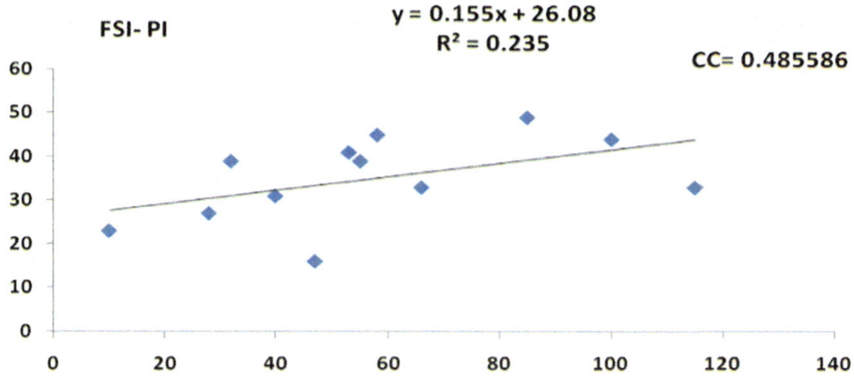

Fig. 2.50 Correlation coefficient and regression analysis of FSI versus the plasticity index for some of the important test pitches of India

2.8.1 Conclusions of Case Study 2

Through the analysis of the physical, chemical, and engineering properties of nearly 76 pitch soil samples (by the IIT-POWAI, Mumbai, through the BCCI in 2007), it is observed that some of the pitch soil samples do not match the desirable quantitative and qualitative proportions of fine swelling and shrinking smectite dominant clay, having non-proportionate particle size distribution (among the sand, silt, and clay fragments).

To achieve the desired ideal pitch surface characteristics under variable weather and climatic conditions, the cricket pitch should have a well-structured, proportionate particle size distribution (50–70% fine clays, 5–20% moderate silt, and 10–30% fine to medium sand) under normal to slightly alkaline pH conditions (full desirable morphological, physical, chemical, mineralogical, and engineering properties desirable for a cricket pitch surface).

The proportion of silt and sand found in some pitch soil samples was nearly 100–300% higher (as observed from Figs. 2.34 and 2.42) than the desirable limits, which is the main major concern and is mainly the sole reason for slow and flat pitches in India.

Most of the examined soil samples are silty clay, silty clay loam, and loamy, sandy, or silty clay loam in texture. The dominance of silty clay in most of the pitch soils was due to the source from which most of the soils had been transported. Most of the soil samples have been excavated from the major river basin areas, ponds, and low-lying submerged areas where silty clay-dominated soils are naturally predominant.

None of the pitch soil samples were examined or evaluated for the quantitative and qualitative nature of the predominant clay minerals present through mineralogical, direct, or indirect methods. There is a weak or moderate correlation between the FSI and clay content or between the different Atterberg limits (Figs. 2.39, 2.40, 2.41, 2.45, 2.46 and 2.47, respectively). It also demonstrates a non-proportionate particle size distribution with a silty clay texture as the dominant pitch soil. Low to moderate values of FSI, LL, and PI (as seen in Figs. 2.48, 2.49 and 2.50; Tables 2.12, 2.13, 2.14 and 2.15) indicate low to moderate swelling and shrinking hydroxy-interlayered vermiculites (HIV) and hydroxy-interlayered smectite (HIS) clay minerals, or non-swelling and shrinking illite or kaolinite as the dominant clay minerals in most pitch soil samples. While examining the direct (FSI, COLE values) and indirect (Atterberg limits) methods of evaluation, the predominance of intermediate clay minerals in the form of Kaolinite, Illite, and a few smectite was noted (Figs. 2.42, 2.43 and 2.44).

2.9 Summary

Previously, some physical and chemical analysis of pitch soils was only considered a litmus test for the preliminary acceptance or rejection of the pitch soil samples before constructing a pitch profile or during the preparation of match wickets. Soils with high clay contents and acceptable desired physical and chemical parameters were generally accepted for laying down the pitch profile. But the recent studies suggest that the predominant clay mineral type present in the pitch soil is much more vital than the high percentage of clay content alone. Pitch soils with high clay contents but little or no swelling and shrinking illite or kaolinite dominant clay minerals are undesirable. Pitch soils, even those with a low clay percentage but having smectite or vermiculite types of swelling and shrinking clay minerals, are mostly preferred to construct a pitch profile.

Clay mineralogy research on cricket pitches (both continental and sub-continental pitches) and its direct effect on cricket pitch behaviour is still in its early stages. Through analysis and interpretation of 76 pitch soil samples (tested by the IIT-Bombay through the BCCI for their physical, chemical, mineralogical, and engineering properties), it was found that some of the soil samples had kaolinite or illite as predominant minerals with a high percentage of clay. However, these soils do not possess the high swelling and shrinking types of soil, thereby resulting in flat and slow pitches. Cricket pitch soils with a low percentage of expanding soils, low Water Holding Capacity (WHC), optimum moisture content (OMC), bulk density (B.D.), and a very weak double diffused layer (DDL) are undesirable because they cannot provide variable pitch behaviour during the game (particularly in test matches). Most Indian cricket pitches have low percentages of clay dominant in kaolinitic or illitic types and a very high percentage of silt and sand contents (as derived from data provided by the Indian Institute of Technology (IIT), Bombay, through the BCCI and various central and state agencies), making them slow, flat, and generally spinner friendly. Kaolinitic or illitic clay minerals with their non-expensive lattice structure lack the plasticity, cohesiveness, swelling, and shrinking capacities and the desired Optimum Moisture Content (OMC) for differential pitch surface density levels. Some South African and Australian pitches (particularly the Perth pitch) favour pace bowlers (the fastest in the world) due to higher levels of sodium-dominated clay (nearly 50–82% with smectite as the major mineral (60–80%) and desirable proportionate sand and silt content. Pitch soils with a high percentage of fine swelling and shrinking smectitic type clay minerals with a predominant silt and sand content are the need of the hour for the Indian subcontinent, and they have the potential to produce miraculous results.

Therefore, in the modern era of pitch soil selection, qualitative and quantitative analysis of the predominant clay mineral type (illite, kaolinite, vermiculite, or smectite) should be considered as a basic preliminary test so as to accept or reject the soil samples before constructing any pitch profile. The type of soil mineralogy and the interlayer cations present influence the physical and chemical behaviour of pitch soils. Apart from different field identification methods, various sophisticated methods like X-ray diffraction, infrared spectroscopy, scanning electron microscopy, and differential thermal analysis are in use to determine the mineralogy of the soils. Other simple and direct methods like the free-sell index (FSI) or free-swell ratio (FSR), the coefficient of linear expansibility (COLE), and indirect methods using Atterberg limits and clay activity can be adopted to analyse and interpret the available predominant minerals. Different continental and sub-continental cricket pitches can be classified based on the analysis and interpretation of dominant minerals as well as cations. Based on these categorizations, their behaviour during the course of the match can be predicted and forecasted narrowly (as every pitch soil with its predominant clay minerals along with the exchangeable cations follows a particular pattern that decides their swelling and shrinking or non-swelling and shrinking surface behaviour at a particular moisture level).

The traditional style of pitch making with the least concentration on pitch soil sampling and its analysis on physical, chemical, mineralogical, and engineering properties before its use in pitch preparation is the common trend in most continental and sub-continental pitches. It was done merely on the basis of permutation and combination and skill developed from generations to generations. Even in the process of pitch soil type selection and its layering concepts (either 3, 4, or 5-layered wickets), deep-rooted grass selection and its uniform growth, rolling concepts, and its schedule at the optimum moisture content for maximum compaction are directly affected by the type and nature of soil used for preparing a pitch. Countries like South Africa, England, Australia, and New Zealand have bouncy and fast pitches only as an outcome of coordinated efforts by the localised cricket associations and engineering and agricultural colleges. Their valuable research and development, based on the experience and skills of the groundskeeper and curator (either on pitch soil selection, layering, grass variety selection and maintenance, rolling techniques and patterns, maximum compaction level at the optimum moisture content, and so on), can only produce fast and bouncy pitches in the international arena in India.

Clay mineralogy is considered the DNA of any pitch soil, which provides the intrinsic template to have most of the desired physical, chemical, or engineering properties for a good pitch. The genes are the basic building block that builds the nature and character of humans, as they have evolved through the process of evolution. Soil type and its basic minerals (illite, kaolinite, montmorillonite, or smectite) are the genes of any pitch soil, which define and govern the basic nature and behaviour of any pitch.

References/Bibliography

Albrecht BA (1996) Effect of desiccation on compacted clays. M.sc. thesis, Department of Civil Engineering, University of Wisconsim, Madison, WI

Aldous DE (2016) Handbook on international turf management. CRC press

Al-Zoubi MS (2008) Swell characteristics of natural and treated compacted clays. Electron J Geotech Eng 13(D)

Baker SW, Cook A, Adams WA (1998) Soil characteristics of first class cricket pitches and their influence on playing performance. J Turf Grass Sci 74:63–79

Baldock JA, Nelson PN (2000) Soil organic matter—handbook of soil science. CRC Press, Boca Raton

BCCI Curators Certification Course Manual (2013) BCCI, Second Edition

Bhattacharya T, Chandran P, Ray SK, Mandal C, Pal DK, Venugopalan MV, Durge SL, Srivastava P, Dubey PN, Kamble GK, Sharma RP, Wani SP, Rego TJ, Ramesh V, Manna MC (2006) Morphological properties of red and black soils of selected benchmark spots in semi arid tropics of India, Global theme on agroecosystems report no. 21, National Bureau of Soil Survey and Land Use Planning, ICAR, Nagpur, ICAR, 2006

Bhattacharyya T, Chandran P, Ray SK, Mandal (Mrs) C, Pal DK, Venugopalan MV, Durge SL, Srivastava P, Dubey PN, Kamble GK, Sharma RP, Wani SP, Rego TJ, Pathak P, Ramesh V, Manna MC, Sahrawat KL (2007) Physical and chemical properties of red and black soils of selected benchmark spots for carbon sequestration studies in semi-arid tropics of India, global theme on agro-ecosystems, Report No.-35, National Bureau of Soil Survey and Land Use Planning, ICAR, Nagpur, ICRISAT Patancheru (A.P) and New Delhi , CRIDA Hyderabad and ICAR New Delhi, 2007

Bowles JE (1979) Physical and geotechnical properties of soils. Wiley, New York

Brady NC, Weil Ray R (2008) Nature and properties of soil, 14th edn. Prentice Hall

Burdet L (1986) Turf cricket pitches. South Australian Cricket Association, Adelaide, p 28

Bureau of Ureau of Indian Standards, Indian Standard Classification and Identification of Soils for General Engineering Purposes, BIS, New Delhi, 1970, (Reaffirmed 1987), IS 1498

Deb S, Sahu SS (2011) Soil mineralogy in relation to physiography in Jumar Sub Watershed of Jharkhand. Agropedology 21(1):1–7

ECB Staff (2007) Recommended guidelines for the construction, preparation and maintenance of cricket pitches and outfields at all levels of the game, England and Wales Cricket Board, Lord's Cricket Ground, London, UK

Ekwue EI, Lall DZ, Stone RJ (2006) Engineering properties of major soils used in cricket pitches in Trinidad. West Indies J Eng 28(2), Technical Paper

Grant CD, Bkackmore AV (1991) Self mulching behavior in clay soils: its definition and measurement. Aust J Soil Res 29:155

Grim RE (1953) Clay mineralogy. Mc Graw-Hill Publ, New York

Gumbs FA, Griffith SM, Shillingford G (1981) The design and physical features of the recently re-laid cricket pitch at the queens park oval, Trinidad. J Sports Turf Res Inst 58:16–20

Harris JR (1961) The crumbling of cricket pitches. The Australian Scientist, 01 April, pp 73–74

Holtz RD, Kovacs WD (1981) An introduction to geotechnical engineering. Prentice-Hall, Upper Saddle River, NJ

IOG/ECB (2006) Management and maintenance of the cricket pitch. The Institute of Groundsmanship, Milton Keynes, UK

Johnson K, Hamilton S (1990) Background to excessive cracking of pitches at the Western Australian Cricket Association Ground, Turf Craft Australia, July 1990, pp 28–32

Kolhe AH, Chandran P, Ray SK, Bhattacharyya T, Pal DK, Sarkar D (2011) Genesis of associated red and black shrink-swell soils of Maharashtra. Clay Res 30(2); Division of Soil Resource Studies, National Bureau of Soil Survey and Land Use Planning, ICAR, Nagpur, ICRISAT Patancheru (A.P.)

Lambe TW (1951) Soil testing for engineers. New York, Wiley

Lambe TW, Whitman VR (1969) Soil mechanics. Willey (SI Version)

Lotse EG, Datta NP, Tomar KP, Motsara MR, Mineralogical composition of some red and black soils of India. Nuclear Research Laboratory, IARI, PUSA, New Delhi, pp 216–226

Marshall TJ, Holmes JW, Rose CW, Soil physics, 3rd edn. Cambridge University Press, New York

McIntyre K, Don (2001) Cricket wickets-science versus fiction, horticulture engineering consultancy, 5-Brimage Palace Kambah, ACT Australia

McIntyre DK, Jackobsen B (1998) Drainage for sports turf and horticulture, Horticultural Engineering Consultancy, Canberra, Australia

Means RE, Parcher JV (1965) Physical properties of soils. Prentice Hall of India (P) Limited, New Delhi

Mehra RK (2014) A textbook of soil science, Indian Council fir Agricultural Research Publication

Mitchell K (2010) Cricket wickets, Copyright ©Queensland Cricket, pp 1–37

Nagaraj TS, Srinivasamurty BR, Vatsala A (1994) Analysis and prediction of soil behavior. Taylor and Francis

NSRI (2001) A guide to better soil structure. National Soil Resource Institute, Cranfield University, Silsoe, UK

Pal DK, Bhattacharya T, Deshpande SB, Sarma VAK, Velayutham M (2000) Significance of minerals in soil environment of India, NBSS Review Series1-ISBN: 81-85460-57-4, September, 2000

Pal DK, Deshpande SB, Velayutham M, Srivastava P, Durge SL (2000) Climate change and polygenesis in Vertisols of the Purna Valley (Maharashtra) and their management, Research Bulletin, Pub. No. 83, National Bureau of Soil Survey and Land Use Planning, ICAR, Nagpur

Parsons S, The aeration of clay soils in cricket, engineering doctorate (Academic year 2008–2012). School of Applied Sciences, Cranfield University

Pearson J (2010) Basic guide to turf cricket pitch preparation, VCA 2010

Prakash K, Sridharan A (2004) Free swell ratio and clay mineralogy of fine-grained soils. Geotech Test J 27(2):220–225

Principles and Practice of Pitch Preparation (1993) United Cricket Board of South Africa, May

Ranjan G, Rao ASR (2002) Basic applied mechanics, 3rd edn. New Age International Private Limited

Rathinaswamy A, Saliha BB (2014) Fundamental of soil science. Scientific Publication

Ray SK, Chandran P, Bhattacharyya T, Durge SL, Pal DK (2003) Layer charge of two benchmark vertisol clays by alkylammonium method. Clay Res 22:13–27

Ray SK, Bhattacharyya T, Chandran P, Sahoo AK, Sarkar D, Durge SL, Raja P, Maurya UK, Pal DK (2006) On the formation of cracking clay soils in West Bengal. Clay Res 25(2):141–152

Raychaudhuri SP, Roy BB, Gupta SP, Dewan ML (1963) Black soils of India, National Institute of Soil Science, Mathura Road, New Delhi

Reddy ES, Sastri KR (2002) Measurement of engineering properties of soils. New Age Publication

Robinson GR, Allam MMM (1998) Effect of clay mineralogy on coefficient of correlation. Tech Bull Clays Clay Minerals, The Clay Mineral Society, Vol. 46, No.5, pp. 596–600

Salini U, Girish MS (2011) Suitability of soil for cricket pitch preparation, College of Engineering Trivandrum, Technical paper at NATCON-2011, Centre for Engineering Research and Development, Kerala, 28–29 Jan 2011

Sarkar D, Nayak DC, Maurya UK, Gajbhiye KS (2008) Mineralogy of benchmark soils of West Bengal. Technical Bulletin, NBSS Pub No-139, National Bureau of Soil Survey and Land Use Planning, ICAR, Nagpur, July, 2008

Sehgal J, Gajbhiye KS, Batta RK, Sharma VAK (1999) Swell-shrink soils (vertisols) of India, 1st edn. Kalyani Publishers

Shimpton P, James L (2009) Optimization of cricket pitch rolling. Ph.D. thesis, Cranfield University, School of Applied Science and Natural Resource Department

Singh DN (2008) Physical, chemical, free swell index and atterberg limit analysis reports of 76 cricket pitch soils under the BCCI, IIT, Mumbai

Singh A (2009) Mordern geotechnical engineering, 3rd edn. CBS Publication

Singh DN, Uday KV (2013) Application of laser microscopy for studying crack characteristics of fine-grained soils. Technical note, Geotech Test J 36(1):1–9

Sivapullaiah PV, Sitaram TG, Rao KSS (1987) Modified free swell index for clays. ASTM Int 10(2):80–85

Sivapullaiah PV, Sitaram TG, Rao KSS (2004) Simple correlation between soil plasticity and expansion potential. Adv Eng Geol Geotech, GE 441

Sridharan A (2014) Soil clay mineralogy and physio-chemical mechanics governing the fine-grained soil behaviour, Fourth IGS-Ferroco Terzaghi Oration: 2014. Ind Geotech J (October–December 2014), pp 371–399

Sridharan A, Rao SM (1988) A scientific basis for the use of index tests in identification of expansive soils. Geotech Test J ASTM 11(3):208–212

Sridharan A, Prakash K (2000) Classification procedures for expansive clays, Institute of Civil, Engineers and Geotechnology, Engineering, Thomas Telford Press Ltd., pp 235–240

Sridharan A, Prakash K (2000) Classification procedures for expansive soils. Proc Inst Civ Engrs Geotech Eng 143:235–240

Sridharan A, Rao SM, Murthy NS (1985) Free swell index of soils: a need for redefinition. Ind Geotech J 15:94–99

Sridharan A, Rao SM, Murthy NS (1986) A rapid method to identify clay type in soils by the free-swell technique. Geotech Test J ASTM 9(4):193–203

Soil Resource Information of Different Agro-Eco Sub-Regions (ASER) of India for crop and 1soil carbon modelling (2016) Technical bulletin, NBSS and LUP Publication 1035, ICAR, Nagpur

Summer EM, Olson CG, Thompson ML, Wilson MA, Handbook of soil science. CRC Press, Baca Raton, London, pp 112–113

Stewart VI, Adams WA (1969) Soil factors affecting the control of pace on cricket wickets. In: The Sports Turf Research Institute (Ed) first International Turf Grass Research Conference Harrogate, pp 533–546

Taiton NM, Klug JR, Edmondson D, Campbell RK, Deventer PWV, and Beer, MJD

TCCB and NCA (1989) Recommendations on the preparation of cricket pitches, Test and county Cricket Board and National Cricket Association, Lords, London

Venkataramaiah C (2006) Geotech Eng. ISBN: 978-93-86649-70-6. New Age International (P) Ltd.,

Wood C (2006) Recommended guidelines for the maintenance of cricket pitches and outfields at all levels of game. England and Wales Cricket Board in association with The Institute of Groundsmanship

Chapter 3
"Cricket Pitch Soil Inventory: Benchmark Soils for Indian Cricket Pitches"

3.1 Introduction—Why Benchmarking of Continental and Subcontinental Pitches?

Through the BCCI, the IIT Powai in Mumbai looked at the physical, chemical, and engineering properties of the cricket pitch soils in all the cricket associations across the country. They found that most of the pitch soils did not have the right amounts of sand, silt, and clay. From Fig. 3.1, we can see that the amount of silt and sand in most pitch soil samples was 100–300% higher than what is desirable. This is the main concern and the main reason why pitches in India and the subcontinent are usually slow and flat. Low to moderate values of the liquid limit plasticity index, the free swell index (FSI), and clay activity (shown and inferred from Chap. 2) show that silt is more common than clay and sand (fine to medium sand) in India and subcontinental silty clay soils.

Even though some state cricket associations use kaolinitic or illitic clays with a high percentage of clay, they do not have high swelling and shrinking capacities. This means that the pitches are slow and flat, and the pitch surface and turf grass do not recover well after a game. Low percentages of clay dominant in non-swelling and shrinking kaolinitic or illitic types and a very high percentage of silt and sand contents (as derived from the data provided by the IIT POWAI, Mumbai, through the BCCI and various central and state agencies) are found in most of the Indian cricket pitches, which makes them slow, flat, and spinner friendly (Fig. 3.1). Pitch soils with the recommended proportionate particle size distribution (PSD) and a Ca^{+2}/Mg^{+2}-dominant smectite type of fine clay are the need of the hour for the Indian subcontinent and can bring the desired magical results.

S. B. Singh et al., *Cricket Pitches*, Springer Transactions in Civil and Environmental Engineering, https://doi.org/10.1007/978-981-99-2913-9_3

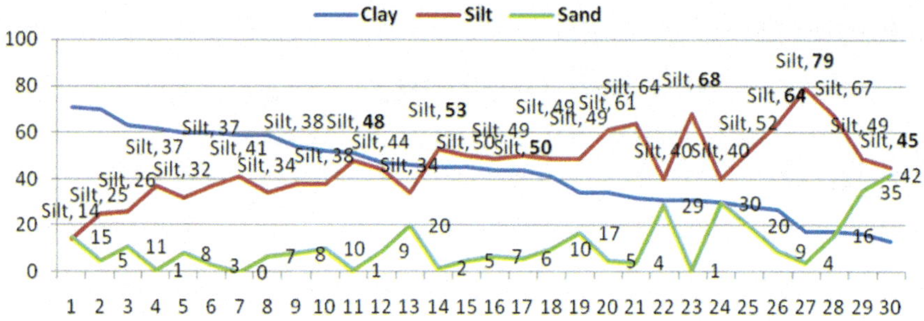

Fig. 3.1 Particle size distribution of soils for cricket pitches of all the cricket associations across the country (India). *Data source* Singh (2007)

Some soil samples contain high proportions of clay (as Ca^{+}/Mg^{+}-dominant swelling and shrinking clays (due to their high liquid limits, plasticity index, free swell index, clay percentage, activity values—as derived from Fig. 3.1) with silt and sand at desirable moderate levels, but the predominant exchangeable cations and minerals have yet to be verified through available mineralogical analysis. Most of the pitch soil samples were not examined for the quantitative and qualitative nature of the predominant clay minerals present. Cricket pitch soils based on recent research on mineralogical analysis are essential to identify the nature and type of clay minerals present before the final selection as a cricket pitch soil.

There is a big difference between the number and quality of pitch materials that are already in use and what the I.C.C.-B.C.C.I. suggests, or between the different types of cricket pitch soils that are already out there and what the scientific institutions have found based on the desirable physical, chemical, mineral, and engineering index properties.

To attain the desired ideal pitch surface characteristics under variable weather and climatic conditions (continental and subcontinental), the cricket pitch should have good structure and a proportionate particle size distribution (50–70% fine Ca^{+2}-dominant fine clays with moderate silt (5–20%) and 10–30% fine to medium sand under normal to slightly alkaline pH conditions (full desirable morphological, physical, chemical, mineralogical, and engineering properties as a cricket pitch soil are provided in the introductory Chap. 1, Table 1.1).

Based on the available climatic and weather variables, 5–30% variations in the physical, chemical, and mineralogical desirable properties of the cricket pitch are considered. Places with a hot, humid climate, and less rainfall (the Indian subcontinent, Australia, South Africa, Zimbabwe, etc.) with high temperatures, relative humidity (RH), solar radiations, etc., can have the privilege of using well-structured Ca^{+2}/Mg^{+2} fine smectites having a very high water holding capacity (WHC). These soils, under conditions of high temperature, relative humidity (RH), and evapotranspiration rates (ET), retain the desirable optimum moisture content (OMC) and the desirable pitch surface structure for a long duration of time (either the third, fourth, or fifth day of the match), providing equal opportunity to all classes of the game.

On the other hand, countries with cooler climatic conditions (England, New Zealand, etc.) with low temperatures and low relative humidity (RH) should have moderate swelling and shrinking clays with moderate silt and sand proportions. Most people prefer silty clay loams, which have an equal amount of fine sand, silt, and clay with a mix of minerals, in cool climates because pitch soils with a lot of fine smectite clay having high water holding capacity (WHC) won't dry out in cool climates. This is one of the major reasons that in spite of having high smectitic fine clay, the English wickets are generally laid on loamy soils.

Both extreme cricket pitch soil conditions, either very low pH or very high pH, should be avoided, as the pitch surface in these undesirable morphological, physical, chemical, and mineralogical conditions will not produce the desired pitch surface behaviour. Pitch soils with higher sand and silt content than the prescribed clay content have non-swelling and shrinking kaolinite-dominant fine or coarse clay with non-proportionate particle size distribution. These types of soils typically produce slow spinner-dominated wickets, and managing pitch surface behaviour during the later stages of the match (3, 4, or 5th day) becomes extremely difficult due to pitch surface structure loss and uncontrolled crumbling of the pitch surface (as shown in Fig. 3.3a). After matches, pitch surface and turf grass recovery is very poor in these types of cricket pitch soils due to Al^{+3}/H^{+} toxicity.

Pitch soils at high pH conditions (> 8.5) with Mg^{+2}/Na^{+} dominant and very high swelling and shrinking smectite clay generate very high undesirable cracks and high surface dry density with non-proportionate PSD (very high clay content with less silt and fine sand, as shown in Figs. 3.2 and 3.3b). These types of cricket pitches provide a narrow surface density, either assisting fast bowlers or the batsman. Spinners have very little room to manoeuvre on these types of pitch surfaces. Due to very high pitch surface density and uncontrolled very wide and deep cracks on the pitch surface, it can be difficult to manage the pitch surface behaviour during the later stages of the matches (3rd, 4th, or 5th day), making it unplayable or injurious for the batsman to play (as shown in Fig. 3.2b).

The Perth and Harvey cricket pitch soils (as shown in Fig. 3.3) having very high comparative clay percentages (82 and 78%) and very high percentages of smectite (> 80%) with very low percentages of recommended silt (6 and 11%) and fine (2 and 7%) sand generally provide narrow pitch behaviour, favouring the fast bowlers (due to very high pitch surface dry density and low wear and tear of the desired pitch surface during the later stages of the match).

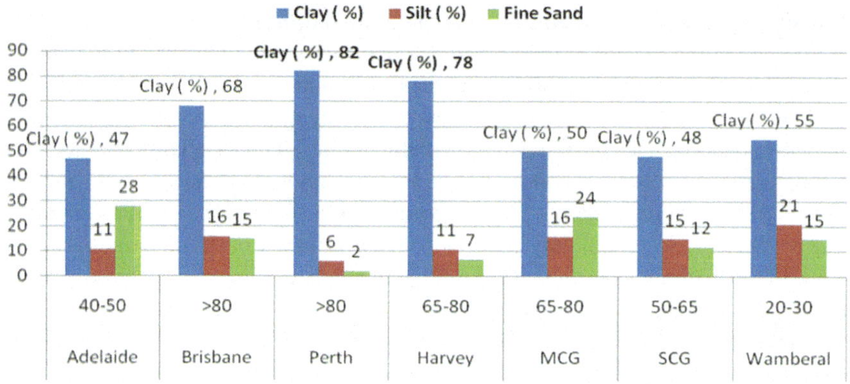

Fig. 3.2 Particle size distribution with predominant clay minerals (in percentage) of cricket pitch soils predominantly used in Australia

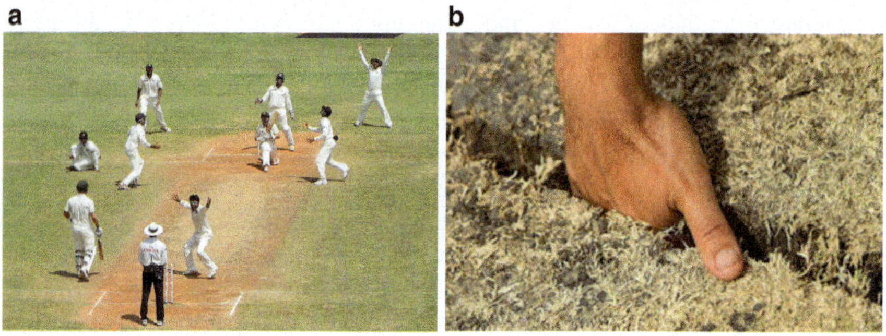

Fig. 3.3 **a**, **b** Extreme undesirable pitch surface behaviour due to non-proportionate clay, silt, and sand proportions or due to very high proportion of kaolinite or smectites-dominant clay minerals in some subcontinental or continental pitches

3.2 The Purpose of Generating a Benchmark for Cricket Pitch Soils in India

1. To make a region-specific data bank (ready reckoner) based on the desired morphological, physical, chemical, mineralogical, and engineering index properties for cricket pitch soils.
2. To narrow down the gap that exists between what it is and what it ought to be. Due to the high proportion of silt and moderate sand fragments in Indian and subcontinental pitches, they are labelled as slow and dead. A wide gap exists between the cricket pitches laid down using traditional or conventional methods and the recently raised stadium pitches, which were raised only after examining

the cricket pitch soil for its desirable morphological, physical, chemical, mineralogical, and engineering index properties. A large variation exists between the quality of pitch soil material being used and the wide range of cricket pitch soils already available in India (based on their physical, chemical, mineralogical, and engineering properties, which are tested and analysed by some of the most reputed institutions in India, such as the ICAR-NBSS&LUP, Nagpur, and IIT Powai). The purpose of benchmarking cricket pitch soils is to synchronise the vast pool of soil profile data and information already available regarding the Indian soils based on their research and findings with the desirable pitch soil index properties before constructing a pitch profile.

3. To predict and analyse the pitch behaviour under different agro-ecological regions based on their textural, structural, physical, chemical, mineralogical, and engineering properties.
4. To pinpoint or classify the different cricket pitches within the different variable micro and macroclimatic variations suitable for different formats of the game. Generally, cricket pitch soils with high swelling and shrinking characteristics in hot and humid conditions favour the desirable pitch variations for Test matches. During the course of the Test matches, high temperatures, high evapotranspiration rates, and high humidity and dew point induce medium-to-wide cracks, resulting in the desired variations day by day or innings by innings.
5. To upgrade and standardise the uniformity and quality of cricket pitch soil materials and the process by which the international stadium pitches are being raised up in India and abroad.
6. Benchmarking of cricket pitch soils will reduce transportation costs as the desirable benchmark pitch soils will be available in the vicinity of cricket stadiums.
7. Benchmarking cricket pitch soils will provide many alternatives, from which curators will be able to select pitch soils based on their suitability and requirements. Generally, a pitch square contains 5–10 wickets constructed from the same type of selected pitch soil material, and all the formats of matches (especially T-20, ODI, Test matches, Ranji matches, etc.) are played on the same pitch square. The desired objectives and variations as laid down for the different formats of the game can also be achieved by having different wickets with different pitch soil materials within the pitch square. Designated pitches with different soil texture and other physical, chemical, and mineralogical parameters can be used within the same pitch square so as to provide desirable variations in pitch behaviour during the different formats of the match (either Test matches, one day international matches, or T-20 matches).
8. A vast pool of available cricket pitch soils (through benchmarking) will enable us to analyse, evaluate, and correlate the impact of textural, structural, physical, chemical, mineralogical, and engineering properties, as well as weather and climatic factors (such as evapotranspiration, relative humidity, temperature, dew, soil moisture, etc.) on the different behaviours of cricket pitches. Based on index-based benchmarking, thousands of such combinations of pitch behaviour can be analysed, and desirable Pitch Behaviour Analysis Index (PBAI), Pitch

Quality Index (PQI), or Pitch Behaviour Forecasting (PBF) methods can be developed. This could help us categorise different pitch behaviours based on benchmarked index soils, as well as analyse and predict complex pitch behaviour more precisely.

3.3 State-Wise Distribution of Benchmarked Soils in India[1]

Recommended proportionate particle size distribution (PSD of fine clay, silt, and fine or medium sand) and the cation exchange capacity (CEC) values have been considered for developing state or nationwide benchmarking for cricket pitch soils in India (full tabular details of the statewide benchmarked soils for cricket pitches in India have been annexed separately and graphically shown as below, from Figs. 3.4, 3.5, 3.6, 3.7, 3.8, 3.9, 3.10, 3.11, 3.12, 3.13, 3.14, 3.15, 3.16, and 3.17). Out of thousands of benchmarked Indian soils for cricket pitches, the top 50 and top 20 have been finally benchmarked based on their highest clay and CEC values, with silt and fine or medium sand under recommended proportions. Various other available recommended physical, chemical, and mineralogical tests (XRD, COLE, FSI, FSR, clay activity, and Atterberg limits, as discussed in Chap. 2) can be considered as an index property range while developing benchmarks for cricket pitch soils.

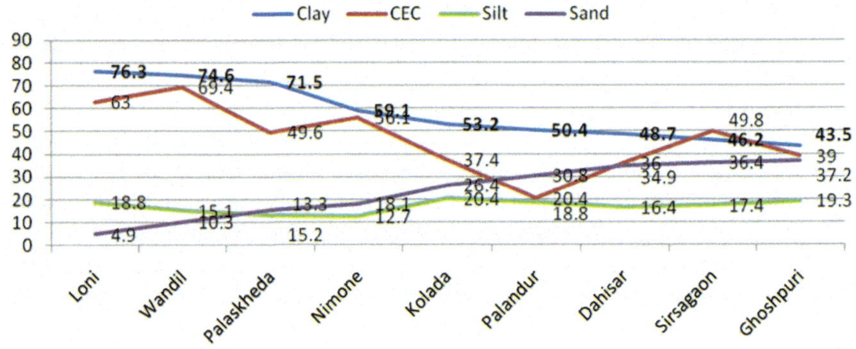

Fig. 3.4 Properties of some benchmark cricket pitch soils of Maharashtra

[1] State-wise distribution of benchmarked soils for cricket pitches in India (based on the desirable index properties of soils) has been annexed to this book as annexure 3.

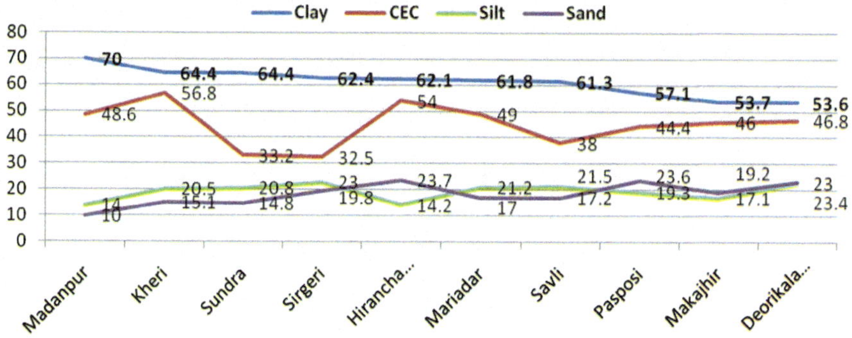

Fig. 3.5 Properties of some benchmark cricket pitch soils of Madhya Pradesh

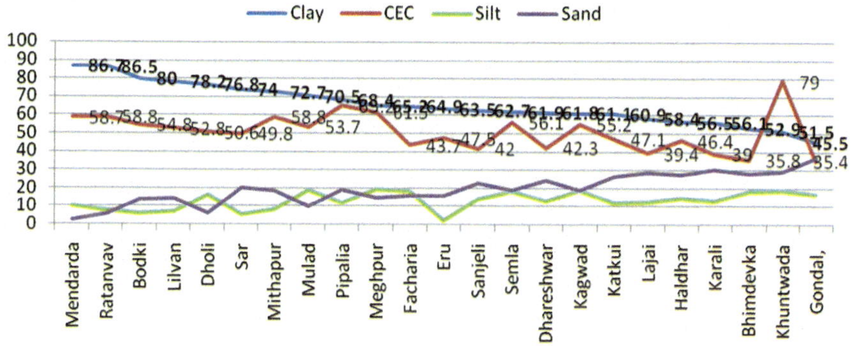

Fig. 3.6 Properties of some benchmark cricket pitch soils of Gujarat

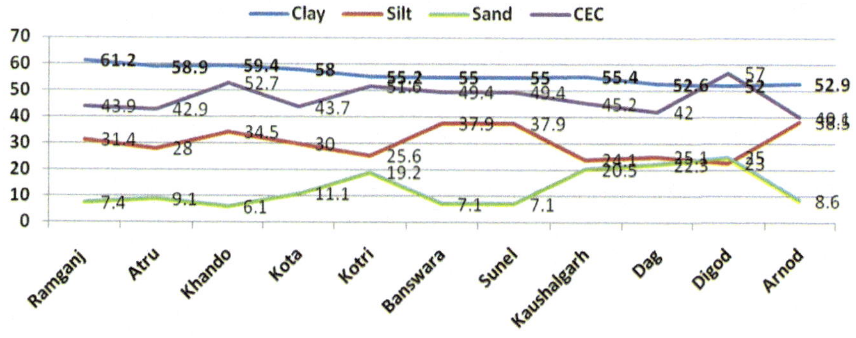

Fig. 3.7 Properties of some benchmark cricket pitch soils in Rajasthan

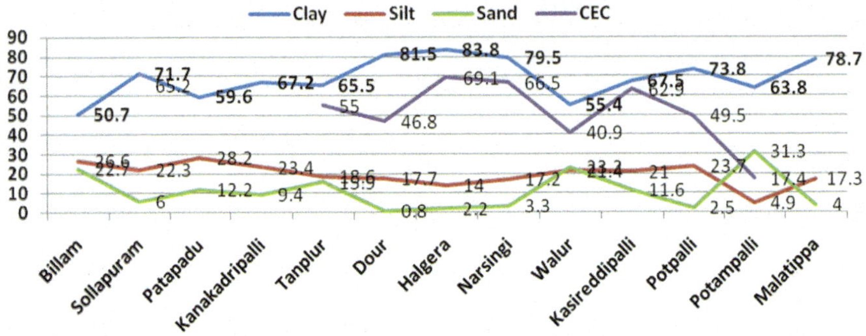

Fig. 3.8 Properties of some benchmark cricket pitch soils in Andhra Pradesh

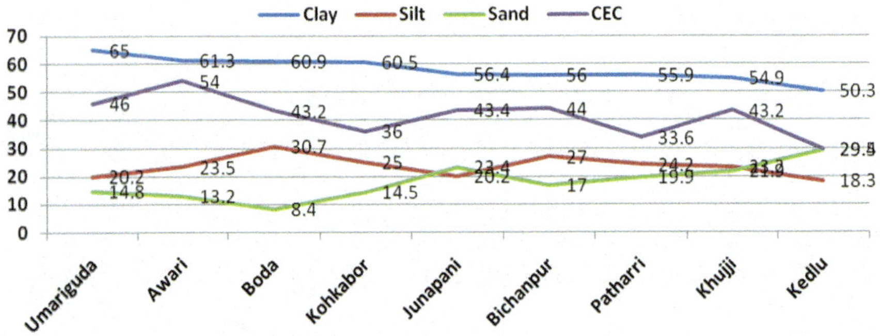

Fig. 3.9 Properties of some benchmark cricket pitch soils in Chhattisgarh

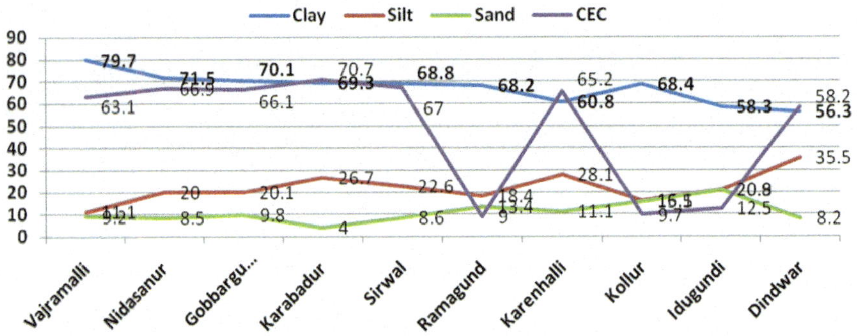

Fig. 3.10 Properties of some benchmark cricket pitch soils in Karnataka

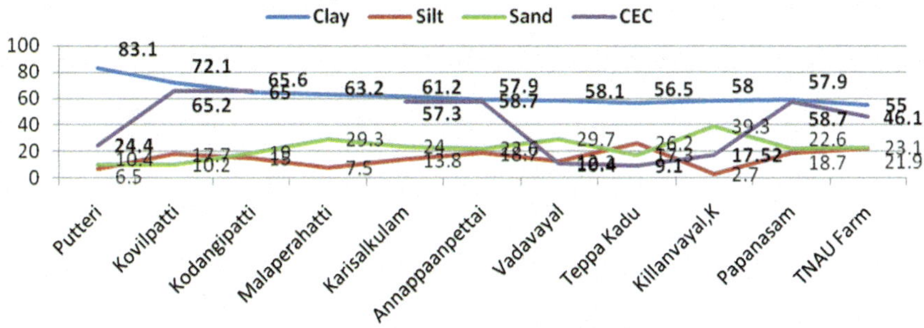

Fig. 3.11 Properties of some benchmark cricket pitch soils in Tamil Nadu

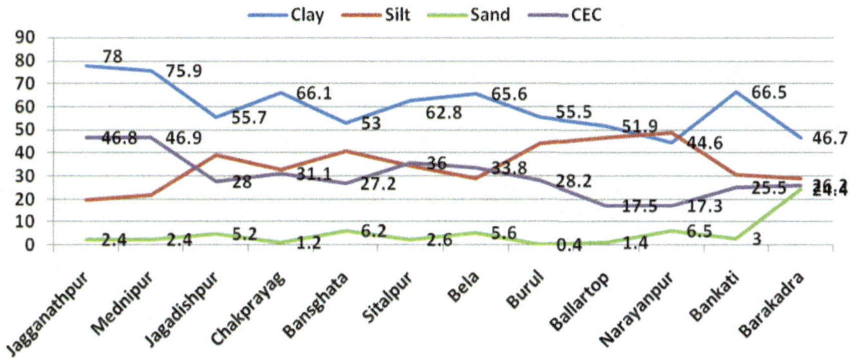

Fig. 3.12 Properties of some benchmark cricket pitch soils in West Bengal

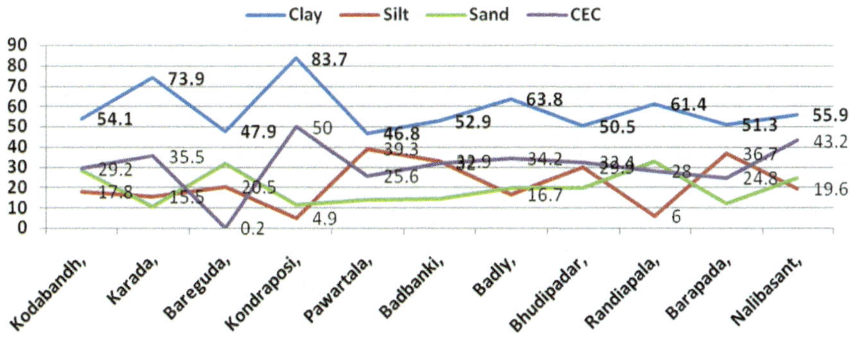

Fig. 3.13 Properties of some benchmark cricket pitch soils in Odisha

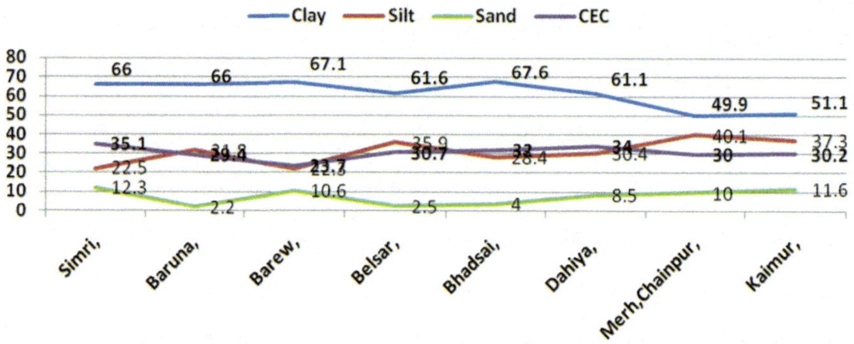

Fig. 3.14 Properties of some benchmark cricket pitch soils in Bihar

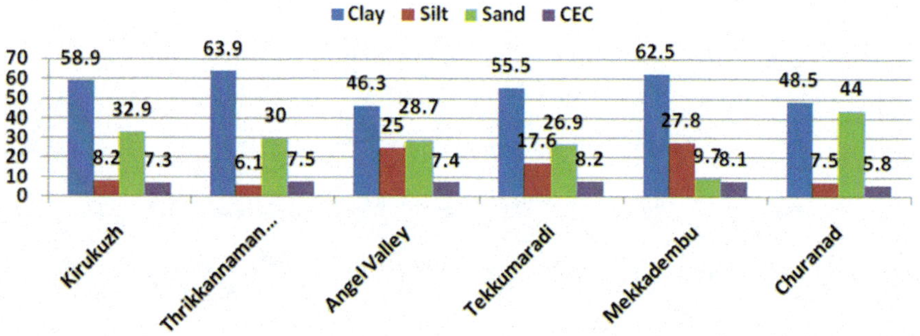

Fig. 3.15 Properties of some benchmark cricket pitch soils in Kerala

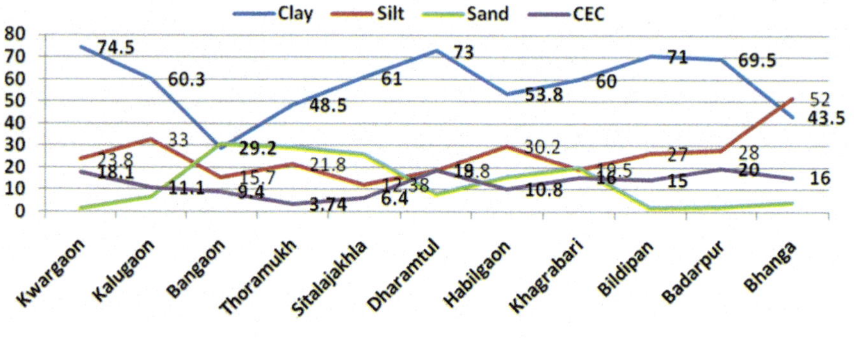

Fig. 3.16 Properties of some benchmark cricket pitch soils in Assam

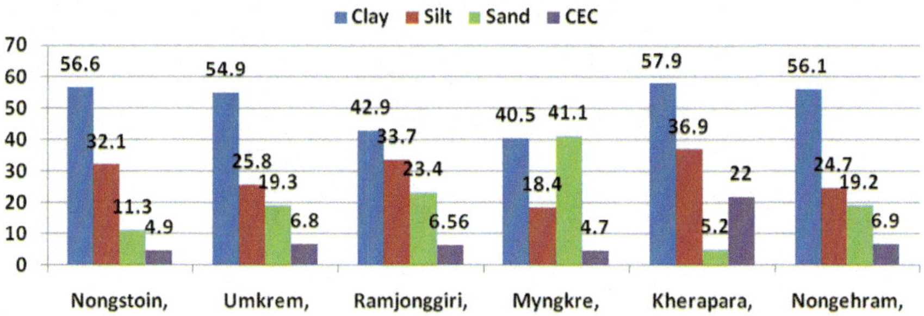

Fig. 3.17 Properties of some benchmark cricket pitch soils in Meghalaya

3.4 Analysis of Selected Top 50 Benchmarked Pitch Soil Samples (State Wise)[2]

See Tables 3.1, 3.2, and 3.3; Figs. 3.18, 3.19, 3.20, 3.21, and 3.22.

[2] Based on desirable index properties (proportionate particle size distribution (PSD) and cation exchange (CEC) values).

Table 3.1 Top 50 selected benchmark soils suitable for cricket pitches in India

S. No.	Locations	Clay	Silt	Sand	CEC	S. No.	Locations	Clay	Silt	Sand	CEC
1	Bodki, Guj	80	6.1	13.9	54.8	26	Pocharam, AP	63.4	12.5	24.1	45.4
2	Vajramalli, Kar	79.7	11.1	9.2	63.1	27	Suklatirth, Guj	62.5	20.3	17.2	51.3
3	Lilvan, Guj	78.2	7.5	14.3	52.8	28	Sirgeri, MP	62.4	23	19.8	32.5
4	Wandil, Maha	74.6	15.1	10.3	69.4	29	Elaram, AP	62.4	22.6	15	48.8
5	Sar, Guj	74	5.9	20.1	49.8	30	Hiranchanda, MP	62.1	14.2	23.7	54
6	Karada, OR	73.9	15.5	10.6	35.5	31	Semla, Guj	61.9	18.7	19.4	56.1
7	Pusalpahad, AP	73.8	16.5	9.7	36.8	32	Dhareshwar, Guj	61.8	13.4	24.7	42.3
8	Mithapur, Guj	72.7	8.7	18.6	58.8	33	Randiapala, OR	61.4	6	32.6	28
9	Kovilpatti, TN	72.1	17.7	10.2	65.2	34	Savli, MP	61.3	21.5	17.2	38 ara>
10	Palaskheda, Maha	71.5	13.3	15.2	49.6	35	Jalia, Guj	60.9	20.3	18.8	49.4
11	Mulad, Guj	70.5	19.2	10.3	53.7	36	Katkui, Guj	60.9	12.3	26.8	47.1
12	Gobbargumpi, Kar	70.1	20.1	9.8	66.1	37	Nimone, Maha	59.1	12.7	18.1	56.1
13	Madanpur, MP	70	14	10	48.6	38	Papanasam, TN	58.7	18.7	22.6	57.9
14	Pipalia, Guj	68.4	12.1	19.5	65.2	39	Pasposi, MP	57.1	19.3	23.6	44.4
15	Giripalli, AP	67.1	18.4	14.5	49.6	40	Haldhar, Guj	56.5	15.1	28.1	46.4
16	Karla, Maha	66.1	23.5	10.4	58.4	41	Junapani, Chattis	56.4	20.2	23.4	43.4

(continued)

Table 3.1 (continued)

S. No.	Locations	Clay	Silt	Sand	CEC	S. No.	Locations	Clay	Silt	Sand	CEC
17	Simri, BI	66	22.5	12.3	35.1	42	Nalibasant, OR	55.9	19.6	24.5	43.2
18	Karabadur, Kar	65.7	23.6	10.7	60	43	Makajhir, MP	53.7	17.1	19.2	46
19	Tanplur, AP	65.5	18.6	15.9	55	44	Kolada, Maha	53.2	20.4	26.4	37.4
20	Meghpur, Guj	65.2	20	14.8	61.5	45	Bhimdevka, Guj	52.9	18.8	28.3	35.8
21	Ranibennur, Kar	65.1	16.8	18.1	60	46	Khuntwada, Guj	51.5	19.2	29.3	79
22	Umariguda, Chattis	65	20.2	14.8	46	47	Bhola, Guj	50.6	21.9	27.5	44.3
23	Kodangipatti, TN	65	15	19	65.6	48	Palandur, Maha	50.4	18.8	30.8	20.4
24	Facharia, Guj	64.9	18.9	16.2	43.7	49	Sirsagaon, Maha	46.2	17.4	36.4	49.8
25	Badly, OR	63.8	16.7	19.5	34.2	50	Gondal, Guj	45.5	17.5	37	35.4

Based on proportionate particle size distribution (PSD) and cation exchange (CEC) values
Source Soil series of India (state wise), ICAR-NBSS-LUP, Nagpur, Maharashtra

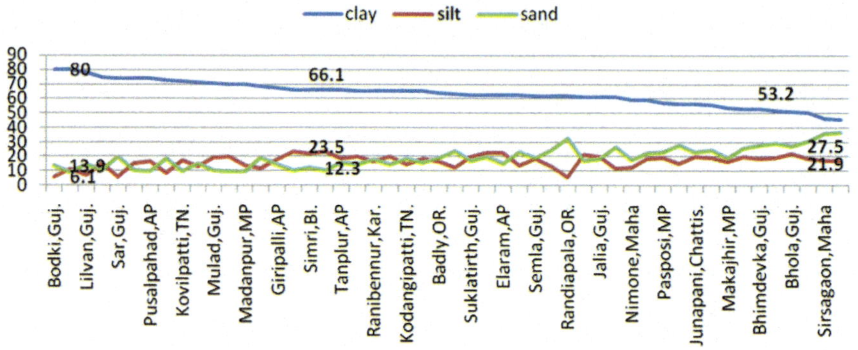

Fig. 3.18 Particle size distribution (PSD) in descending order of clay percentage of top 50 selected benchmarked soils suitable for pitches in India

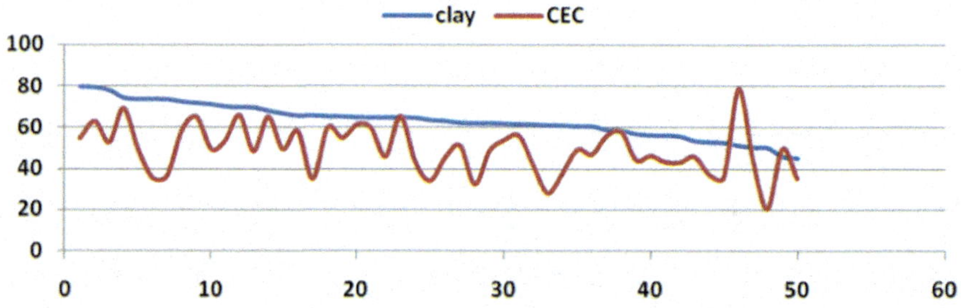

Fig. 3.19 Clay percentage and cation exchange capacity (CEC) of selected top 50 benchmarked smectite-dominant swelling and shrinking clay soils suitable for pitches in India

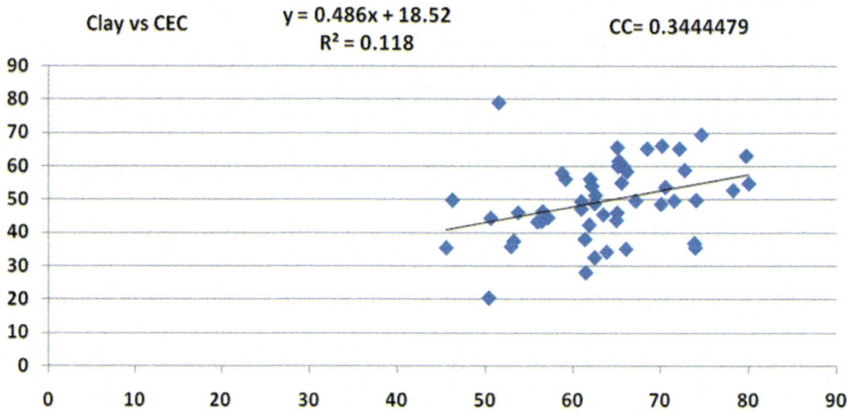

Fig. 3.20 Strong correlation in between the clay percentage and cation exchange capacity (CEC) of selected top 50 benchmarked smectite-dominant swelling and shrinking clay soils suitable for pitches in India

Table 3.2 Selected benchmark soils (top 20) based on highest clay content (in descending order) and within the desirable range of silt and sand proportions

S. No.	Location	Clay	Silt	Sand	CEC
1	Mendarda, Guj	86.7	10.7	2.6	58.7
2	Ratanvav, Guj	86.5	7.9	5.6	58.8
3	Halgera, AP	83.8	14	2.2	69.1
4	Kondraposi, OR	83.7	4.9	11.4	50
5	Putteri, TN	83.1	6.5	10.4	24.4
6	Tornal, AP	81.8	17	1.2	70
7	Dour, AP	81.5	17.7	0.8	46.8
8	Bodki, Guj	80	6.1	13.9	54.8
9	Vajramalli, Kar	79.7	11.1	9.2	63.1
10	Narsingi, AP	79.5	17.2	3.3	66.5
11	Malatippa, AP	78.7	17.3	4	
12	Lilvan, Guj	78.2	7.5	14.3	52.8
13	Singarbugda, AP	78	17.1	4.9	65
14	Jagganathpur, WB	78	19.6	2.4	46.8
15	Dholi, Guj	76.8	16.5	6.6	50.6
16	Loni, Maha	76.3	18.8	4.9	63
17	Kalegaon, Maha	76.2	21	2.8	57.5
18	Mednipur, WB	75.9	21.7	2.4	46.9
19	Singarbuguda, AP	75.4	17.5	7.1	70
20	Wandil, Maha	74.6	15.1	10.3	69.4

High values of CEC indicate presence of swelling and shrinking smectite-dominant clay mineral fine clays

Source Soil series of India (state wise), ICAR-NBSS&LUP, Nagpur, Maharashtra

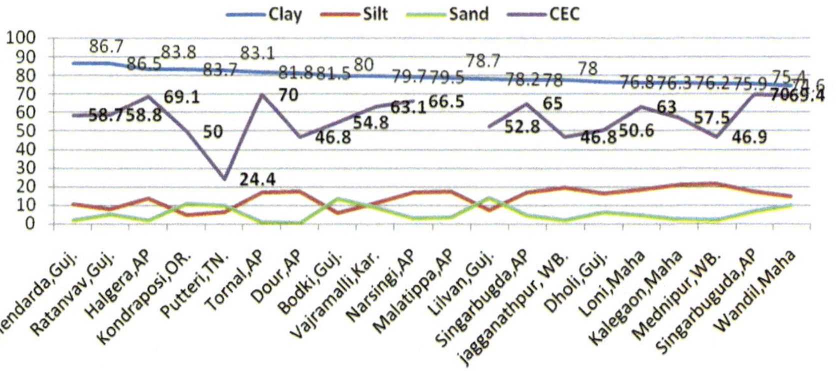

Fig. 3.21 Selected benchmark (top 20) soils based on highest clay content (in descending order) with the desirable range of silt and sand proportions. High values of CEC indicate presence of swelling and shrinking smectite-dominant clay mineral fine clays

Table 3.3 Selected benchmark soils (top 20) based on highest CEC values (in descending order) and within the desirable range of clay, silt, and sand proportions

S. No.	Location	CEC	Clay	Silt	Sand
1	Khuntwada, Guj	79	51.5	19.2	29.3
2	Vihigaon, NE	71.3	63.6	31.9	4.5
3	Tornal, AP	70	81.8	17	1.2
4	Singarbuguda, AP	70	75.4	17.5	7.1
5	Wandil, Maha	69.4	74.6	15.1	10.3
6	Halgera, AP	69.1	83.8	14	2.2
7	Sirwal, Kar	67	68.8	22.6	8.6
8	Nidasanur, Kar	66.9	71.5	20	8.5
9	Narsingi, AP	66.5	79.5	17.2	3.3
10	Gobbargumpi, Kar	66.1	70.1	20.1	9.8
11	Kodangipatti, TN	65.6	65	15	19
12	Pipalia, Guj	65.2	68.4	12.1	19.5
13	Sollapuram, APr	65.2	71.7	22.3	6
14	Karenhalli, Kar	65.2	60.8	28.1	11.1
15	Kovilpatti, TN	65.2	72.1	17.7	10.2
16	Singarbugda, AP	65	78	17.1	4.9
17	Bhatumbra, Kar	64.4	66.1	29.5	4.4
18	Bhoirepathar, Maha	63.9	53.5	30	16.5
19	Masala, Maha	63.9	60.8	29.4	9.8
20	Hulihalli, Kar	63.5	62	26.6	11.4

High values of CEC indicate presence of swelling and shrinking smectite-dominant clay mineral fine clays

Source Soil series of India (state wise), ICAR-NBSS-LUP, Nagpur, Maharashtra

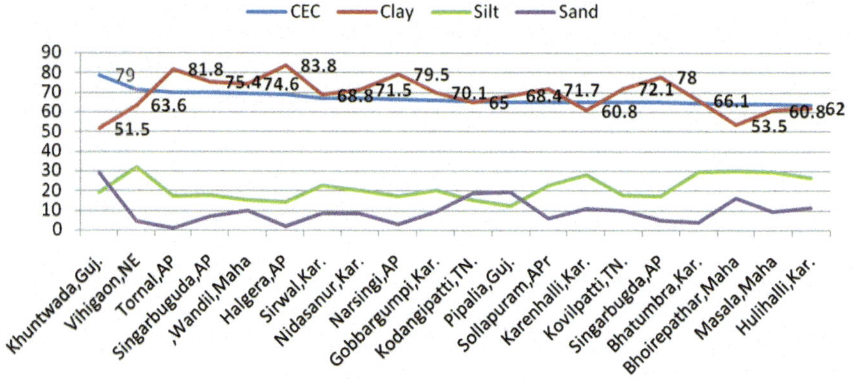

Fig. 3.22 Selected benchmark soils (top 20) based on highest CEC values (in descending order) and within the desirable range of clay, silt, and sand proportions. High values of CEC indicate presence of swelling and shrinking smectite-dominant clay mineral fine clays. *Source* Soil series of India (state wise), ICAR-NBSS-LUP, Nagpur, Maharashtra

3.5 Critical Analysis of Selected Benchmarked Pitch Soil Samples[3]

The desirable variations in cricket pitch behaviour are not only attained by cricket pitch soils with the highest clay content and smectite as the dominant clay mineral or with the highest non-swelling and shrinking kaolinite-dominant minerals having non-proportionate particle size distribution, but also by pitch soils with good flocculated structure with recommended particle size distribution (PSD) and high smectite fine clays in a desirable pH range (neutral to slightly alkaline conditions). The ideal pitch behaviour can be achieved by creating benchmarked soil data for cricket pitches based on the existing climatic and weather variables and the suggested morphological, physical, chemical, and mineralogical index properties.

Though the clay percentage and their proportion of sand and silt in Table 3.4 and Fig. 3.23 are in consonance with the desirable particle size distribution ranges, they get disqualified when we consider the CEC of these soils. Low CEC values indicate the presence of fine clay minerals that don't swell or shrink and are mostly made of kaolinite at very low pH ranges and with silt and fine sand in the disproportionate amounts. These types of pitch soils are predominant in the north-eastern states and some parts of Kerala, which have high rainfall patterns (as depicted in Figs. 3.15, 3.16, and 3.17).

Clay CEC (soil CEC*100/clay%) below 16 is considered dominantly kaolinitic, which may not be suitable soils to achieve the objectives and variations desired in cricket pitches. Generally, these types of soils having CEC below 16 and are kaolinitic in mineralogical composition and thus lack the desirable swelling and shrinking capacity to generate variations in pitch behaviour, and after-match recovery of the pitch surface and the turf grass is very poor. They also lack the optimum moisture content (OMC) desirable to generate a good pitch surface structure and density, which results in a powdery or crumbled pitch surface (during later stages of the match, especially the 3rd, 4th, or 5th day of the match), making them spinner friendly and with low unpredictable bounce.

[3] Based on proportional particle size distribution (PSD), physical, chemical, mineralogical, and engineering index-based properties.

Table 3.4 Benchmarked soils in ascending order of CEC values

S. No.	Location	Clay	CEC	Silt	Sand
1	Thoramukh, Assam	48.5	3.74	21.8	29.7
2	Myngkre, Meghalaya	40.5	4.7	18.4	41.1
3	Nongstoin, Meghalaya	56.6	4.9	32.1	11.3
4	Churanad, Kerala	48.5	5.8	7.5	44
5	Sitalajakhla, Assam	61	6.4	12.4	26.62
6	Ramjonggiri, Meghalaya	42.9	6.56	33.7	23.4
7	Umkrem, Meghalaya	54.9	6.8	25.8	19.3
8	Nongehram, Meghalaya	56.1	6.9	24.7	19.2
9	Kirukuzhi, Kerala	58.9	7.3	8.2	32.9
10	Angel Valley, Kerala	46.3	7.4	25	28.7
11	Thrikkannamangal, Kerala	63.9	7.5	6.1	30
12	Kollur, Karnataka	66.7	8	18.9	14.4
13	Mekkadembu, Kerala	62.5	8.1	27.8	9.7
14	Tekkumaradi, Kerala	55.5	8.2	17.6	26.9
15	Ramagund, Karnataka	68.2	9	18.4	13.4
16	Teppa Kadu, T. N.	56.5	9.1	26.2	17.3
17	Bangaon, Assam	29.2	9.4	15.7	31.1
18	Kollur, Karnataka	68.4	9.7	16.1	15.5
19	Giriappanamane, Kar	53.6	10.1	18.6	27.8
20 ara>	Karuvakurchi, TN	58.3	10.2	1.4	40.8

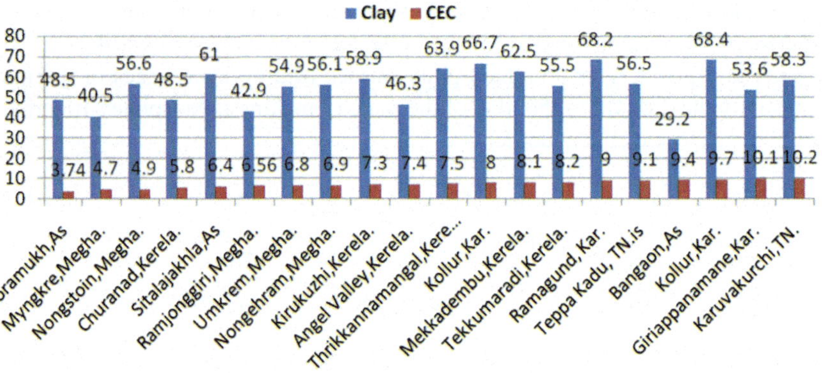

Fig. 3.23 Selected soil samples with high clay contents but having low CEC indicating the presence of non-expanding non-active and undesirable clays such as kaolinites

The same analogy as non-proportionate particle size distribution with very high clay percentage and smectite minerals (as in the states of Maharashtra, Madhya Pradesh, Gujarat, Rajasthan, Karnataka, Andhra Pradesh, Odisha, etc.) can be correlated with the cricket pitch soils used for raising the Perth and Harvey soils in Australia. Though the soil sample from Purna Valley, Maharashtra (as depicted in Figs. 3.24, 3.25, and 3.26), or from the states of Maharashtra, Madhya Pradesh, Gujrat, Rajasthan, Karnataka, Andhra Pradesh, and Odisha, contains a very high proportion of clay percentage (as Ca^{+}/ Mg^{+}-dominant smectite as dominant clay minerals due to their high CEC, COLE, and VSP values, as derived from Fig. 3.24, 3.25 and 3.26 and from the state-wise benchmarking table, as annexed as annexure 3), this indicates the presence of very high swelling and shrinking smectite-dominant mineral fine clays.

But due to the presence of a non-proportionate particle size distribution (of clay, silt, and sand proportions) with a very high percentage of silt (about 30%, as observed in Figs. 3.4 and 3.27, Andhra Pradesh, Odisha, etc.), these soils can be correlated with the cricket pitch soils used for raising the Perth and Harvey soils in Australia. Though the soil sample from Purna Valley, Maharashtra (as depicted in Figs. 3.24, 3.25 and 3.26), or from the states of Maharashtra, Madhya Pradesh, Gujrat, Rajasthan, Karnataka, Andhra Pradesh, and Odisha, contains a very high proportion of clay minerals (as Ca^{+}/Mg^{+}-dominant smectite as dominant clay) due to their high CEC, COLE, and VSP, these types of soils are generally not suitable for accepting them for benchmarking or using them for constructing a pitch profile.

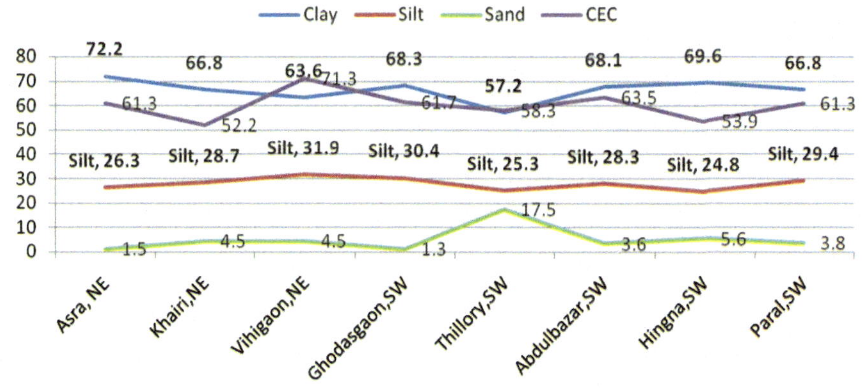

Fig. 3.24 Non-proportionate particle size distribution with CEC of bench marked soils of Purna Valley, Amrawati, Maharashtra

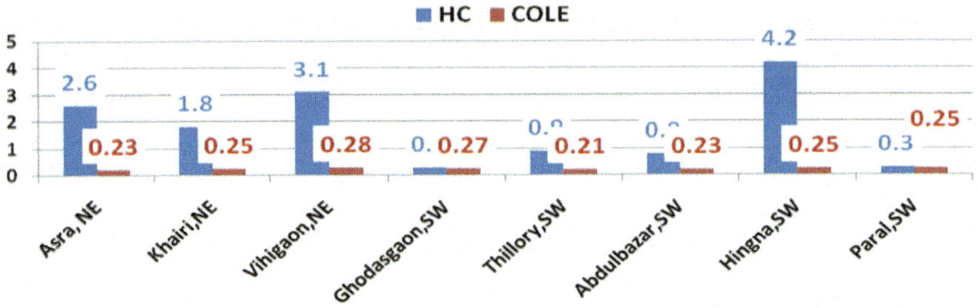

Fig. 3.25 Hydraulic conductivity (HC) and coefficient of linear extensibility (COLE) values of benchmarked soils of Purna Valley, Amrawati, Maharashtra

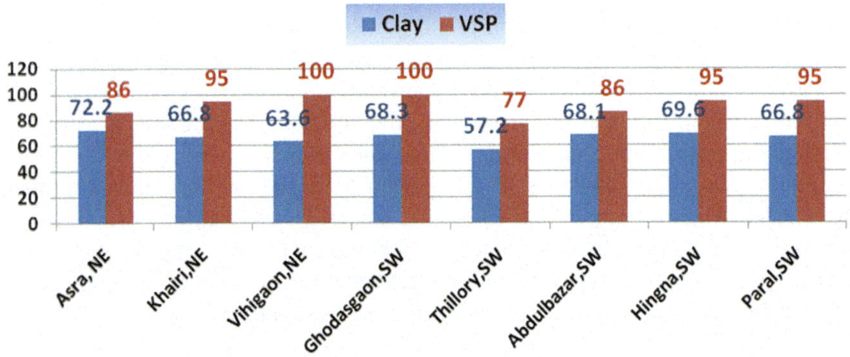

Fig. 3.26 Clay percentage and volumetric shrinkage percentage (VSP) values of bench marked soils of Purna Valley, Amrawati, Maharashtra

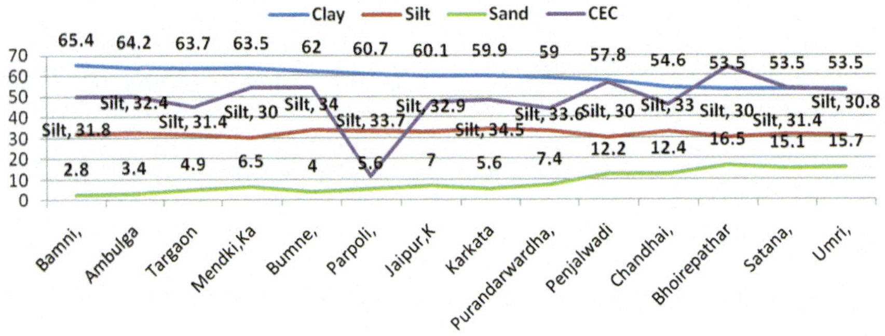

Fig. 3.27 Non-proportionate particle size distribution of some of the bench marked soils from the state of Maharashtra

3.6 Summary

The desirable pitch soil quality or standards with variations in their behaviour during the game (especially Test matches) can only be achieved through proper selection, analysis, and use of the active swell-shrink soil behaviour with proportionate particle size distribution and having the physical, chemical, and mineralogical properties in the defined index range. For constructing any cricket pitch, much more concentration and effort are confined to textural, physical, and chemical analysis only. Research on cricket pitch clay mineralogy and various engineering properties is still in its early stages for constructing any pitch profile or its preparation methods. It is worth noting that neither the Indian pitch soils nor their predominant soil mineralogical type or composition have been investigated, nor has any scientific study or research been conducted to assess their impact on the behaviour of cricket pitches.

You can get different types of pitch not only by using a pitch soil with the right minerals or the most clay, but also by making sure the clay, silt, sand, organic matter, and other ingredients are in the right amounts. Every major (clay, silt, sand, turf grass, air, moisture) or minor (organic humus, soil pH levels, predominant exchangeable cations, etc.) pitch soil component in defined proportions, along with the available pitch preparation methods and techniques, and existing continental or subcontinental climatic and weather variables (temperature, relative humidity, evapotranspiration rates, dew, etc.), are equally important and have a specific predetermined role to play in defining the desirable pitch behavioural patterns. It goes without saying that the proportion and quantity of silt and sand (as recommended) in the pitch soil are equally important, as their proportionate ratio provides the desired broad-based pitch behaviour during the various phases of the match.

Out of the total benchmark soils based on desirable morphological, physical, chemical, and mineralogical index properties, about 50 of them have been selected strictly based on their recommended proportionate particle size distribution and physical, chemical, and mineralogical index properties. Some of the benchmark soils have the highest clay content or the highest CEC but lack the desired proportion of clay, silt, and sand. Further, the best 20 pitch benchmark soil samples were selected based on their favourable mineralogy, proportionate particle size distribution (PSD), and other physical and chemical properties that form a useful combination to achieve the desired pitch behaviour objectives.

So, it's important that we choose benchmark soils that have the best mix of physical, chemical, and mineralogical properties based on their recommended proportionate particle size distribution and physical, chemical, and mineralogical index properties, so that competitive cricket pitches can be made to make the game of cricket better.

Based on the scientifically researched and approved index properties, the benchmarking of the cricket pitch soil can be done, which will help us build a nationwide data base of all the different types of soils desired in constructing the pitch profile. Based on these indexed and benchmarked cricket pitch soils, we can make the Pitch Behaviour Analysis Index (PBAI) and Pitch Behaviour Forecasting (PBF), which will help us narrowly categorise, classify, understand, analyse, and predict the complex pitch behaviour.

References/Bibliography

BCCI Curators Certification Course Manual (2013) BCCI, Second Edition

Bhattacharya T, Chandran P, Ray SK, Mandal C, Pal DK, Venugopalan MV, Durge SL, Srivastava P, Dubey PN, Kamble GK, Sharma RP, Wani SP, Rego TJ, Ramesh V, Manna MC (2006) Morphological properties of red and black soils of selected benchmark spots in semi arid tropics of India, Global Theme on Agro-ecosystems Report no. 21, National Bureau of Soil Survey and Land Use Planning, ICAR, Nagpur, ICAR, 2006

Bhattacharyya T, Chandran P, Ray SK, Mandal (Mrs) C, Pal DK, Venugopalan MV, Durge SL, Srivastava P, Dubey PN, Kamble GK, Sharma RP, Wani SP, Rego TJ, Pathak P, Ramesh V, Manna MC, Sahrawat KL (2007) Physical and chemical properties of red and black soils of selected benchmark spots for carbon sequestration studies in semi-arid tropics of India, Global theme on agro-ecosystems, report no.- 35, National Bureau of Soil Survey and Land Use Planning, ICAR, Nagpur, ICRISAT Patancheru (A.P) and New Delhi , CRIDA Hyderabad and ICAR New Delhi

Brady NC, Weil Ray R (2008) Nature and properties of soil, 14th edn. Prentice Hall

Challa O, Gajbhiye KS, Velayutham (2008) Soil series of Maharashtra, Technical bulletin, NBSS Pub. 79, National Bureau of Soil Survey and Land Use Planning, ICAR, Nagpur and Department of Agriculture, Govt. of Maharashtra

Dexter AR (1998) Advances in characterization of soil structure. Soil Till Res 11:199

ECB Staff (2007) Recommended guidelines for the construction, preparation and maintenance of cricket pitches and outfields at all levels of the game, England and Wales Cricket Board, Lord's Cricket Ground, London, UK

Haris JR (1978) Maintenance of soil structure under playing turf. In: Emerson WW, Bond RD, Dexter AR (eds) Modification of soil structure. Wiley

Ivapullaiah PV, Sitaram TG, Rao KSS (1987) Modified free swell index for clays. ASTM Int 10(2), pp 80–85

Kolhe AH, Chandran P, Ray SK, Bhattachatyya T, Pal DK, Sarkar D (2011) Genesis of associated red and black shrink-swell soils of Maharashtra. Clay Res 30(2); Division of Soil Resource Studies, National Bureau of Soil Survey and Land Use Planning, ICAR, Nagpur, ICRISAT Patancheru (A.P.)

Lotse EG, Datta NP, Tomar KP, Motsara MR, Mineralogical composition of some red and black soils of India, Nuclear Research Laboratory, IARI, PUSA, New Delhi, pp 216–226

Mahapatra SK, Singh SP, Ram G, Gajbhiye KS (2004) Soil series of Delhi state, Technical bulletin, Pub. No. 112, National Bureau of Soil Survey and Land Use Planning, ICAR, Nagpur

Means RE, Parcher JV (1965) Physical properties of soils. Prentice Hall of India (P) Limited, New Delhi

Mehra RK (2014) A textbook of soil science. Indian Council for Agricultural Research Pub

Mitchell K (2010) Cricket wickets. Copyright ©Queensland Cricket, pp 1–37

Mitchell PH, Graham J, Williams DJ (1972) Cracking in drying soils. Can Geotech J 29:263–277

Natrajan A, Reddy PSA, Sehgal J, Velayutham M (1996) Soil resources of Tamil Nadu for land use planning. NBSS Pub. No-46, National Bureau of Soil Survey and Land Use Planning, ICAR, Nagpur and Department of Agriculture, Coimbatore

Nayak DC, Dipak S, Velayutham M, Soil series of West Bengal, Technical bulletin, Pub. No-89, National Bureau of Soil Survey and Land Use Planning, ICAR, Nagpur, NBSS Regional Centre, Kolkata and Department of Agriculture, West Bengal

NSRI (2001) A guide to better soil structure. National Soil Resource Institute, Cranfield University, Silsoe, UK

Pal DK, Bhattacharya T, Deshpande SB, Sarma VAK, Velayutham M (2000) Significance of minerals in soil environment of India, NBSS Review Series1. ISBN: 81-85460-57-4, September, 2000

Pal DK, Deshpande SB, Velayutham M, Srivastava P, Durge SL (2000) Climate change and polygenesis in vertisols of the Purna Valley (Maharashtra) and their management, Research bulletin, Pub. No. 83, National Bureau of Soil Survey and Land Use Planning, ICAR, Nagpur

Parsons S, The aeration of clay soils in cricket, engineering doctorate (Academic year 2008–2012), School of Applied Sciences, Cranfield University

Prasad Shiva CR, Reddy RS, Sehgal J, Velayutham M (1998) Soils of Karnataka—their kinds, distribution, characterization and interpretation for Optimizing Land Use, NBSS Pub. No-47, National Bureau of Soil Survey and Land Use Planning, ICAR, Nagpur and Department of Agriculture and State Soil Survey, Karnataka

Rathinaswamy A, Saliha BB (2014) Fundamental of soil science. Scientific Publication

Raychaudhuri SP, Roy BB, Gupta SP, Dewan ML (1963) Black soils of India, National Institute of Soil Science, Mathura Road, New Delhi

Reddy RS, Budhilal SL, Kumar Ramesh SC, Naidu LGK (1996) Benchmark soils of Andhra Pradesh, Technical bulletin, Pub No-128, National Bureau of Soil Survey and Land Use Planning, ICAR, Nagpur and Department of Agriculture, Govt. of Andhra Pradesh

Reddy RS, Budhilal SL, Kumar Ramesh SC, Naidu LGK (1996) Soil series of Medak series, Andhra Pradesh, Technical bulletin, Pub No-127, National Bureau of Soil Survey and Land Use Planning, ICAR, Nagpur and Department of Agriculture, Govt. of Andhra Pradesh

Rose CW (1966a) Agricultural physics. Pergamon Press, Oxford

Sahdev CB, Lal T, Rana KPC, Sehgal (1995) Soil series of Haryana for optimizing land use, Soils of India Series 3, NBSS Pub. No-44, National Bureau of Soil Survey and Land Use Planning, Department of Agriculture, Haryana

Sahoo AK, Dipak S, Gajbhiye KS (1996) Soil series of Bihar. Technical Bulletin, Pub. No. 98, National Bureau of Soil Survey and Land Use Planning, ICAR, Nagpur and RAU, Pusa, Bihar

Sahoo AK, Dipak S, Gajbhiye KS, Sah KD (1998) Soil series of Orissa, Technical bulletin, Pub No. 119, National Bureau of Soil Survey and Land Use Planning, ICAR, Nagpur and Directorate of Soil Conservation (Survey), Orissa

Salini U, Girish MS (2011) Suitability of soil for cricket pitch preparation, College of Engineering Trivandrum, Technical paper at NATCON-2011, Centre for Engineering Research and Development, Kerala, 28–29 January

Sarkar D, Nayak DC, Maurya UK, Gajbhiye KS (2008) Mineralogy of benchmark soils of West Bengal, Technical bulletin, NBSS Pub No-139, National Bureau of Soil Survey and Land Use Planning, ICAR, Nagpur

Sehgal J, Gajbhiye KS, Batta RK, Sharma VAK (1999) Swell-shrink soils (Vertisols) of India, 1st edn. Kalyani Publishers

Sohan L, Tapas B, Jagdish P (1990) Benchmark swell-shrink soils of India. Morphology, characteristics and classification, National Bureau of Soil Survey and Land Use Planning, ICAR, Nagpur

Sohan L, Deshpande SB, Sehgal J (1990) Soil series of India, National Bureau of Soil Survey and Land Use Planning, ICAR, Nagpur

Sharma JP, Giri JD, Shyampura RL, Gajbhiye KS (1999) Soil series of Gujarat, Technical bulletin, NBSS Pub. 120, National Bureau of Soil Survey and Land Use Planning, ICAR, Nagpur and Department of Agriculture, Govt. of Gujarat, ISBN: 81-85460-99-X

Shyampura RL, Singh SK, Singh RS, Jain BL, Gajbhiye KS (1981) Soil series of Rajasthan, Technical bulletin, Pub. No 96, National Bureau of Soil Survey and Land Use Planning, ICAR, Nagpur and Department of Agriculture, Govt. of Rajasthan and CAZRI, Jodhpur, ISBN: 81-85460-75-2

Sidhu GS, Walia CS, Tarsem L, Rana KPC, Sehgal (1995) Soil series of Punjab for optimizing land use, soils of India Series 2, NBSS Pub. No-45, National Bureau of Soil Survey and Land Use Planning, Department of Agriculture, Punjab

Singh DN (2007) A report on analysis of soils for cricketing pitches of all the cricket associations across the country, Submitted to The Board of Control for Cricket in India (BCCI), Mumbai, Department of Civil Engineering, Indian Institute of Technology Bombay, Mumbai

Singh RS, Utpal B, Dipak S, Butte PS, Gajbhiye KS (1999) Soil series of Meghalaya, Technical bulletin, Pub. No. 121, National Bureau of Soil Survey and Land Use Planning, Department of Agriculture, Meghalaya. ISBN: 81-85460-49-3

Soil Resource Information of Different Agro- Eco Sub-Regions (ASER) of India for crop and soil carbon modelling (2016) Technical Bulletin, NBSS and LUP Publication 1035, ICAR, Nagpur

Soil Series of Kerala (1996) Technical bulletin, Pub. No-136, National Bureau of Soil Survey and Land Use Planning, ICAR, Nagpur and Department of Agriculture, Kerala

Tamgadge DB, Gajbhiye KS, Velayutham M, Kaushal GS (1996) Soil series of Madhya Pradesh, Technical Bulletin, Pub No 78, National Bureau of Soil Survey and Land Use Planning, ICAR, Nagpur and Department of Agriculture, Govt. of Madhya Pradesh

Tamgadge DB, Gajbhiye KS, Pandey GP (2010) Soil series of Chhattisgarh State, Technical bulletin, Pub. No 85, National Bureau of Soil Survey and Land Use Planning, ICAR, Nagpur and Department of Agriculture, Govt. of Chhattisgarh State

Vadivelu S, Sen TK, Bhaskar BP, Baruah U, Sarkar D, Maji AK, Gajbhiye (1999) Soil series of Assam, Technical bulletin, Pub. No. 101, National Bureau of Soil Survey and Land Use Planning, ICAR, Nagpur and Department of Agriculture, Assam, ISBN: 81-85460-50-7

Wani S, Pralhad R, Johan O, Yousef T (2006) Global theme on agro-ecosystems, ISAT Patancheru (A.P.), CRIDA Hyderabad and Indian Institute of Soil Science Bhopal, Report No-21

Chapter 4
Effect of Soil Chemical Properties on Cricket Pitch Behaviour

4.1 Introduction

Soil chemical properties that directly or indirectly affect cricket pitch behaviour are pH, predominant exchangeable cations, and organic humus. No other single chemical soil characteristic is more important in determining the chemical environment of cricket pitch soil and pitch grass than its pH. Soil pH is largely controlled by soil colloids and their associated exchangeable cations, which directly affect the swell-shrink behaviour or crumbling behaviour of the pitch profile and hence control the pitch surface density levels during the course of the match. Though these soil colloids and predominately exchangeable cations are present in minuscule amounts, their direct or indirect effect on cricket pitch behaviour is quite significant.

4.2 What is Soil pH?

Soil pH is a measure of the acidity or alkalinity of the soil. It is also called a soil reaction. Soil reaction is a term used to indicate the acid–base reaction in soils (Tisdale and Nelson 1975; Brady 1974, 1980). Commonly expressed in terms of soil pH, the terms "acid," "neutral," and "basic" (ranging from pH 0 to 14) are derived and expressed as the negative logarithm of the H^+ ion concentration ($-\log_{10}[H^+]$) (Table 4.1).

S. B. Singh et al., *Cricket Pitches*, Springer Transactions in Civil and Environmental Engineering, https://doi.org/10.1007/978-981-99-2913-9_4

Table 4.1 Most common classes of soil pH according to the United States Department of Agriculture (Soil Survey Staff, 2002)

Terms	pH range
Ultra-acid	< 3.5
Extremely acid	3.5–4.4
Very strongly acid	4.5–5.0
Strongly acid	5.1–5.5
Moderately acid	5.6–6.0
Slightly acid	6.1–6.5
Neutral	6.6–7.3
Slightly alkaline	7.4–7.8
Moderately alkaline	7.9–8.4
Strongly alkaline	8.5–9.0
Very strongly alkaline	> 9

Brady and Weil (1996) considered the soil pH as a master variable that affects most of the soil properties. They reported that pH will not only affect the chemical and biological properties of soils but also the physical properties, including air and water movement in the soil mass. This movement, retention, and release of pitch soil moisture and air within and just above the pitch profile (due to the colloidal behaviour or predominant exchangeable cations and minerals, soil pH, or weather variables) generate sand and regulate the desired compression levels and bulk density of the pitch surface during the course of the match or during the pitch preparation.

4.2.1 *Acid Soils*

An acidic soil is common in all regions where precipitation is high enough to leach appreciable quantities of exchangeable cations (Ca^{+2}, Mg^{+2}, K^{+}, and Na^{+}) from the surface layer of the soil, thus leading to the exchange complex being dominated by H^{+} ions.

Two adsorbed ions, hydrogen and aluminium, are largely responsible for soil acidity. The mechanism by which these two cations exert their influence depends on the degree of soil acidity and on the source and nature of the soil colloids. Available aluminium becomes soluble in strongly acidic soils (pH less than 5) and is either tightly bound by organic matter or exists in the form of aluminium (Al^{+3}) or aluminium hydroxy cations Al $(OH)^{+2}$. The negative charge of soil colloids, which at low pH values are dominant permanent charges associated with 2:1 type silicate clay minerals, adsorbed these exchangeable ions on the clay micelle in preference to other cations.

In moderately acid soils having pH values between 5 and 6.5, aluminium and hydrogen compounds also account for H^{+} ions present in soil solution, but are held to the clay by different mechanisms. The percentage base saturations in moderately acid soils are slightly higher than in strongly acid soils. Aluminium ions can no longer exist as ions because they have been converted to aluminium hydroxy ions. Some of the aluminium hydroxy ions are adsorbed and act as exchangeable cations. As such, they may be in equilibrium within the soil solution in a similar way as the Al^{+3} ions are in strongly acidic soils. In moderately acid soils, the small amount of readily exchangeable hydrogen contributes to soil acidity in the similar manner as shown for strongly acidic soils. Furthermore, as pH rises, some hydrogen atoms that were bound tenaciously through covalent bonding by organic matter, Fe and Al oxides, and 1:1 type clay at low pH are now subject to release as H^+ ions.

4.2.2 *Classification of Soil Acidity*

Generally, there are three types of soil acidity: (1) active (soil solution) acidity; (2) salt-replaceable (exchangeable) acidity; and (3) residual acidity.

1. Active soil solution acidity is a measure of the H^+ ion activity in the soil solution at any given time. However, the quantity of $[H^+]$ ions owing to active acidity is very small as compared to the quantities of exchangeable and residual acidity. For example, only about 2 kg of calcium carbonate would be required to neutralise the active acidity in a hectare-furrow slice of soil at pH 4 and 20% moisture.
2. Salt-replaceable (exchangeable) acidity: this type of acidity is primarily associated with the exchangeable aluminium hydroxy ions and hydrogen ions that are present in the largest quantities in strongly acidic soils. These ions can be released into the soil solution by an unbuffered salt such as KCl. In moderately acid soils, the quantity of easily exchangeable aluminium and hydrogen is quite limited. At a given pH value, exchangeable acidity is generally highest for smectites, intermediate for vermiculites, and lowest for kaolinite.
3. Residual acidity is that which remains in the soil after active and exchangeable acidity has been neutralised. Residual acidity is generally associated with aluminium hydroxyl ions and with hydrogen and aluminium atoms that are bound in non-exchangeable forms by organic matter and silicate clays. When lime is added to soil, the pH rises, and the aluminium hydroxy ions transform into uncharged gibbsite.

The residual acidity is commonly far greater than either the active or salt-replaceable acidity. Conservative estimates suggest that the residual acidity may be 1000 times greater than the soil solution or active acidity in sandy soil and 50,000 or even 100,000 times greater in clayey soil, which is high in organic matter.

4.3 Neutral to Alkaline Soils

Alkalinity occurs when there is a comparatively high degree of saturation with base-forming cations. The presence of calcium, magnesium, and sodium carbonates also results in a preponderance of hydroxyl ions over hydrogen ions in the soil alkalinity. Potassium carbonate (K_2CO_3), the OH-ions greatly enhance alkalinity. When salts of strong bases and weak acids, such as Na_2CO_3, K_2CO_3, and $MgCO_3$, dissolve, they undergo hydrolysis and develop alkalinity. Under such conditions, the soil is alkaline. Alkaline soils are, of course, characteristics of arid and semi-arid regions. Because of the lack of extensive leaching in arid and semi-arid soils, the level of base-forming cations is quite high; as a result, the pH is frequently 7 or higher. Positively charged aluminium and aluminium hydroxy ions are absent, and hydrogen ions are extremely low. Adsorbed Ca^{+2} and Mg^{+2} ions dominate the exchange complex, although concentrations of Na^+ and K^+ ions are generally higher in acidic soils. Organic matter is lower in dry soils than in soils in humid regions. The CEC of soils in dry areas is commonly as high as or higher than that of comparable acid soils. This is because the pH-dependent negative charges on humus, 1:1 type minerals, allophane, and Fe and Al oxides are more effective at higher pH levels.

4.4 Saline and Sodic Soils

In saline soils, salts accumulate in some surface soils of arid and semi-arid regions because there is insufficient rainfall to flush them from the upper soil layers. The salts are primarily chlorides and sulphates of calcium, magnesium, sodium, and potassium. The sources of these salts are the weathering of rocks and minerals, rainfall, groundwater, and irrigation. Once deposited or released in the soil, the salt is brought to or near the surface by upwardly moving water, which then evaporates, leaving the salts behind. In sodic soils, pH is raised by the preponderance of sodium ions in the exchange complex and the precipitation of calcium ions as $CaCO_3$. The soil solution is dominated by carbonates and bicarbonates of Ca^{2+}, Mg^{2+}, and Na^+ ions.

Fig. 4.1 Deposition of $CaCO_3$ (in powered form) on the ped and at the subsoil from where they have been excavated (from Balangir, Odisha)

The presence of calcium carbonate is very common in cricket soils that have been transported from arid or semi-arid regions with low rainfall, high temperatures, and high evapotranspiration rates. As shown in Fig. 4.1, Balangir's black-coloured, smectite-rich fine-grained soil is generally used for pitch construction in major districts of Odisha (Balangir, Sambalpur, Bhuvneshwar, and Cuttack) and in some parts of Jharkhand, Uttar Pradesh, and Assam, where it has been transported for pitch construction on a trial basis.

Carbonate precipitation (in powdered or small, whitish-yellow granules) is generally induced by either a lowering of CO_2 or by evaporation (as observed in Figs. 4.1 and 4.2). Thus, water loss through evapotranspiration (ET) is considered the primary mechanism in the precipitation of pedogenic carbonates (Arkley 1963; Scholz 1971; Rabenhorst et al. 1984). Though it affects the ET, temperature also plays an important role in controlling water flow in the soil (Arkley 1963; Ahmad 1978). Arid climates are found to be the most important factor in the formation of calcareous sodic soils (Pal et al. 1999).

When there is a sufficient amount of calcium ions in the soil solution due to the weathering of calcium-bearing minerals and the lowering of pressure carbon dioxide (PCO2) by evaporation of the soil solution and precipitation of calcite in soils, pedogenic carbonates (PC) form (Cerling 1984; Srivastava 2002; Pal et al. 2003). In thin sections, they can be observed as coatings and nodules that have diffused boundaries with the ground mass.

Fig. 4.2 Precipitated $CaCO_3$ and whitish grey/whitish-yellow granules embedded within the peds (Ca^{+2}-dominant fine smectite from Balangir black soil, Odisha, India) used for pitch profile construction at JCA cricket complex, Ranchi, Jharkhand, India

4.4.1 Saline Soils

These soils contain concentrations of neutral soluble salts sufficient to interfere seriously with the growth of most plants. The EC of a saturated extract of the soil solution is more than 4 dS/m. The ESP is less than 15, and the pH is usually less than 8.5. Saline soils are called white alkali soils because of surface incrustation. Although the Na^+ ion concentration may be somewhat higher than those of Ca^{+2} and Mg^{+2}, the SAR is less than 13 for saline soils. These soils can be reclaimed by merely leaching out the salts, an important consideration in their management.

4.4.2 Saline-Sodic Soils

These soils contain appreciable quantities of sodium ions that seriously affect most of the plants. The ESP is greater than 15, and the EC of a saturated extract is more than 4 dS/m. The pH is commonly 8.5 or less, and the SAR is at least 13 in these sodic soils, though they do not contain a great deal of soluble salts. The detrimental effects of these soils on turf grass are not only due to the toxicity of Na^+, HCO_3^-, and OH-ions but also due to the reduced water infiltration and aeration. The high pH is largely due to the hydrolysis of sodium carbonate.

Sodic soils are usually in an unsatisfactory physical condition for crop growth due to the deflocculating (clay dispersion) influence of sodium, but they are desirable conditions for cricket pitches, as these types of soils with high water holding capacity (WHC due to hydration of Na^+ -dominant smectitic clay), compressibility, surface dry density, and less air voids (due to deflocculation and hydration of sodium/magnesium-dominant clay usually generates high pace and bounce on turf pitches. The only problem with these types of pitch soils (sodic soil) is that the regeneration of pitch and turf grass after the match (on an overly compacted surface) is very poor.

The ESP of sodic soils is decidedly more than 15, and the SAR is more than 13. Because the concentration of neutral soluble salts decreases with increasing hydroxide ions and pH, the adsorbed sodium is free to hydrolyse. The pH is always above 8.5, often rising to 10 or more. The EC of the saturated extract is normally less than 4 dS/m.

Because of the extreme alkalinity resulting from the high sodium content, the surface of the sodic shrink-swell soils is usually discoloured by the dispersed humus carried upward by capillary water and deposited when it evaporates. Hence, the name "black alkali" (also known as "solods," "solonetz," or "slick spots") has been used to describe these soils.

4.4.3 Soil Group Identification Based on ESP and ECe Values

(1) Saline soils, (2) Saline alkali, and (3) Non-saline alkali (sodic soils).

1. The saline soils have an ECe > 4 mm-Hg/cm at 25 °C and ESP < of 15%. Dispersion of saline soils occurs at ESP = 15%. The soil pH is ordinarily less than 8.5. Because of the presence of excess salts and a low amount of Na^+ ions in the exchange position, these soils are usually in a flocculated state, and their permeability is considered to be equal to or higher than that of the two other soils.
2. The saline-alkali soils are soils with ECe > 4 mmHg/cm and ESP > 15%. These soils have both free salts and exchangeable Na^+. As long as excess salts are present, the soil is flocculated, and the pH is normally less than or equal to 8.5. When the soils are leached, the free salt content decreases, and the soil reaction may become strongly alkaline (pH > 8.5) because of the hydrolysis of exchangeable Na^+.
3. Non-saline-alkali soils, also known as sodic soils, have an ECe of 4 mmho/cm and an ESP of greater than 15%. Most of the Na^+ is in exchangeable form, and very small amounts of free salts are present in the soil solution. The soil pH ranges from 8.5 to 10.

4.5 What Controls the Soil pH?

There are various sources that cause a soil to be acidic or alkaline in nature. In general, the pH is influenced by mineralogy, climatic variables (rainfall, temperature, relative humidity, evapotranspiration, etc.), and parent material weathering. The presence of aluminium and hydrogen ions in soil increases its acidity, whereas calcium and other base-forming cations (particularly sodium) increase its alkalinity.

Colloidal factors affecting soil pH are

A. The proportion of cation exchange capacity (CEC) that is satisfied by base-forming cations.
B. The nature of micelle.
C. The kind of adsorbed bases.
D. The level of soluble salts in the soil solution.

A. Per cent base saturation is the percentage of the CEC that is satisfied by the base-forming cations.

$$\%\ \text{base saturation (BS)} = \text{Sum of exchangeable cations (cmol/kg)}/\text{CEC(cmol/kg)}$$

A low percentage base saturation means acidity, whereas a percentage base saturation of 50–80 will result in neutrality or alkalinity. In general, in humid regions, soils dominated by silicate clays and humus are acidic if their base saturation percentage is below 80. When such soils have a base saturation percentage of 80 or above, they are usually neutral or alkaline.

2 Nature of the micelle: there is considerable variation in the pH of different types of colloids at the same percentage base saturation. This is due to differences in the ability of the colloids to furnish hydrogen ions to the soil solution. For example, the dissociation of adsorbed H^+ ions from the smectites is much higher than that from the hydrous Fe and Al oxide clays. Consequently, the pH of soils dominated by the smectites is appreciably lower than that of the oxides at the same percentage base saturation. The dissociation of adsorbed hydrogen from 1:1 silicate clays and organic matter is intermediate between that from smectites and that from hydrous oxides.
3 Kind of adsorbed base-forming cations: the comparative qualities of each of the base-forming cations present in the colloidal complex are another factor influencing soil pH. Sodium being a monovalent cation, soils with high sodium saturation have much higher pH values than those dominated by calcium and magnesium ions, which are divalent cations. Thus, at a base saturation percentage of 90, the presence of calcium, magnesium, potassium, and sodium ions in the ratio 4:1:1:9 would certainly result in a higher pH than if the ratio were 10:3:1:1. In the first instance, the sodium-calcium complex is dominated by the sodium; in the second, calcium is dominant.
4 Neutral salts in solution: the presence in the soil solution of neutral salts, such as the sulphates and chlorides of sodium, potassium, calcium, and magnesium, has a tendency to increase the acidity of the H^+ ion in solution and, consequently, decrease the soil pH.

4.6 Measuring Soil pH

Soil acidity and alkalinity are measured in units of pH. The pH scale is from 0 (most acidic) to 14 (most alkaline), and a pH of 7 is neutral. Soil pH can be measured in the field using a test kit or by sending a sample to a laboratory for more accurate results. The standard depth of sampling is 10 cm. Soil pH in the field can be measured using a simple test kit based on a colour-card method available from agricultural supply stores called the Raupach soil pH kit. The kit gives the soil pH on the water scale and should be used only as a guide to soil pH.

Analysis in a laboratory provides the most accurate measurement of soil pH. The two main laboratory methods used to determine the soil pH are calcium chloride and water. The standard method of measuring soil pH is in calcium chloride. An air-dry soil sample is mixed with five times its weight of a dilute concentration (0.01 M) of calcium chloride ($CaCl_2$), shaken for 1 h, and the pH is measured using an electrode.

The results are usually expressed as pH ($CaCl_2$). Soil pH in distilled water is used in place of 0.01 M calcium chloride, and the results are expressed as pH (w). The pH ($CaCl_2$) test is the more accurate of the two pH tests, as it reflects what the plant experiences in the soil. The values of pH ($CaCl_2$) are normally lower than pH (w) by 0.5–0.9. A useful, but not consistently accurate, conversion is to subtract 0.8 from the pH (w) to obtain a pH ($CaCl_2$) value.

4.7 Measuring Salinity and Alkalinity

Soil pH is not only the best indicator to be used for the characterisation of saline, alkaline, or sodic soils. The characteristics of soils should be assessed by the combination of their properties and not solely by their pH. The saline soils have a pH of 8.5 or lower. The sodic soils may possess a pH even higher than 10, but some of the soils may be neutral whereas others are acidic in reaction.

The predominance of exchangeable cations (Ca^{+2}, Mg^{+2}, or Na^{+}) at various saline or alkaline pH levels may not be the only criteria used to classify and distinguish cricket pitch behaviour. Kamphorst and Bolt (1976) reported that sodification does not necessarily yield a rise in soil pH levels. Many sodic soils are neutral in reaction, whereas a number of solonetzic soils are even acidic in reaction. In sodic soils with a neutral soil reaction, the Na salts are neutral salts such as Na–Cl.

As given in Table 4.2 and Fig. 4.3, Ca^{+2}/Mg^{+2}/H^{+}-dominant fine smectite soils of the Perth, Harvey and the Wamberal cricket pitch soils having high cation exchange capacity (CEC), base saturation (BS), water holding capacity (WHC), moderate exchangeable sodium percentage (ESP), dispersion index (DI) and low pH and electrolyte content (EC) values are an exception to the categorization of cricket pitch soils based on considering soil pH as an index property. In these soils from Australia, flocculation with good pitch soil structure and compressibility is observed even at low pH (4.5–5.9) due to predominance of Ca^{+2} and Mg^{+2}ions. One of the major reasons that the Perth wicket provides maximum pace and bounce as well as good after-match pitch surface and turf grass recovery is the highly flocculated soil structure with Ca^{+2} and Mg^{+2}-dominant fine smectite clay.

Table 4.2 Typical chemical properties of soils used in few cricket pitches [Perth, Harvey, and Sydney Cricket Ground (SCG)] in Australia

Soil name	Perth	Harvey	SCG	Wamberal
Clay (%)	82	78	48	55
pH	5.8	4.5	5.9	5.5
EC	1.13	1	0.56	1.02
CEC	67.1	64	30.1	29.4
Ca^{+2}	11.8	7.6	9	1.59
Mg^{+2}	24.6	22.7	1.8	2.2
Na^{+}	6	6.8	1.2	1.1
H^{+}	24.2	26.3	17.4	25.2
ESP	8.9	10.6	4	3.7
DI	10	12	4	10

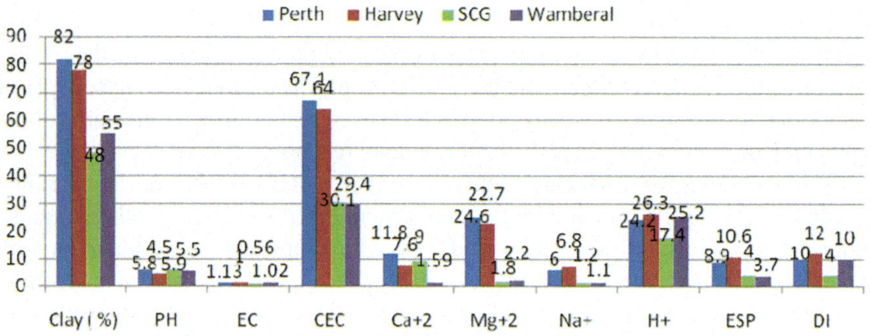

Fig. 4.3 Clay proportion, pH values (very strongly acidic to moderately acidic), CEC, ESP and predominant exchangeable cations of soils used in few cricket pitches [Perth, Harvey, Sydney Cricket Ground (SCG)] in Australia. *Source* McIntyre Keith and Don, Cricket Wickets—Science versus Fiction, 2001, p. 68

To distinguish saline and sodic soils from other soils, the US Salinity Laboratory (Richards 1954) proposed to use as criteria soluble salt and exchangeable Na^+ content. These parameters are expressed in terms of (1) electrical conductivity (EC_e) for salt content, and (2) for exchangeable salt Na^+ content, exchangeable sodium percentage (ESP). The salinity of soils is characterised by measuring the ECe in millimhos per centimetre (mmho/cm) of the saturated soil extract, and the same is obtained by suction and filtration of a water-saturated soil paste.

The exchangeable sodium percentage is calculated as

$$ESP = \left(\text{exchangeable}\,Na^+/CEC\right) * 100$$

Salinity: the salt content of the soil is estimated by methods based on the ability of the salt to conduct electricity. Laboratory measurements of the electrical conductivity

(EC) of the soil solution extracted from a saturated sample of soil give an indication of the salt levels. The salinity is commonly expressed in terms of the EC, which is measured in deci-Siemens per metre (dS/m).

Sodium status: two methods are used to characterise the sodium status of these soils. The exchangeable sodium percentage (ESP) identifies the degree to which the exchange complex is saturated with sodium. ESP levels of > 15 generally yield pH values of 8.5 and above. Higher ESP levels may bring the pH even higher, to 10.

ESP is complemented by a second, more easily measurable characteristic, such as the sodium adsorption ratio (SAR), which gives information on the comparative concentration of Na^+, Ca^{+2}, and Mg^{+2} in soil solutions.

$$\text{It is calculated as SAR} = \left(Na^+\right)/\left\{1/2\left(Ca^{2+} + Mg^{2+}\right)\right\}^{1/2}$$

where Na^+, Ca^{+2}, and Mg^{+2} are the concentrations (in $cmol_c kg^{-1}$) of the sodium, calcium, and magnesium ions in the soil solution

The sodicity hazard of irrigation water is usually evaluated by SAR (Table 4.3).

Table 4.3 Quality class of irrigation water

SAR	Quality class
< 10	Safe
10–20	Moderate
> 20	Unsafe

Using EC, ESP, and SAR characteristics and soil pH, soils affected by salt are classified as saline, saline-sodic, and sodic (Table 4.4).

Table 4.4 Classification of normal soils compared to acid, saline, sodic, and saline-sodic soils

	pH	EC (dS/m)	SAR	ESP
Normal	6.5–7.2	< 4	< 13–15	< 15[a]
Acid	< 6.5	< 4	< 13–15	< 15
Saline	< 8.5	> 4	< 13–15	< 15
Saline- sodic	< 8.5	> 4	> 13–15	> 15
Sodic	> 8.5	< 4	> 13–15	> 15

[a]*Source* Brady and Weil (2008, p. 244)

In this respect, the electrical conductivity (ECe) is frequently used as an index for salanisation hazard. The hazard of salanisation is considered as low if irrigation water used has an ECe < 0.75 mmho/cm (Richards 1954; Taylor and Ashcroft, 1972) (Table 4.5).

Table 4.5 Hazard of salinisation at different ECe range

Salinity hazard	ECe values (mmho/cm)
Low	< 0.75
Medium	0.75- 1.5
High	1.5 to 3.0
Very high	> 3

Richards (1954), Taylor and Ashcroft (1972)

In arid regions, salinisation is a natural phenomenon. Therefore, the changes for salinisation are considered very high if water with ECe = 3.0 or higher is used for irrigation over many years, even for non-saline soils (Fig. 4.4).

Table 4.6 Chemical analysis for some of the important cricket pitch soils of India—IIT Bombay, [in ascending order (pH value)]

S. No.	Sample location	EC	pH	TDS
	Low salinity—EC e < 0.75 mmhos/cm			
1	TDSA Stadium, Tinsukia, Assam	0.2938	5.64	14.76
2	Golaghat, Goalpara, Assam	0.3211	5.67	15.84
3	JSCA International Cricket Stadium, Ranchi	0.357	6.09	17.84
4	Keenan Stadium, Jamshedpur	0.393	6.42	19.68
5	KCA Stadium, Thalassery, Kerala	0.4011	6.21	20.11
6	N.F.R.S.A, Assam	0.4041	6.2	20.08
7	DDSA, Dibrugarh, Assam	0.4181	6.56	20.91
8	St. Paul's College Ground, Kalamassery, Kerala	0.4689	5.36	23.35
9	PTG Cricket Ground, Agartala	0.4745	5.24	23.78
10	Birsa Munda Cricket Stadium, Chaibasa, Jharkhand	0.5342	6.43	26.71
11	Nurlamin Stadium, Nagaon, Assam	0.5804	6.65	28.65
12	KCA Stadium, Perinthalmanna, Kerala	0.6866	6.28	34.32
	Medium salinity—EC e-0.75–1.5 mmhos/cm			
13	Wankhade Stadium, Mumbai	0.7631	5.76	34.32
14	Ravenshaw University Ground, Cuttak	0.7682	6.19	38.08
15	MBB Stadium, Agartala	0.7851	6.48	39.23
16	Nehru stadium (Soil-1), Gauhati, Assam	0.8748	6.19	43.81
17	Barabati Stadium Ground, Cuttak and Sambalpur, Odisha	0.9072	6.62	45.47
18	Nimapada Stadium, Cuttak, Odisha	1.04	7.08	52.16
19	Lucknow, U.P.	1.043	7.13	52.16
20	Sagar Cricket Ground, Sagar, MadhyaPradesh	1.052	6.69	52.6
21	Nagpur, Maharashtra	1.121	6.44	56
22	BITS, Pilani, K.K Birla Goa Campus Cricket Ground, Goa	1.121	6.44	56

(continued)

(continued)

S. No.	Sample location	EC	pH	TDS
23	Jawaharlal Nehru International Stadium, Kochi, Kerala	1.202	6.16	54.09
24	Mangaldai Sports Association, Assam	1.246	6.24	62.39
25	Nehru stadium (Soil-2), Gauhati, Assam	1.274	6.49	63.64
26	Luhnu Cricket Stadium, Bilaspur, Himachal Pradesh	1.282	6.9	64.55
27	Sonwar, Srinagar, J and K	1.296	7.45	67.35
28	Mohali, Chandigarh	1.366	7.56	61.6
	High salinity—EC e-1.5 to 3.00 mmhos/cm			
29	Mohan Meakins Cricket Ground, Ghaziabad, Uttar Pradesh	1.556	7.5	77.92
30	Porvorim Ground, Goa	1.584	6.86	79.04
31	Sardar Patel Gujarat Stadium, Motera, Ahmedabad, Gujarat (Brown soil)	1.597	6.94	79.89
32	Bidanashi Stadium, Cuttak, Odisha	1.621	6.62	81.11
33	Railway Sports Promotion Board (RSPB), Delhi	1.672	7.26	75.27
34	Capt. Roop Singh Stadium, Gwalior, Madhya Pradesh	1.679	7.43	84.12
35	Indira Stadium, UNA, Himachal Pradesh	1.76	6.93	88.2
36	HPCA Cricket Stadium, Dharmshala, Himachal Pradesh	1.766	6.94	88.37
37	JNV Pekhubela, UNA, Himachal Pradesh	1.784	6.97	89.06
38	Shivaji Stadium, Ratnagiri, Maharashtra	1.789	6.95	90.02
39	Bhamashala Park, Meerut, Uttar Pradesh	1.804	6.75	90.14
40	Silchar, Assam	1.885	6.29	94.33
41	SCA Ground, Khanderi, Saurashtra, Gujarat	1.945	7.28	94.46
42	Subrata Roy Sahara Stadium, Gahunje, Pune, Maharashtra	1.949	7.28	97.4
43	KSD Shambhag School Ground, Satara, Maharashtra	2.025	6.81	101.2
44	Anant Kanhere Golf Maidan, Nasik, Maharashtra	2.025	7.22	101.1
45	Atal Bihari Vajpayee Cricket Stadium, Amtar, Himachal Pradesh	2.026	7.07	100.3
46	Sardar Patel Gujarat Stadium, Motera, Ahmedabad, Gujarat (Reddish soil)	2.031	7.3	91.54
47	RCA Academy Ground, Jaipur, Rajasthan	2.047	7.51	92.31
48	PYC Hindu Gymkhana, Pune, Maharashtra	2.116	7.14	95.17
49	Poona Club Ground, Pune, Maharashtra	2.219	7.7	100.6
50	Emerald Heights School, Indore, Madhya Pradesh	2.226	7.01	109.7
51	Race course ground, Saurashtra, Gujarat	2.396	7.3	118.9
52	Kamla Club Ground, Kanpur, Uttar Pradesh	2.403	6.82	122.7
53	Fort Maidan, Palakkad, Kerala	2.501	6.28	124.6
54	Railway Cricket Ground, Saurashtra, Gujarat	2.509	7.33	124.9

(continued)

(continued)

S. No.	Sample location	EC	pH	TDS
55	YSR-ACA-VDCA Stadium, Vishakapattnam, Andhra Pradesh	2.729	6.99	137.1
56	VCG, Rajiv Gandhi International Stadium Hyderabad, Telangana	2.844	6.8	142.2
	Very high salinity > 3 mmhos/cm			
57	Bandra Kurla Complex Stadium Mumbai, Maharashtra	3.016	6.53	150.4
58	Holkar Stadium, Indore, Madhya Pradesh	3.022	7.63	151.1
59	MA Chidambaram Stadium, Chennai, Tamilnadu (Black Clay)	3.034	7.07	151.9
60	M Chinnaswamy Stadium, Bangalore, Karnataka	3.039	7.91	137.6
61	Palam Ground (B), New Delhi	3.179	7.45	158.6
62	SMS Stadium, Rajasthan	3.253	7.09	161.9
63	Deccan Gymkhana, Poona, Maharashtra	3.295	7.18	164.2
64	Rewa Cricket Ground, Rewa, Madhya Pradesh	3.878	7.63	195.8
65	Eden Gardens Ground (Kolkata), Salt Lake, Kalyani	4.222	6.49	211.2
66	Kandivili, Mumbai, Maharashtra	4.267	5.99	212.6
67	FerozshahKotla Ground, Delhi	4.494	7.75	225.1
68	Motibaug, Baroda, Gujarat	6.498	7.14	324.1
69	Alembic, Baroda, Gujarat	7.002	7.86	316.5
70	GDA, Ghaziabad, Uttar Pradesh	12.12	7.92	606.2
71	Bhiwani, Haryana	16.06	7.12	809.2

Note EC = Electrical conductivity (S = micro-Siemens/cm), TDS = total dissolved solids; ppm = parts per million
Data source: Singh, D.N. (2008)
n

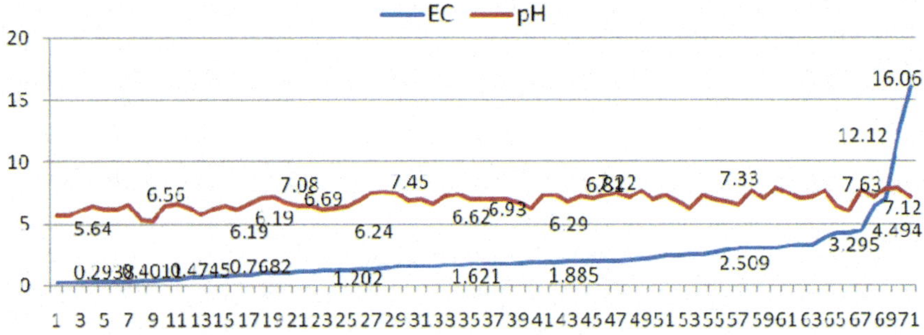

Fig. 4.4 Comparative analysis of ECe (ascending order) and pH values of some of the major cricket pitch soils of India (as tabulated in Table 4.5)

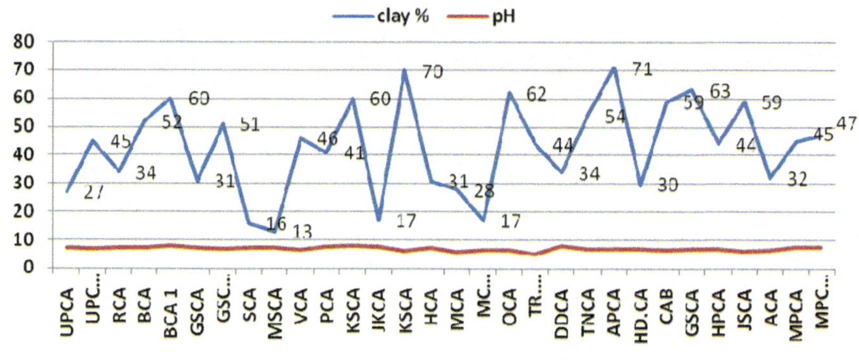

Fig. 4.5 Variation of per cent clay content and pH values of some Test cricket pitch soils in India

4.8 Chemical Analysis and Interpretation for Some of India's Most Important Cricket Pitch Soils

Most of the cricket pitch soil samples (as shown in Figs. 4.4 and 4.5 and Table 4.6) collected during national and international matches in India are found to be in the desirable pH range, either in the neutral (6.6–7.3) or slightly alkaline (7.4–7.8) range.

Cricket pitch soil samples for the state of Assam, Meghalaya, Mizoram, Manipur, Nagaland, Sikkim, Arunachal Pradesh, and Kerala which are prepared with 1:1 kaolinitic-dominant clay, formed under very high rainfall, strongly to slightly acidic conditions, high in organic matter content, low base saturation (showing H^+/Al^{+3}-dominant exchangeable cations generating honey comb structure with high air voids generating high BD values and compressibility but are with very poor grass growth and after-match recovery of the pitches and the turf grasses (as shown in Table 4.6).

4.9 The Mechanism Through Which the Cricket Pitch Soil pH Affects the Pitch Behaviour at Different pH Ranges is as Follows

1. Availability of nutrients: soil pH influences the solubility of nutrients. It also affects the activity of micro-organisms responsible for breaking down organic matter and most chemical transformations in the soil. The types and population densities of micro-organisms change with pH. A pH of 6.6–7.3 is favourable for microbial activities that contribute to the availability of nitrogen, sulphur, and phosphorus in soils. Soil pH affects the availability of several plant nutrients.

Soils that have a pH below 5.5 generally have a low availability of calcium, magnesium, and phosphorus. At this low pH, the solubility of aluminium, iron, and boron is high and low for molybdenum. At low pH, turf grass growth and after-match recovery are poor due to the non-availability of basic essential plant nutrients and high toxic levels of H^+ and Al^{+3}.

A pH range of 6–7 is generally most favourable for plant growth because most plant nutrients are readily available in this range. At pH 7.8 or more (slightly alkaline), calcium and magnesium are abundant. Molybdenum is also available if it is present in the soil minerals. High pH soils may have inadequate availability of iron, manganese, copper, zinc, and especially phosphorus and boron.

The availability and non-availability of different essential basic minerals at different cricket pitches and soil pH levels should be considered before the application of suitable fertilisers. Most pesticides, insecticides, and fertilisers are labelled for specific soil conditions. If the pH of the soil exceeds the allowable range, pesticides/insecticides and fertilisers may become ineffective, change to an ineffective form, or fail to degrade as expected, causing problems for the following cricketing seasons.

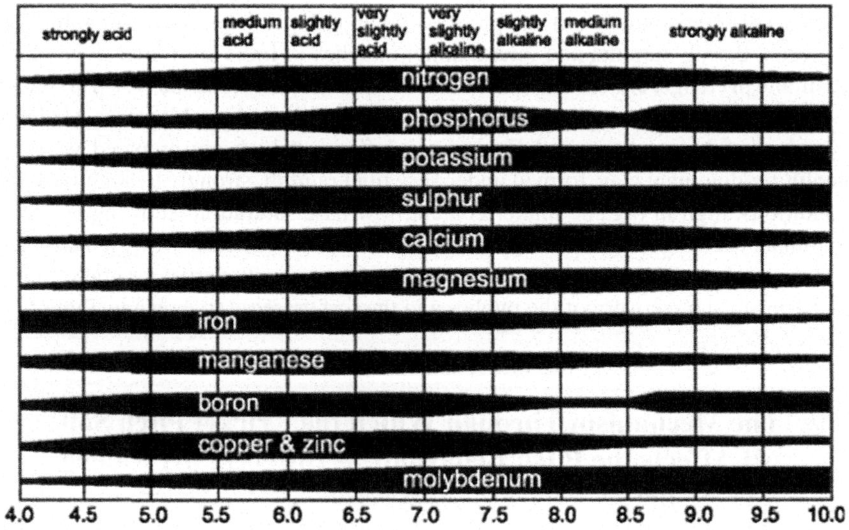

Fig. 4.6 Relationship between soil pH and the availability of plant nutrients. The wide portions of the bands indicate the zones of greatest microbial activity and the most readily availability of nutrients

2. Soil pH at different ranges regulates the soil structure, compressibility, desirable pitch surface density, swelling, and shrinking capacity of the pitch profile by regulating the predominant minerals, exchangeable ions on the clay micelle (cations and anions), soil solutions, double diffused layer (DDL) shape, size, and number adsorbed on the clay micelle, cation exchange capacity (CEC), base

saturation levels (BS), water holding and retention capacity (WHC), pore space, and desirable density of the pitch profile.

The pH of cricket pitch soil should be kept at desirable levels (slightly acidic to moderately alkaline), because extreme levels of soil pH (either strongly acidic, strongly alkaline, or sodic) have a negative impact on the desirable surface structure and density, and regeneration of turf pitch grass and the pitch surface after the match is poor. A pH range in the cricket pitch profile directly affects the pitch surface density, which is dependent upon the clay mineralogy, soil structure, and the pH of the pore fluid. Laboratory studies show that the pH of soils will alter the engineering properties of soils such as hydraulic conductivity (Gajo Maines 2007) and plasticity properties (Sridharan et al. 2000).

In the area of pitch making and construction, very little research-based information is available on the direct or indirect influence of soil pH variation on cricket pitch soil compression levels with different dominant clay minerals in them. However, the information available in the literature on the influence of pH on the plasticity and engineering properties of soils can provide a good base for understanding its influence on the behaviour of cricket pitches.

In recent years, sufficient efforts have been made to understand the influence of pH conditions on soil plasticity (Kamon et al. 1997), strength (Gajo and Maines 2007; Wang and Siu 2006; Graychev and Sassa 2013; Gatchev and Towhata 2013), and compressibility (Imai et al. 2006; Gratchev and Towhata 2011a). These studies have clearly brought out that several processes could occur due to soil–water–acid interaction at low pH, including:

A. The dissolution of chemical compound or organic matter.
B. Changes in thickness of DDL.
C. Formation of pH-dependent charges on the edges of clay particles. It is noted that these processes are common for pH ranging from3 to 6.

A. The dissolution of chemical compounds or organic matter: it has been recognised that acid can dissolve cementation agents such as calcium carbonates and ferric oxides (Mitchell 1993), altering the structure of natural clays. When acids come in contact with naturally consolidated clays, they may destroy or dissolve the chemical bonds or cementation between clay particles, forming relatively loose structures (Imai et al. 2006). The formation of loose structures during leaching can be seen in the higher initial void ratios of the samples with a lower pH.

Results of laboratory tests conducted on kaolinitic soils indicate that clay with open-flocculated structures exhibits greater permeability (Mitchell 1993; Sridharan and Prakash 1999; Sridharan et al. 1988; Gratchev and Sassa 2009). The formation of an open-flocculated structure is expected to affect the compressibility of clay as well, increasing the soil's capacity to compress under load application.

As depicted in Fig. 4.7, high liquid limits at low pH are also generally observed in most of the kaolinite-dominant pitch soils in high rainfall areas.

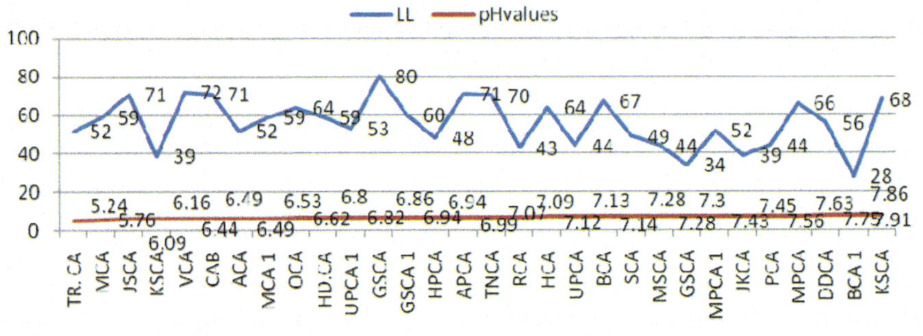

Fig. 4.7 pH and liquid limits of some test cricket pitch soils in India

Under low pH range, kaolinite-dominant clay with open flocculated structure and larger air voids initially generates larger compression during initial stage of rolling of the pitch surface, but regeneration of these over-compacted pitch surface and the turf grass after the match is very poor. These type of kaolinite-dominant pitch soils at low pH range needs longer time for the pitch and grass to recover and a need for the use of mechanical means to decompact the over-compacted pitch surface during initial stage of rolling and to aerate the pitch surface after the match due to their non or limited swelling capacity. Generally, it is due to predominance of non-swelling and shrinking kaolinitic clay with high organic matter content at low pH conditions. This phenomenon is generally observed on the cricket pitches of north-east Indian states and in the state of Kerala which have slightly acidic to strongly acidic conditions of the soil formed under high rainfall condition.

2 Changes in thickness of DDL: a high concentration of acid in the pore water also may affect the thickness of the DDL of clay particles (Van Olphen 1991). As the pH of pore fluids drops from neutral to acidic, H^+ ions replace the commonly found exchangeable cations in the clay, resulting in an increase in DDL (Gajo and Maines 2007). Due to their superior position in the Hofmeister or Lyotropic series (which describes how readily some cations are adsorbed at a given concentration), H^+ ions would likely replace the commonly found exchangeable cations such as Na^+, Ca^{+2}, Al^{+3}, or Fe^{+3} (Sridharan et al. 1986). From clay colloidal chemistry (Van Olphen 1991), it is known that this process would lead to an increase in DDL thickness, resulting in a greater compressibility of soil.

However, as the amount of H^+ ions in the pore fluid still increases (that is, the pH becomes more acidic), the DDL is expected to decrease (Van Olphen 1991). At some point, the amount of H^+ in the pore fluid will eventually reach the concentration at which the DDL will collapse (Mitchell 1993), a process that will affect the compressibility of soil. Mitchell (1993) noted that the process of altering the DDL would have a significant impact on soils with montmorillonite or smectite (but be negligible for kaolinite soils), while changes in particle orientation caused by chemicals should have a dominant effect on the properties of kaolinite soils.

3 Formation of pH-dependent charges on the edges of clay particles: in an acidic medium, the charge on the edges of the clay particles will become increasingly positive due to the adsorption of H^+ ions, resulting in flocculated, face-to-edge (F-E) structures (Dolinar and Trauner 2007). Although this mechanism can occur in soils with different clay mineralogy, it appears to have the most influence on kaolinitic soils due to the size and structure of kaolinitic particles (Mitchell 1993; Ma and Eggleton 1999).

The mechanism that governs the compressibility of kaolinite and illite at low pH can be related to the pH-dependent charge of the kaolinite edges (Mitchell 1993). Van Olphen (1991) noted that when the pH is less than 5.5, the charge on the edges of kaolinite particles becomes increasingly positive, a process that leads to the formation of a more open, flocculated structure. The existence of such a structure was experimentally verified by Wang and Siu (2006) and Dolinar and Trauner (2007). This possibility was evidenced by the SEM views of the microfabric. Thus, it is logical to assume that soil with such an open microfabric would be more compressible, producing a soil structure with a high compressibility index (CI) when the pH of the pore fluid drops below 5.5. By preventing the interlayer expansion (swelling and shrinking capacity) and blocking the exchange sites (with Al^{+3} or H^+ ions) in an acidic environment, it affects the desirable compression levels of the pitch surface.

At low pH, aluminium ions dominate the exchange sites and break the clay micelle fabric of 2:1 smectites, converting them into non-swelling 1:1 kaolinitic clay minerals.

The pitch surface shows desirable variations through cracking or crumbling patterns in neutral to slightly alkaline cricket pitch soils with Ca^{+2}, Mg^{+2}, predominant swelling and shrinking fine clay at different moisture levels, and a well-flocculated structure. Swelling and shrinking clays with high WHC and retention capacities provide good pitch surface density with consistent pace and bounce. These types of pitches express their behaviour through the rate, intensity, and size of the cracks and the flocculated structure on the pitch surface. At these pH ranges (neutral to slightly alkaline), in the cricket pitch soil organic humus along with $CaCO_3$ acts as a bridge or acts as a binding agent and provides a good soil structure by gelling the primary particles (clay, silt, and clay) together, thereby generating good pace and bounce on the pitch surface.

4.10 Management and Reclamation of Problematic Cricket Pitch Soils

1. The best suitable soil pH range to achieve the desired pitch surface behaviour during the different levels of the match is the slightly acidic to moderately alkaline range. Even if the pitch soil is in an undesirable pH range (highly acidic to

strongly alkaline or sodic levels), the pitch surface behaviour can be controlled and managed by adopting the following practices:

Reclamation and management of saline and sodic soils are based mainly on proper irrigation and drainage, on the exchange of Na^+ for Ca^{+2} on the exchange complex, and on the use of salt-tolerant turf grass varieties on cricket pitches.

Generally, 3–4 in. of the cricket pitch surface are saturated (as shown in Fig. 4.8) with portable water during the initial phase of pitch preparation, and frequent watering is done on the pitch square to maintain the pitch turf grass. Before the application of available portable water for these purposes, irrigation water must be examined for its important chemical properties like pH, predominant exchangeable cations (Na^+, Mg^{2+}, K^+), or anions (Cl^-, CO_3, HCO_3, ECe, ESP, SAR, etc.).

Fig. 4.8 Photograph showing irrigation of pitch as a part of pitch preparation before a match. **a** Dr. Ambedkar stadium, Baramati, Maharashtra, Ranji match, 07–13 February 2020. **b** DRIEMS ground, Cuttack, Odisha, Ranji match, and 23–27 January 2020

ESP, SAR increases the cricket pitch soil pH from neutral to undesirable strongly alkaline, saline, or sodic levels, which adversely affects the pitch surface's compressibility, swelling and shrinking capacity, and hydraulic conductivity (loss of soil structure due to dispersion of Na^+ or Mg^{+2}-dominant smectite clay), resulting in very poor after-match recovery of the over-compacted pitch surface. Salts accumulating in the root zone affect the turf grass by osmotic suction and restrict the availability of soil water to the turf grass, thereby resulting in poor turf grass growth, after-match grass recovery, and also pitch surface recovery.

2. To decompact and aerate the over-compacted pitch surface at saline or alkaline pH levels, leaching of excess salts by irrigation is generally conducted. A proper drainage method and the use of irrigation water with the desired salt quality are necessary. Of late, 4 in. of sandy loam soil and another 4 in. of river-washed sand are laid as a drainage layer below the 8–12 in. of top pitch surface clay soil (a two- or three-layered wicket profile). This subsurface pitch profile subsoil drain is typically installed to flush out undesirable excess salts accumulated or developed within the root zone of grass turf (generally 125–150 mm depth) as well as the draining downward of dissolved salts in either gravitational or capillary water. In some cricket pitches with multiple pitch surface layers or without a proper subsoil drainage layer, there is severe deposition and accumulation of these undesirable salts on the pitch surface or within the pitch profile, resulting in undesirable pitch behaviour.
3. Even some incorrect pitch preparation methods, such as rolling the pitch surface with heavy rollers at high moisture levels, lead to the deposition of excess salts on the pitch surface. Heavy rolling at high moisture squeezes the excess moisture from the micropores, and the salts are deposited on the surface after the surface soil moisture evaporates at high temperature and humidity. Whitish grey or whitish black spots or concentric whitish rings on the pitch surface are generally observed when the wet pitch surface in wet conditions is rolled with heavy rollers.
4. Adopting different pitch preparation methods and techniques (watering, rolling, turf grass management, etc.). Most continental or subcontinental cricket pitches use generalised methods and techniques for watering, rolling, and turf grass management. Based on the curator's experience or existing style of pitch maintenance and match pitch preparation, these methods are mostly suitable for pitch surfaces in the neutral to slightly alkaline pH range. As such, now very little research has been done on cricket pitch preparation methods under very adverse pH levels (either under very acidic or strongly alkaline conditions). The pitch surface behaves entirely differently at different pH levels, and hence, the pitch preparation methods under adverse and undesirable pH levels need to be developed accordingly.
5. Use of pH-tolerant turf grass varieties: under Indian subcontinental weather conditions, Bermuda and *Seashore Paspalum* are the best salt-tolerant (under alkaline to saline pH levels) varieties of grass that are suitable to be used as turf grass on cricket pitches.

4.11 Different Pitch Preparation Methods and Techniques Based on the Different pH Levels

A. Low pH range: very strongly acidic (4.5–5.0) to slightly acidic (6.1–6.5)

Pitch surface conditions include pitch surface with Al^{+3} (toxicity) or H^{+}, kaolinite-dominant fine clay with non-swelling and shrinking capacity, flocculated soil structure, poor to very poor after-match pitch surface and turf grass recovery, crumbling pitch surface when dry (due to low water holding and retention capacity), providing high dry density, but in a very narrow moisture range during the match.

Pitch preparation methods and techniques: these pitches with low water holding and retention capacity should be watered frequently during match pitch preparation to maintain the desired moisture and pitch surface density levels. Limited or optimum rolling with light to medium rollers is generally recommended during match pitch preparation, as these types of non-swelling and shrinking soils lack the natural ability to decompact and regenerate after the match. On these low pH pitch soils, limited use of heavy rollers at desirable optimum moisture levels is generally required. The pitch profile with poor recovery and regeneration rate should have as many pitches as possible within the pitch square with large boundaries, so that after every match, the pitch can be rotated and provided with sufficient pitch surface and grass recovery time to regenerate the grass of the used pitch.

2 Pitch Surface Conditions: pitch surface with slightly acidic to moderately alkaline Ca^{+2}-dominant fine swelling and shrinking smectite clay, high water holding and retention capacity (due to hygroscopic and capillary water), good flocculated and aerated soil structure (due to Ca^{+2}-clay-humus bonding), providing a wide range of desirable pitch surface density in a wider pitch surface moisture range.

Recommended pitch preparation methods and techniques: soils within this type of pH range (slightly acidic to moderately alkaline) provide a wide range of moisture and good flocculated pitch surface structure to work towards the desired pitch surface density during the pitch preparation. Optimum watering and rolling (with different roller weights) are desirable as, after a match, the natural recovery and regeneration of the pitch surface and turf grass is very good.

3 Pitch Surface Conditions: very strongly alkaline to sodic soils (Na^{+}/Mg^{+2}-dominant fine smectite), high water holding and retention capacity (due to hydration of Na^{+}-dominant clay), low air voids generating very high dry density on drying, dispersed or deflocculated soil structure generating very poor after-match grass recovery.

Recommended pitch preparation methods and techniques: optimum rolling and maximum drying of the pitch surface are recommended as the over-compacted pitch surface during match pitch preparation requires longer time to regenerate the turf grass.

4.12 Cricket Pitch Soil Behaviour Based on the Predominant pH Range and Exchangeable Cations

See Table 4.7.

Table 4.7 Cricket pitch soil behaviour at different pH range

S. No.	pH	pH range	Dominant cations	Pitch surface behaviour
1	Moderately acidic to slightly acidic	Below 5.6–6.5	Al^{+3} or H^{+}	Generally, kaolinite-dominant fine clay has low CEC, WHC, base saturation (BS), SAR, ESP, and EC values The high initial compressibility is caused by high air voids and clay flocculation at low pH levels. However, high Al concentrations, though beneficial for flocculation of clay, are harmful for turf grass growth and regeneration. Regeneration of grass and the pitch surface is poor due to the unavailability of basic nutrients and the presence of Al^{+3} at toxic levels. Ped loses its structure in a highly acidic environment due to the dissolution of clay, humus, and $CaCo_3$ binding
2	Neutral to moderately alkaline	6.6–8.4	Ca^{+2}	Generally smectite–illite-dominant swelling and shrinking fine clay with high exchangeable calcium, base saturation levels, EC, moderate CEC, SAR, and ESP Ca^{+2}-dominant fine smectite clays have good flocculation (stable soil aggregate and structure due to clay-calcium-humus bridge), which powers negatively charged clay micelles and reduces the toxic effect of high Al^{+3} concentrations. The regeneration of grass and the pitch surface during and after the match is good
3	Strongly alkaline or sodic soil	Above 8.5	Mg^{+2} or Na^{+}	Generally, smectite-dominant fine swelling and shrinking clay has high CEC, WHC, ESP, SAR, and low EC, HC, and air voids, resulting in a pitch surface with a very high dry density Very high dry density due to deflocculation (dispersion of clay particles in Na^{+}-dominant saline conditions) and hydration of clay particles soil structure is lost, and the soil is dispersed, resulting in sticky soil when wet and very hard soil when dry. High WHC and low air voids result in a dense pitch surface. Regeneration of grass and the pitch surface is poor due to low porosity (air voids), dispersion of clay particles, and the presence of slats at toxic levels

4.13 Classification of Cricket Pitch Soils Based on the pH Range and Predominant Exchangeable Cations of Subcontinental and Continental Pitches

Condition 1—Moderately low (5.6–6.0) to slightly acidic (5.6–6.5) pH acidic conditions (as depicted in Figs. 4.9, 4.10, 4.11, 4.12, 4.13, and 4.14), dominant Al^{+3}/H^{+} cations, kaolinitic clays which restricts the swelling and shrinking behaviour by breaking down the 2:1 smectites clay to 1:1 non swelling clays which reduces the SAR, CEC, BS, WHC, BD at desired OMC—e.g. most of the cricket pitch soils of north-eastern states (Assam, Meghalaya, Manipur, Mizoram, Nagaland, Arunachal Pradesh, Sikkim, and Tripura) and Kerala with very high rainfall comes under this category (Table 4.8).

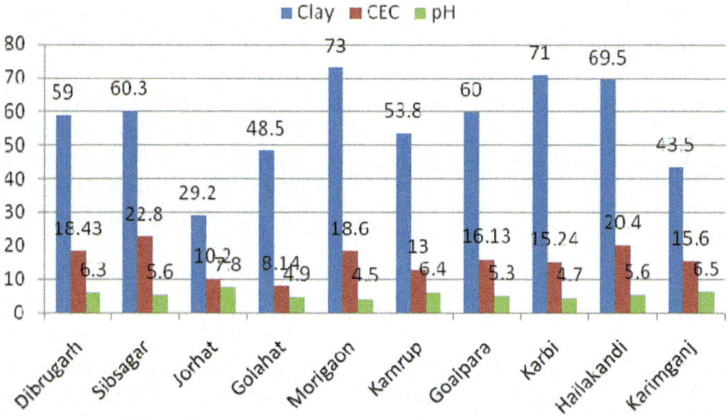

Fig. 4.9 Variation of clay percentage, CEC, and pH of some selected cricket pitch benchmark soils from Assam

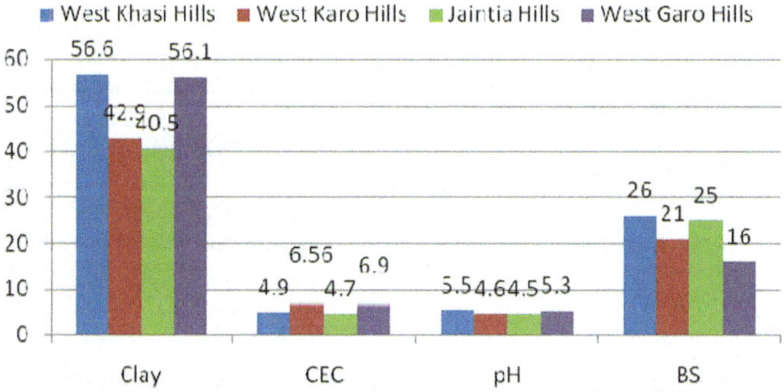

Fig. 4.10 Clay percentage, CEC, pH values, and of some selected benchmark soils from Meghalaya

Fig. 4.11 View showing the soil profile and the puffy/spongy ball in moulded state of the ICAR, Krishnagiri farm house soil sample, Imphal

Table 4.8 Particle size distribution, soil organic clay (SOC), pH, cation exchange capacity (CEC) and base saturation (BS) of Krishnagiri farm house soil sample, Longol hills, ICAR Research complex for NEH, Imphal

Horizon	Depth (cm)	Sand	Silt	Clay	SOC	pH	CEC	BS %
A	0–15	2.5	48.8	48.7	2.34	4.9	14.7	27
Bw1	15–40	2.9	46.3	50.8	1.64	4.5	12.8	11
Bw2	40–71	4.5	52.3	43.2	1.71	4.6	12.4	11
Bw3	71–99	2	41.4	56.6	0.97	4.7	12	10
Bw4	99–127	3	47.8	49.2	0.93	4.1	11.7	11
Bw5	127–150	1.9	41	57.1	0.35	4.9	10.6	9

a

b

Fig. 4.12 **a** Very poor after-match recovery of pitch turf grass and the pitch surface due to low pH levels of the pitch square of the Luwangpokpa Cricket Stadium (LCS), Imphal and **b** Over-compacted pitch surface with very poor after-match pitch surface and grass (Rye grass) recovery at Polo ground, Shillong, Meghalaya

Most of the soils that are available in the north-east sister states are blackish-grey, light grey, red, yellowish red, or brownish red in colour, but their low values of cation exchange capacity (CEC) and base saturation (BS) values (as given in Table 4.4) indicate the non-swelling types of silty clays with kaolinite as a dominant mineral. Generally, red/yellowish red/brownish red soils are found in well-drained areas like the hilly areas, and black/grey/light grey soils are found in the submerged valleys with poor natural drainage. With a humid subtropical climate and heavy rainfall areas, most of the soils in these regions have high organic matter content, very low pH levels with Al^{+3} toxicity, and non-swelling and shrinking kaolinitic silty clays, which make them infertile and lack the basic nutritive value to support the growth and regeneration (after-match recovery) of grasses on the pitches. This is one of the main reasons why the cricket turf grass and the pitch surface recovery after the match are very poor on these types of soils (as shown in Fig. 4.12a, b). The pitches from these types of non-swelling and shrinking kaolinitic silty clays with very high organic matter contents (as shown in Figs. 4.11 and 4.12a, b), low water holding capacity, low pH, and very poor grass growth generally result in pitches with slow pace and bounce that crumble in the later stages of the game.

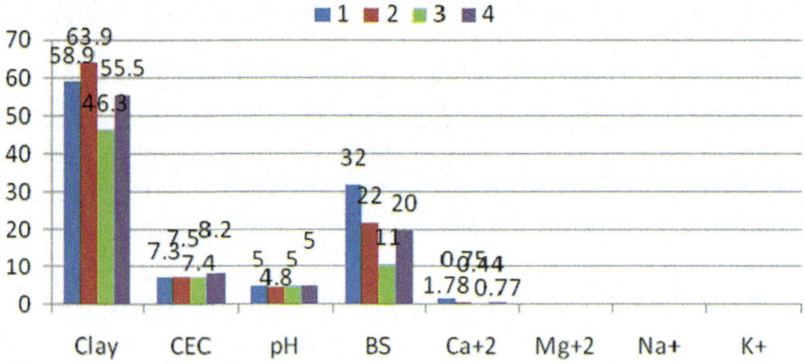

Fig. 4.13 Variation of clay content, CEC, pH, and predominant exchangeable cations of some selected cricket pitch benchmark soils from Kerala

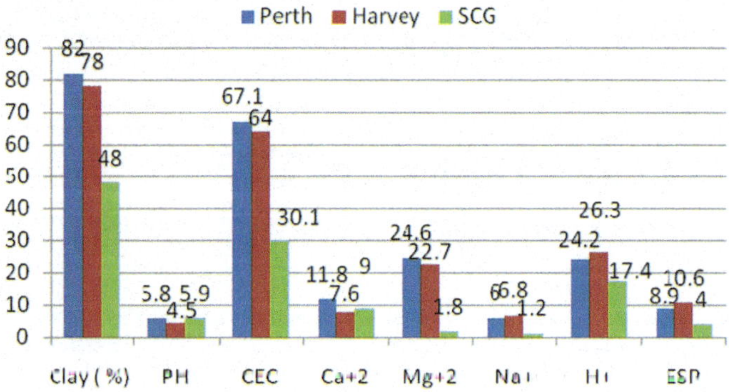

Fig. 4.14 Clay proportion, pH values (very strongly acidic to moderately acidic), CEC, ESP, and predominant exchangeable cations of some of the pitch soils predominantly used in Australian pitches. *Source* McIntyre Keith and Don, Cricket Wickets—Science versus Fiction, 2001, p. 68

Condition 2—Neutral (6.6–7.3) to slightly alkaline (7.4–7.8) pH (as depicted in Figs. 4.15 and 4.16), 2:1 calcium/magnesium-dominant fine smectites, silty clay, high CEC, BS, EC, WHC, good flocculated structure with good grass growth and after-match recovery generating moderate to high pitch surface dry density, e.g., soils from Balangir (Odisha), Eden gardens (West Bengal), Chinnaswami (Karnataka) stadium.

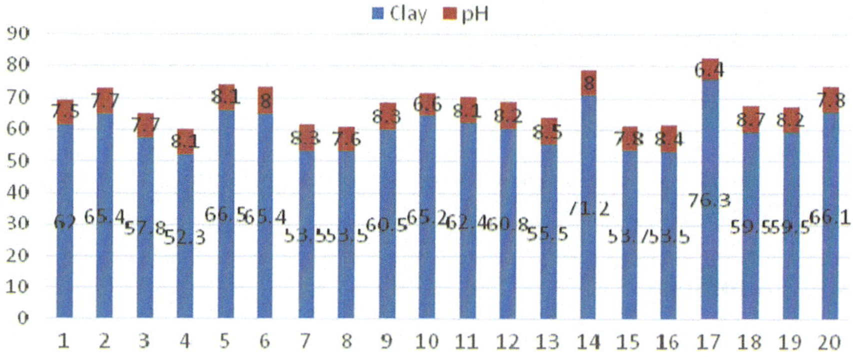

Fig. 4.15 Variation of clay content and pH of some selected benchmark soils of Maharashtra

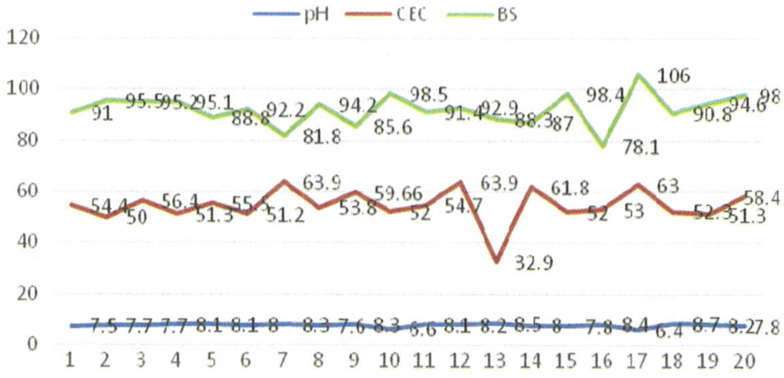

Fig. 4.16 Variation of clay, base saturation and pH of some selected benchmark soils of Maharashtra

Condition 3—moderately (7.9–8.4) alkaline—(as depicted in Fig. 4.12) having Ca^{+2}, Mg^{+2}-dominant fine smectite clay with high CEC, EMP, Ca^{+2}/Mg^{+2} ratio very low EC, HC, ESP, SAR, generating good flocculated soil structure (Fig. 4.17).

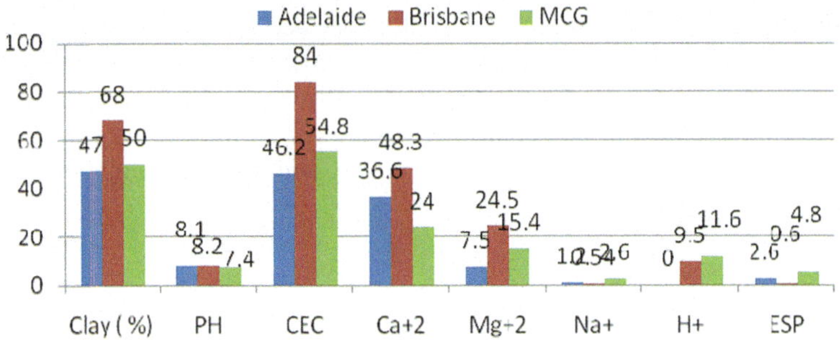

Fig. 4.17 Per cent clay proportion, moderately (7.9–8.4) alkaline pH values (slightly alkaline to moderately alkaline), CEC, ESP and predominant exchangeable cations of some of the pitch soils predominantly used in Australian pitches. *Source* McIntyre Keith and Don, Cricket Wickets—Science versus Fiction, 2001, p. 68

4.14 Summary

Soil pH (acidic, neutral, alkaline, or saline) with predominating exchangeable cations (flocculating Al^{+3}, H^{+} in acidic conditions, Ca^{+2}, Mg^{+2} in neutral to moderately alkaline levels, or deflocculating Na^{+} cations in non-saline, alkaline sodic soils) provides different pitch surface density and variations at a given moisture level. Soil pH, by regulating the predominant exchangeable ions (Al^{+3} or H^{+} in acidic conditions, Ca^{++}, Mg^{++}, or Na^{+} in alkaline, saline, or sodic conditions), controls the size and amount of diffuse double layer (DDL), water holding and retention capacity (WHC), and pitch behaviour by controlling the soil structure, porosity, hydraulic conductivity, surface density, swelling or shrinking capacity, or crumbling patterns of the pitch profile. At different pH levels, cricket pitch soils with the dominant colloidal minerals (smectite, illite, or kaolinite) with organic humus and dominant exchangeable cations (H^{+}, Al^{+3}, Ca^{+2}, Mg^{+2}, Na^{+}, or K^{+}) control the swelling and shrinking of the pitch profile and provide the desired variable densities by the loss or gain of moisture levels on the pitch surface or within the pitch profile.

The desirable pH range for the cricket pitch profile should be neutral (6.6–7.3), slightly alkaline (7.4–7.8), or moderately alkaline (7.9–8.4), as these pH ranges generate good flocculated pitch structure with desirable pitch surface density, compressibility, pace, and bounce. In addition to this, after-match recovery of the pitch surface and turf grass is also good. Soils in the low pH range are generally non-swelling and less shrinking kaolinite-dominant fine clay, with low CEC, WHC, BS, and compressibility, but with poor after-match pitch surface and turf grass recovery. These types of pitch soils are predominantly found in areas with high rainfall and good drainage. Cricket pitch soils with a pH range of moderate to high are smectite-dominant fine clay with high swell-shrink behaviour, CEC, BS, WHC, high pitch surface density, and good pitch surface and turf grass after-match recovery. These types of pitch soils are predominant in areas with low rainfall, poor drainage, high temperatures, high relative humidity, and high evapotranspiration rates.

Extreme soil pH conditions (either low pH 5.6 or high pH > 8.5) are not desirable for cricket pitches because they affect the swelling and shrinking capacity (desirable pitch surface density/compression levels) and after-match regeneration of the pitch soil surface or turf grass is very poor or takes much longer than the pitch surface's usual recovery rate under a desirable pH range. On cricket pitches, salt-tolerant turf grass varieties such as Bermuda and Seashore paspalum should be used because they survive and recover well in alkaline to saline pH levels. Under acidic pH conditions, centipede grass is suitable for pitch turf grass.

Cricket pitch soils with an extreme pH range (either strongly acidic or very strongly alkaline) can be reclaimed by using the pitch soil from benchmarked soil samples. Benchmarked cricket pitch soils in Maharashtra and parts of Madhya Pradesh, Gujarat, Rajasthan, Karnataka, Odisha, and West Bengal have 2:1 swelling and shrinking type and within the desirable pH range (neutral to slightly alkaline pH range), high CEC, BS (showing calcium, magnesium as dominant exchangeable cations), low EC, ESP, SAR generating flocculated structure with high pitch surface dry density/compressibility and good pitch.

Considering the intensity and direct impact on cricket pitch behaviour, soil pH along with the predominant exchangeable cations should be considered as a preliminary test before the final selection and layout of any new wicket profile or during the stage of pitch preparation before the match.

Cricket pitches behave differently at different pH levels due to their variability in compression levels, WHC, density levels, structure, swelling, shrinking, or crumbling patterns, or in the growth or regeneration of the pitch surface and the turf grass. Based on the direct and indirect impact of soil pH range on cricket pitch behaviour, different research-based pitch preparation methods and techniques (different rolling, pitch watering schedules and techniques, turf pitch grass management, etc.) should be evolved so as to achieve the desired pitch surface behaviour at different pH levels. This will help us in analysing, predicting, and comprehending the complex nature of pitch behaviour narrowly and at the desired benchmarked levels.

References/Bibliography

Bolt G (1956) Physicochemical analysis of the compressibility of pure clays. Géotechnique 6(2):86–93

Brady NC, Weil Ray R (2008) Nature and properties of soil, 14th edn. Prentice Hall

Chen J, Anandarajah A, Inyang H (2000) Pore fluid properties and compressibility of kaolinite. J Geotech Geo-Environ Eng ASCE 126(9):798–807

Gratchev I, Sassa K (2009) Cyclic behavior of fine grained soils at different pH values. J Geotech Geol Eng ASCE 135(2):271–279

Gratchev I, Towhata I (2009) Effect of acid contamination on the geotechnical properties of marine soils in Japan, Proc ISOPE-2009 Osaka, 19 the International Offshore (Ocean) and Polar Engineering Conference. Japan 2:151–155

Gratchev I, Towhata I (2011) Compressibility of natural soils subjected to long term acid contamination. Environ Earth Sci 64:193–200

Gratchev I, Sassa K, Osipov V, Fukuoka H, Wang G (2007) Un-drained cyclic behavior of bentonite sand mixtures and factors affecting it. Geotech Geol Eng 25:349–367

Lambe TW (1953) The structure of inorganic soils. Proc ASCE. 79(Separate no 315), p 13

Lambe TW (1958) The structure of compacted clay. Proc ASCE 84(SM2):718

Lambe TW, Whitman RV (1983) Soil mechanics. Wiley Eastern, New Delhi

Leonards GA, Altschaeffl AG (1964) Compressibility of clays. J Soil Mech Found Div ASCE 90(SM5):133–155

McIntyre K, Don (2001) Cricket wickets—science vs fiction, Horticulture Engineering Consultancy, 5-Brimage Palace Kambah, ACT Australia, 2001 Edition

Mitchell JK (1960a) The application of colloidal theory to the compressibility of clays. In: Proceedings of the seminar on interparticle forces in clay water electrolyte systems, C.S.I.R.O., Melbourne, Australia, pp 2.92–2.97

Mitchell JE (1960b) The applicability of colloidal theory to the compressibility of clays. Interparticle forces in clay-water-electrolyte systems, Commonwealth Scientific and Industrial Research Organization, Australia, pp 92–98

Mitchell JK (1993) Fundamentals of soil behavior, 2nd edn. Wiley, New York, p 435

Olsen RE (1974) Shearing strengths of Kaolinite, Illite and Montmorillonite. J Geotech Engg Div ASCE 100(GT 11):1215–1229

Olsen RE, Mesri G (1970) Mechanisms controlling compressibility of clays. J Soil Mech Found Div, ASCE 96 (SM6):1863–1878

Prasad CRV, Reddy PHP, Murthy VR, Sivapullaiah PV (2018) Swelling characteristics of soils subjected to acid contamination. Sci Direct Soils Found 58:110–121

Sridharan A (1991) Engineering behavior of soils—a fundamental approach (IGS Annual Lecture Dec. 1990). Ind Geotech J 21:1–136

Sridharan A, Rao GV (1973) Mechanisms controlling volume change of saturated clays. Geotechnique 23(3):359–382

Sridharan A (2001) Engineering behavior of clays: influence of mineralogy. Invited Lecture. In: Proceedings of international workshop on chemo mechanical couplings in clays from nano-scale to engineering applications, Maratea, Italy, pp 03–28

Sridharan A (2005) On swelling behavior of clays. Key note Lecture. In: Proceedings of international conference on problematic soils, Eastern Mediterranean University, North Cyprus, vol 2, pp 499–516

Sridharan A, Rao AS (1982) Mechanisms controlling the secondary compression of clays. Geotechnique 32(3):249–260

Sridharan A, Rao AS, Makan SK (1983) Shear strength behavior of expansive clays. In: Proceedings of the 7th Asian regional conference on SM and FE, Haifa, Israel, pp 81–83

Sunil BM, Nayak S, Shriharj S (2006) Effect of pH on the geotechnical properties of laterite. Eng Geol 85(1–2):197–203

Tan KH, Principles of soil chemistry, 3rd edn. CRC Press, pp 357–394

Chapter 5
Cricket Pitch Soil Water—An Elixir for Pitch Behaviour

5.1 Introduction

Soil moisture is very important in assessing its response to loading, either from an engineering point of view or for sports-related performance. Cricket pitches do have turf, and the growth of grass needs soil moisture. So, for the proper growth and maintenance of turf-related cricket pitches, an understanding of the soil moisture needed for plant growth and its sustenance is needed. So, the information about soil moisture related to its environs from an agricultural point of view has to be understood.

The amount of water required by plants depends upon their physiological demand and the capacity of the soil and environment to supply water to plants. Generally, plants require water on a continuous basis, depending upon the climatic conditions and availability of water. The water in soil is stored by inorganic as well as organic components present in it. Soil water is a vital component of soil and regulates the various soil properties as well as the health of the flora and fauna in the soil. Thus, it is important to know the different forms of soil water, some of which are available to plants, whereas some forms are unavailable to plants for their growth.

Water present in the void spaces of the soil mass is called soil water. The term "soil moisture" refers to the portion of subsurface water that fills the voids in the soil above the ground water table. Soil water in a cricket pitch profile may be in the form of "free water," "gravitational water," or "held water." The "free water" or "gravitational water" is free to move through the pore space of the soil mass under the influence of gravity; the "held water" is held in the proximity of the surface of the soil grains by a certain force of attraction.

S. B. Singh et al., *Cricket Pitches*, Springer Transactions in Civil and Environmental Engineering, https://doi.org/10.1007/978-981-99-2913-9_5

5.2 Types of Soil Moisture Available Within a Pitch Profile

Soil water is classified into four types: (i) gravitational water, which is held very loosely and is not readily available to plants; (ii) capillary water, which is held with such energy by soil particles that most of it is available to plants; (iii) hygroscopic water, which is held very tightly in soil and is not readily available to plants; and (iv) film moisture (Table 5.1; Figs. 5.1. and 5.2).

Gravitational water

"Gravitational water" is the water in excess of the moisture that can be retained by the soils. It translocates as a liquid and can be drained by the gravitational force. It is capable of transmitting hydraulic pressure. This type of water occupies the larger soil pores (macropores) and moves down readily under the force of gravity. Water in excess of field capacity (– 0.1 to – 0.3 bar and upwards) is termed gravitational water. Gravitational water is of limited use to plants because it is present in the soil for a short period and occupies the larger pores, reducing soil aeration. Drainage of gravitational water from the cricket pitch profile is generally required for optimum grass growth and flushing out entrapped undesirable salts due to the heavily compacted pitch soil profile. To facilitate this, a drainage layer consisting of 4 in. of sandy loam and 4 in. of river-washed sand is provided just beneath the 8 in. of upper playing surface so as to drain gravitational water from the pitch profile complex.

Gravitational water can be subdivided into (a) free water (bulk water) and (b) capillary water.

(a) Free water (bulk water) has the usual properties of liquid water. It moves at all times under the influence of gravity or because of a difference in hydrostatic pressure.
(b) Capillary water, which is in a suspended condition, is held by the forces of surface tension within the interstices and pores of capillary size in the soil. Capillary water is held in the capillary pores (micropores) and retained on the soil particles by surface forces (– 0.1 to – 0.31 bar). It is held so strongly that gravity cannot separate it from the soil particles. The molecules of the capillary water are free and mobile and are present in a liquid state. Due to this reason, it evaporates easily at room temperature. Though it is held firmly by the soil particle, plant roots are able to absorb it. Capillary water is therefore known as available water for the growth of plants, and in the context of a cricket pitch, it is for the growth of pitch grass. This is the reason why it is desirable to have a uniformly spread and deeply rooted grass on the pitch surface, so that the available capillary water is lost by evapotranspiration and helps in having a uniformly deep, dry, and compacted pitch profile.

"Held water" is that water that is held in soil pores or void spaces because of certain forces of attraction. It can be further classified as (a) structural water and (b) adsorbed water. Sometimes, capillary water may be said to belong to this category of held water since the action of capillary forces will be required to come into play in this case.

(a) Structural water is chemically combined as a part of the crystal structure of the mineral in the soil grains. This water cannot be separated by any means. Even drying at 105°–110 °C does not affect it. Hence, structural water is considered part and parcel of the soil grains.
(b) Adsorbed water comprises (1) hygroscopic moisture and (2) film moisture.

1. Hygroscopic water

This type of water is held tightly on the soil colloidal surface (potential value lower than − 31 bars) and is essentially in vapour form. Hygroscopic water is held so tenaciously by the colloidal fraction that the plants cannot absorb it. Unlike capillary water, which evaporates easily at atmospheric temperature, hygroscopic water is unavailable for the growth of grass and cannot be separated from the soil unless it is heated. This is the reason why clayey soil (high colloidal content) with a high smectitic or vermiculitic type is required for pitch soil, as it contains a higher percentage of tightly adsorbed hygroscopic water with a formidable diffuse double layer (DDL) and has optimal moisture content (OMC), which is desired to have higher and more variable compaction levels in a cricket pitch compaction levels. Hygroscopic moisture refers to very thin moisture films that form around mineral grains. This form of moisture is dense and surrounds the surfaces of the individual soil grains as a very thin film (as shown in Figs. 6.1 and 6.2). The soil particles derive their hygroscopic moisture not only from water but also from the atmospheric air by the physical force of attraction of unsatisfied ionic bonds on their surfaces. The weight of the oven-dried soil sample, when exposed to the atmosphere, will increase to a limit depending upon its maximum hygroscopicity, which in turn depends upon the temperature and relative humidity of the air, and the characteristics of the soil grains. Coarse-grained soils have relatively low hygroscopic moisture due to their low "specific surface," or surface area per unit volume. The average hygroscopicity of sand, silt, and clay is 1%, 7%, and 17%, respectively; the high value of clay is because of the very small grain size and high specific surface area. The hygroscopic moisture film is known to be bound rigidly to the soil grains with an immense force of up to 10,000 atmospheres. The closer the soil moisture is attracted to the surface of the grain, the more it is densified. These physical forces are now established to be electrochemical in nature. Hygroscopic moisture is affected neither by gravity nor capillary forces and would not move in liquid form. Ordinarily, it cannot be evaporated. However, hygroscopic moisture can be removed by oven drying at 105°–110 °C. Moisture in this form has properties that differ considerably from those of liquid water: hygroscopic moisture has a greater density, a higher boiling point, greater viscosity, greater surface tension, and a much lower freezing point than ordinary water. Hygroscopic moisture has a pronounced effect on the cohesion and plasticity characteristics of a clay soil.

2. Film moisture forms on the soil grains as a result of aqueous vapour condensation; this is attached to the surface of the soil particle as a film upon the layer of the hygroscopic moisture film. This film is also held by molecular forces of

high density, but not as high as in the case of the hygroscopic moisture film. Migration of film moisture can be induced by the application of external potential, such as thermal or electric potential; the migration will be from points of higher temperature or higher potential to points of lower temperature or lower potential. Film moisture does not transmit external hydrostatic pressure. It migrates rather slowly (Figs. 5.1 and 5.2; Table 5.1).

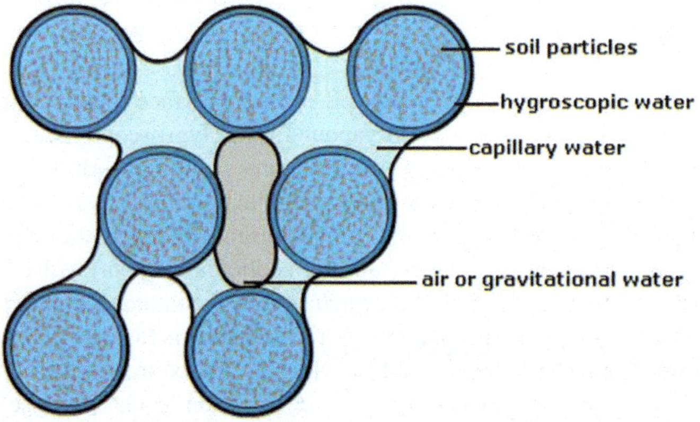

Fig. 5.1 Types of soil moisture available within the cricket pitch profile

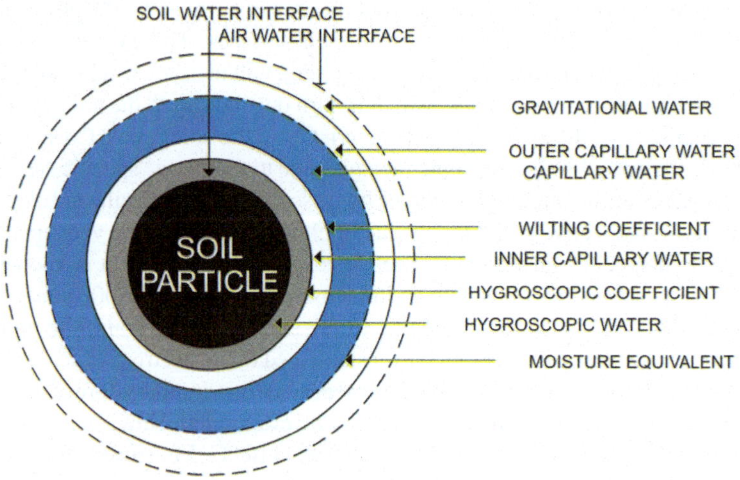

Fig. 5.2 Types of soil moisture adhered with the clay micelle

Table 5.1 Soil-water potential and moisture condition for various types of water held in soil

S. No.	Soil-water potential (bar)	Moisture condition	Type of water
1	0	Saturation (maximum water holding capacity)	Gravitational water
2	0 to − 1/3 (33)[a]	Saturation to field capacity	Gravitational water drains
3	1/3 to − 15 (1500)	Field capacity to wilting point	Capillary water (available water for plants)
4	15 to − 31 (3100)	Air dry	Capillary water + hygroscopic water (not available water for plants)
5	> − 31 (3100)	Oven dry	Hygroscopic water

[a]*Note* Units given in parenthesis are in kPa

5.3 Factors Affecting the Amount of Plant-Available Soil Moisture

The retention and movement of water in soils, its uptake and translocation in plants, and its loss to the atmosphere are energy-related phenomena. The more strongly water is held in the soil, the more heat energy is required. In other words, if water is to be removed from a moist soil, work has to be done against adsorptive forces. Conversely, when water is adsorbed by the soil, a negative amount of work is done. The movement is from a zone where the free energy of water is high (a standing water table) to one where the free energy is low (dry soil). This is called the soil-water energy concept. The free energy movement of the adsorbed water molecules is significantly reduced by matric (soil solid) force, also known as the matric potential, due to the attraction of the soil solids for water. This free energy influences the amount of soil moisture that the plant can adsorb under conditions of sufficient moisture (field capacity) and moisture stress (wilting point). On the other hand, the osmotic (soil solute) force in soil water represents the attraction of various ions in soil solute, which controls the free energy of soil solution, also known as the osmotic potential.

5.3.1 Energy Concepts in Soil-Water Systems

The presence of salts in soils, either from applied fertilisers or as naturally occurring compounds, can influence soil-water uptake. Osmotic potential in soil solutions tends to reduce the range of available moisture in such soils, increasing the amount of water left in the soil at the time the plants wilt permanently (wilting coefficient). For soils high in salts, the total moisture stress will include the matric potential as well as the osmotic potential of the soil solution. Matric and osmotic potentials are negative and reduce the free energy levels of the soil water, which are termed suction or tension. The force of gravity acts on soil water, which tends to pull the water downwards. This force is always positive. The difference between the energy states of soil water and pure free water is known as soil-water potential.

Total water potential (Pt) is the sum of the contributions of gravitational potential (Pg), matric potential (Pm), and osmotic potential (Po).

$$\mathrm{Pg} + \mathrm{Pm} + \mathrm{Po} = \mathrm{Pt}.$$

5.3.2 *Factors Affecting Gravitational Water*

1. Soil texture plays an important role in controlling the rate of movement of gravitational water. The flow of water is proportional to the size of the particles. The larger the particle size, the faster the flow or movement. Because of the larger size of the pores, water percolates more easily and rapidly in sandy soils than clayey soils.
2. Soil structure also affects gravitational water. In a platy structure, the movement of gravitational water is slow, and water stagnates in the soil. Granular and crumby structure helps to improve gravitational water movement. In clayey soils with a single-grained structure, the gravitational water percolates more slowly. If the clayey soils form aggregates, the movement of gravitational water improves.

5.3.3 *Factors Affecting Capillary Water*

Four main factors that are responsible for variations in the amount of capillary water are

1. Surface tension: an increase in surface tension increases the amount of capillary water.
2. Soil Structure: the finer a soil's texture, the more capillary water it holds. This is mainly due to the greater surface area and large number of micropore spaces. A platy structure contains more water than a granular structure.
3. Organic matter helps increase the capillary capacity of soil. Undecomposed SOM is generally porous, having a large surface area, which helps in holding more capillary water. The humus in colloidal form has a very high WHC due to its electronegative charge, high CEC, and large surface area.
4. Clay minerals and the types of ions associated with them also affect capillary water as different minerals attract water molecules in varying capacities. Moreover, the type of cation in the exchange complex of a particular mineral would govern the intensity with which water is held by the soil particles, which greatly determines the amount of capillary water available for plants as well as for cricket pitch making.

5.3.4 *Factors Affecting Hygroscopic Water*

Hygroscopic water is absorbed on the colloidal surface of the clay and organic humus by the dipole orientations of the water molecules. The amount of hygroscopic water is inversely proportional to the size of the soil particles. The smaller the particle, the greater the amount of hygroscopic water it adsorbs. Fine-textured soils like clay contain more hygroscopic water than coarse-textured soils. Not only the amount of clay but also the mineralogy of clay influences the amount of hygroscopic water. Clay minerals with montmorillonitic, vermiculitic, and other properties absorb more water than kaolinitic or illitic clay minerals. Further, as stated above, the type of cationic complex associated with a particular clay mineral would also determine the hygroscopicity of the water.

5.4 Soil-Water Relationships

The three classes of water (gravitational, capillary, and hygroscopic) are very broad and do not accurately represent the soil-water relationship that exists under field conditions. Though the maximum capillary capacity represents the maximum amount of capillary water that a soil holds, the entire amount of capillary water is not available for use by plants. A part of it, as its lower limit approaches the hygroscopic coefficients, is not available for use by plants. Hence, two more constants, field capacity and wilting coefficients, are considered for expressing the soil–plant–water relationships (Fig. 5.3).

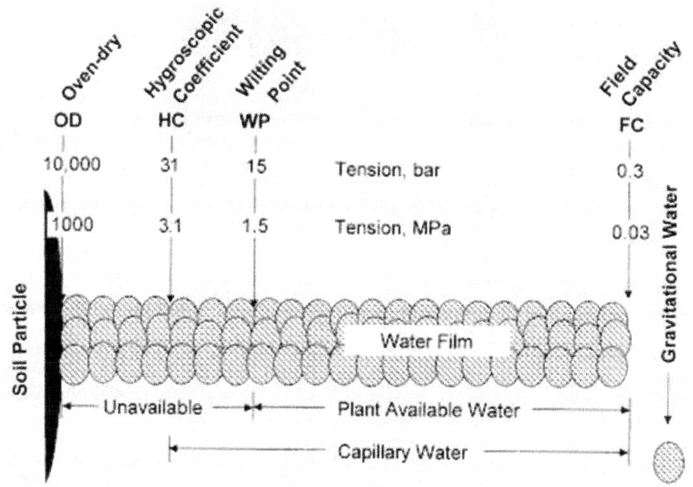

Fig. 5.3 Soil moisture constants at various tensions

5.5 Type of Water Movement in Soil

Water movement in soil may occur in the following manners:

1. Infiltration: this refers to the downwards movement or entry of water into the soil profile, which is primarily affected by the condition of the surface soil. A compacted surface limits infiltration, whereas soil surfaces with vegetative covers have more infiltration compared to bare soils. Warm soils absorb more water than cold ones. Coarse surface texture, granular structure, and high soil organic matter (SOM) in the surface soil all help to increase infiltration (Fig. 5.4). Infiltration rates are generally lower in moist soils than in dry soils (Fig. 5.4).

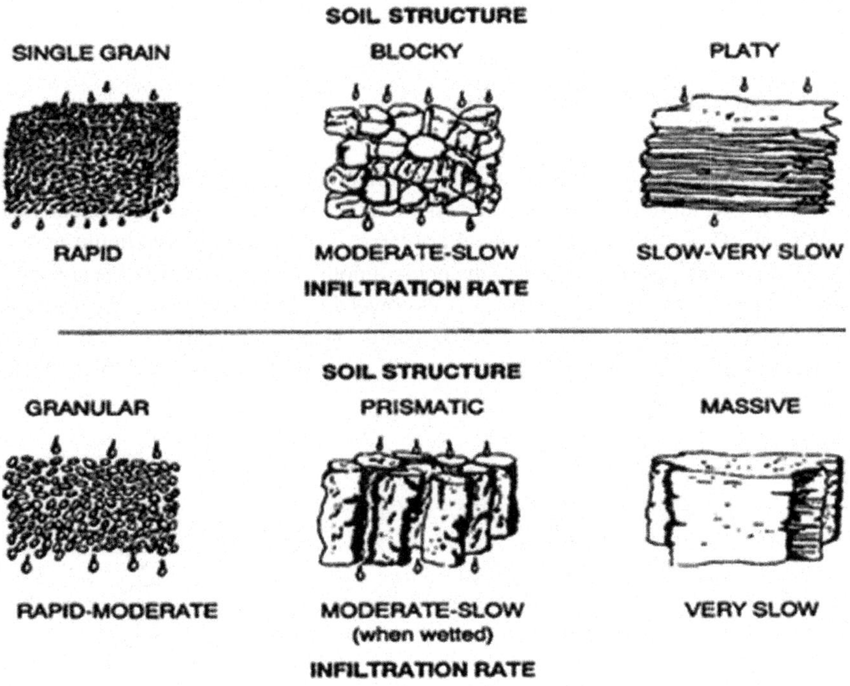

Fig. 5.4 Relationship between soil structure and infiltration rate of water in soils

2. Percolation: the movement of water through a column of soil is called percolation. It generally recharges the groundwater and carries plant nutrients down, often out of reach of plant roots. Sandy soils permit greater percolation in comparison with clayey soils. Vegetation and a high water table reduce percolation losses.

3. Permeability: the relative ease with which water moves within the soil. The permeability is basically dependent on the pore size distribution in the soil (Fig. 5.5). The larger the number of macropores, the greater the permeability. The pore size distribution is greatly determined by the extent of aggregation in the soil. The more aggregates there are, the more non-capillary pores there are. The permeability decreases with depth as the subsoil layers become more compact and reduce the macropores. Normally, permeability decreases with increasing fine texture, but the extent of aggregation in fine-textured clayey soils may override the effects of texture. A fine-textured soil that cracks during dry weather at first allows rapid water movement; later, the cracks swell and close, thereby drastically reducing the water movement. Irrigation water with a high sodium content can cause soil dispersion (compaction) and thus reduce permeability (Fig. 5.5).

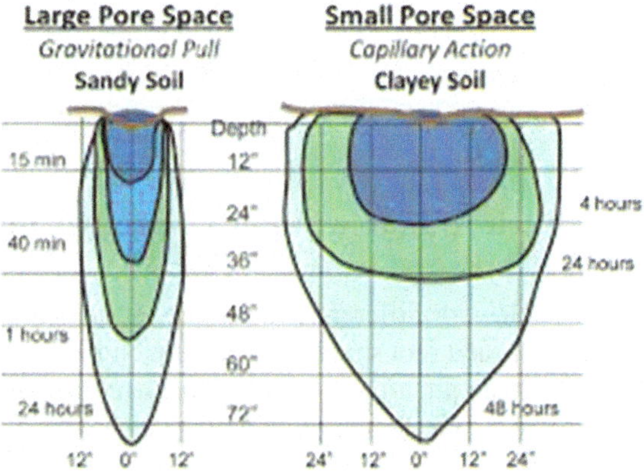

Fig. 5.5 Comparative rates of irrigation water movement into a sandy loam and a clay loam. Note the much more rapid rate of movement in the sandy loam, especially in a downwards direction. *Source* Cooney and Peterson (1955)

5.6 Soil-Water Movement in Cricket Pitch Profile

Three types of movement within the soil are recognised: saturated flow, unsaturated flow, and water vapour movement.

1. Saturated water flow is the condition of the soil when all the micro and macropores are filled with water. The direction of flow is from a zone of higher moisture potential to a lower one. Generally, water percolates downwards into the lower layers. But horizontal flow also occurs less rapidly in comparison with vertical flow. Horizontal movement is much more evident in clay soil, whereas vertical movement is much more evident in sandy loam.
2. Unsaturated flow is the condition when soil pores contain some air as well as water. The macropores are mostly filled with air. Water movement under these conditions is very slow compared to that which occurs when the soil is saturated.

Movement will be from a zone of low suction (thick moisture film) to one of high suction (thin moisture film). The forces responsible for this suction are (1) the attraction of soil solids for water (adhesion) and (2) the capillaries. The movement of water takes place because of the difference in water potential. The portion of soil that is moist has a low water potential, while that which is dry has a high water potential. Moisture moves from a high potential to a low potential, so water rises in the capillary tube.

3. Water vapour movement is of two types: (a) internal movement, in which the change from the liquid to the vapour state takes place within the soil, that is, in the soil pores and (b) external movement, in which the phenomenon occurs at the land surface and the resulting vapour is lost to the atmosphere by diffusion and convection (surface evaporation). The diffusion of water vapour from one area to another in soils does occur. Water vapour moves from high vapour pressure (generally in moist soils) to low pressure (generally in dry soils). The actual amount of water vapour in a soil at optimum moisture for plant growth is surprisingly small, at one time being 10 kg in the upper 15 cm of a hectare of soil. This compares with 375,000 kg of liquid water in the same soil volume.

5.7 Pitch Available Water Capacity/Content (PAWC)

Every pitch with different soil texture, structure, layers, clay quantity and type, grass shoots and root spread and growth, climatic and weather variations just above and within the pitch profile, etc., does have different pitch available water content (PAWC), which results in different pitch behaviour.

Generally, PAWC depends upon the daily diurnal weather variations available on and within the pitch profile (in the form of evapotranspiration rates, relative humidity, temperature, dew point, etc.), the type and quantity of clay minerals, and the exchangeable cations (especially the Ca^{+2} levels, Al^{+3} and H^{+}, E.S.P., E.M.P., etc., which regulate the availability, movement, and retention of held moisture within the pitch profile.

Judicious retention and release of the PAWC during the pitch preparation process and during the match defines the pitch behaviour by providing the desired pitch surface density during the match pitch preparation or at the time of the match (either available in capillary, hygroscopic, or film moisture in the liquid or vapour phase).

The curator's skill lies in capturing and retaining the PAWC during the pitch preparation process, which is the most important component in achieving the desired target densities at various stages of the match.

PAWC on the playing surface and within the pitch profile is generally regulated, controlled, and maintained by

1. Choosing fine Ca^{+2}/Mg^{+2}-dominant fine smectite-dominant clay soils with high water holding and retention capacity.

2. Having a 2- or 3-layered pitch profile so as to flush out the undesirable excess gravitational water from the drainage layer and also to check the upward movement (towards the pitch surface) of undesirable water due to the capillary rise.
3. The judicious use of rolling techniques and schedules.
4. Turf grasses (as it increases the shear and tensile strength of the pitch profile and regulates the moisture levels within the pitch profile through ET).
5. Use of hessian cloth before (to capture and retain the desired moisture levels at the time of rolling) and after rolling (to destress the grass and reduce the moisture loss through ET).
6. Manual brushing (so as to increase the ET rate).
7. Use of cut grass on the bare patches on the pitch to regulate the ET rate bare patches on the pitch to regulate the ET rate.
8. Use of soil conditioners or additives in the form of common salt, gypsum, etc., to regulate the soil structure and PAWC.

5.8 The Science Behind the Art of Pitch Soil Moisture Retention, Movement, and Its Loss

Water contained in the soil (in micro and macropores and adsorbed on a colloidal surface) is known as soil moisture. Not all the water that soils can hold is available to plants. Much of the water remains in the soil as a thin film. Soil water contains dissolved salts and makes up the soil solution, which is important as a medium for supplying nutrients to growing plants. However, the water in the smallest pores is so close to solid particles that it is strongly attracted to and held on the particle surfaces. This water may be so strongly held that plant roots cannot pull it away. Consequently, not all soil water is available to plants.

Depending on the type of soil, 1/6 to 1/2 of the water retained in the soil can be utilised by the plants. However, capillary water (in the liquid phase) and hygroscopic water (in the vapour phase) are generally not available to plants. Interestingly, these unfavourable forms of water for plants are a boon for cricket pitch profiles (during pitch preparation and matches) because they provide the desired adsorbed and entrapped OMC for achieving the maximum compaction level and controlling and regulating the behaviour of the pitch during matches (especially Test matches) through the slow and consistent release of pitch soil moisture in both liquid and vapour forms.

When the pitch profile is fully irrigated and all the micro and macropores are completely filled with water, it is said to be in a "saturated state." After the application of water, there will be a relatively continued downwards movement of some water due to gravity. After 2 to 3 days, this rapid downwards movement will become negligible. The soil is then said to be at its field capacity. At this time, water has moved out of the macropores, and its place has been taken by air. It is the capacity of soil to retain water against the downwards pull of gravity. At this stage, water is held with a force of about 1/3 of an atmosphere, which is readily available to plants and micro-organisms.

Moisture movement will continue to take place, but the rate of movement (unsaturated flow) is slow because it is now due primarily to capillary forces, which are effective only in micropores. As the moisture contents fall (due to evapotranspiration), a point is reached when the water is so firmly held by the soil particles that plant roots are unable to draw water and start to wilt. This stage of the soil moisture relationship is known as the "wilting point," and the percentage amount of water held by the soil at this stage is known as the "wilting coefficient." It represents the point at which the soil is unable to supply water to the plants. Water at a wilting coefficient is held with a force of 15 atmospheres. Between the wilting percentage (15 atmospheres) and the field capacity (1/3 atmosphere), the water is available to plants (Fig. 5.6).

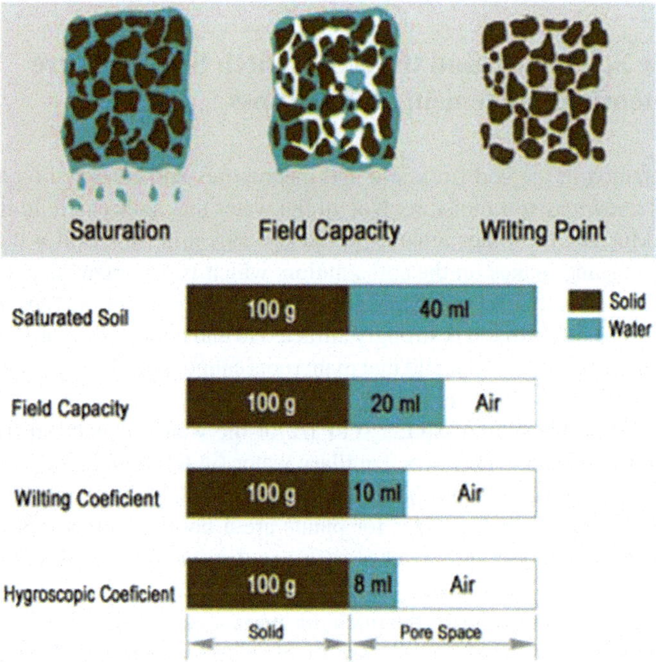

Fig. 5.6 Volume of water and air associated with 100 g of a well-granulated silt loam at different moisture levels The top bar shows the situation when a representative soil is completely saturated with moisture. Water will soon drain out of the larger macropores. The soil is then said to be at field capacity. Plants will remove moisture from the soil quite rapidly until they wilt. When permanent wilting of the plants occurs, the moisture is said to be at the wilting coefficient. There is still considerable moisture in the soil, but it is held too tightly to permit its absorption by plant roots. A further reduction in moisture content due to the hygroscopic coefficient is illustrated in the bottom bar. The water is held very tightly at this point, primarily by soil colloids. *Source* United States Department of Agriculture (USDA) Engineering Field Manual, 2013

Hygroscopic coefficient is the maximum amount of hygroscopic water absorbed by 100 g of dry soil under saturated conditions of humidity (50% RH) and temperature (25%). The tension is equal to a force of 31 atmospheres (Table 5.2). As the soil moisture is lowered below the wilting point, the water molecules that remain are very tightly held, mostly being adsorbed by the colloidal surface. This stage is approximated when the atmosphere above a soil sample is essentially saturated (98% relative humidity) with water vapour and equilibrium is established. The water is held so tightly (at − 31 bars) that much of it is considered non-liquid and can only move in the vapour phase. The moisture content at this point is termed the hygroscopic coefficient. Soils high in colloidal materials will hold more water under these conditions than will coarse soils.

Table 5.2 Volumetric water content at field capacity and hygroscopic coefficient for three representative soils and the calculated capillary water

	Volume %		
	FC	HC	Capillary water
Soil type	(− 0.3 bar)	(− 31 bar)	(col. 1–col. 2)
Sandy loam	12	3	9
Silt loam	30	10	20
Clay	35	18	17

F.C. field capacity, *H.C.* hygroscopic coefficient
Note The clay soil retains most water at the field capacity, but much of that water is held tightly in the soil at − 31 bars potential by soil colloids (HC)

5.9 Process of Wetting and Drying a Cricket Pitch Profile During Pitch Preparation

Depending upon the pitch soil structure, texture, and mineralogy, its hydraulic conductivity, previously available moisture content in the pitch profile, soil organic matter (SOM) content, available salts in the soil solutions, and climatic and weather parameters (such as temperature, relative humidity, and radiation) available just above the pitch profile complex and weather forecast (at least two weeks before and during the match days), the pitch soil profile is saturated slowly and uniformly (as shown in Figs. 5.7 and 5.8) up to 100 mm (especially during evening hours to reduce and avoid evaporative losses) during the initial phase of the pitching process, especially during the evening hours to reduce and avoid depending upon the pitch soil mineralogy, water holding capacity, texture, structure, infiltration, and hydraulic conductivity, the pitch profile is left overnight for 6–12 h so as to flush out the gravitational water available within the pitch profile through the subsoil drains or the drainage layer of sandy loam and river-washed sand just lying beneath the 8 in. of pitch clay surface.

Cricket pitch soils that are dominated by swelling and shrinking smectitic fine clay with high water holding capacity and low hydraulic conductivity require more time (generally 12 h or more) before the first rolling session (when the pitch surface is in plastic consistency). Cricket pitch soils with kaolinitic non-swelling and shrinking minerals or mixed clay minerals with a high proportion of fine or medium sand-dominated soils with low water holding capacity and high hydraulic conductivity require less time (less than 6 h) to reach the appropriate plastic limits so that the first rolling sessions can begin. Considering the low water holding capacity, high hydraulic conductivity, and availability, these types of cricket pitch soils require less frequent watering. Water based on the available moisture test analysis (in the capillary or hygroscopic water of the pitch profile) through the thumb feel method or through spikes, the pitch surface is rolled with a light roller (straight followed by diagonal or cross rolling) so as to seal and entrap the available moisture within the pitch soil profile (Figs. 5.7 and 5.8).

Fig. 5.7 Fully saturating the pitch surface (at least 4 in.) at the initial stages of pitch preparation (before rolling) with hosepipe

Fig. 5.8 Sprinkling of 10–20 L of water after medium to heavy rolling (to compensate the daily evapotranspirative loss of moisture from the pitch surface and to check and regulate the initiation of cracks on the pitch surface)

This process aids in trapping the desired available optimum moisture content (OMC) within the pitch profile to achieve the maximum compaction level via the various roller weights and schedules, as well as reducing excessive pitch soil moisture losses via evapotranspiration. Further, the pitch profile is rolled with the predetermined roller weight and schedules to obtain the maximum compaction levels. The available soil moisture in the liquid phase exists as capillary water, and the unavailable water in vapour form exists as hygroscopic water, which acts as a lubricant for rearrangement of soil particles to achieve the desired compactness necessary for fast and bouncy pitches. Further, available water in the pitch profile is in the form of capillary water in the micropores and some hygroscopic water adsorbed electrostatically on the colloidal micelle (forming the diffused double layer) and within the internal multi-layered lattice, which is possibly squeezed out by the help of rollers and evapotranspiration (through evenly grown pitch turf grass).

It is also well established (through research and development) that irrespective of roller weight, it has a compactive effect only up to 100 mm (which is nearly 1/3 of the total pitch profile) of the upper pitch profile and only helps in squeezing out the available water in the macro and micropores of the pitch profile. However, nearly 2/3 of the pitch profile beneath the playing surface is dried or desiccated only through transpiration from uniformly distributed, deeply rooted pitch grasses.

5.10 Relationship Between Soil Moisture and Pitch Behaviour

Generally, pitches on the first and second days of Test matches have some capillary water and more hygroscopic water and film moisture. Initially, this favours fast bowlers. As the pitch moisture, in the form of hygroscopic water, gets depleted, it favours the batsman on the third and fourth days. On the fourth and fifth days, in the absence of hygroscopic moisture, the pitch surface cracks or crumbles on desiccation, favouring the spinners.

Even unavailable hygroscopic water adsorbed on the clay micelle affects pitch behaviour during the game (especially Test matches) by slow release of soil moisture in the form of vapour (referred to as thermo-osmosis or sweating). Thermo-osmosis, or the sweating of entrapped, tightly adsorbed hygroscopic water within the pitch profile, is due to the diurnal (day-night) differential thermal variations occurring above and below the pitch profile and generally to the use of thick, heavy covers over the playing pitch surface (after the match days and during the night).

The gases accumulate between the internal surface of the pitch cover and the pitch surface, which increases the temperature and enhances the evapotranspirational losses of pitch profile moisture. This hygroscopic water in the vapour phase is generally observed as an accumulation of water droplets (as shown in Fig. 5.10) just below the pitch cover inner layer during the morning hours (when the covers are put off), which is dependent upon the temperature and is attained when the pitches are covered to release the energy through the latent heat of water. Even unavailable hygroscopic water adsorbed on the clay micelle influences pitch behaviour during the game (especially in Test matches) through the slow release of soil moisture in the form of vapour (called thermo-osmosis or sweating). This hygroscopic water in the vapour phase is generally observed as an accumulation of water droplets (as shown in Fig. 5.10) just below the pitch cover inner layer during the morning hours (when the covers are put off), which is dependent upon the temperature and is attained when the pitches are covered to release the energy through the latent heat of water. Even unavailable hygroscopic water adsorbed on the clay micelle effects the pitch behaviour during the course of the game (especially for the Test matches) by slow release of soil moisture in the form of vapour (called thermo-osmosis or sweating). Thermo-osmosis, or the sweating of entrapped, tightly adsorbed hygroscopic water within the pitch profile, is due to the diurnal (day-night) differential thermal variations available above and below the pitch profile and generally to the use of thick, heavy covers over the playing pitch surface (after the match days and during the night). Though the soil moisture in vapour form is present in a lesser amount within the pitch profile, both affect the swelling and shrinking capacities of the pitch soil colloidal complex to a large extent, thereby affecting and regulating the pitch behaviour to a large extent as the game proceeds. This is the reason why the dry and desiccated pitches on the fourth and fifth days of Test matches provide a grip favouring the spinners (Figs. 5.9 and 5.10).

Fig. 5.9 **a**, **b** Dr. Sanjay Kumar Ray, Principal Scientist and Head, Regional Centre, ICAR-NBSS&LUP, Jamuguri Road, Borbheta, Jorhat, Assam along with Dr. S. B. Singh (Curator, BCCI) and Abei, Curator, Nagaland Cricket Association, observing the method of covering the pitch surface and impact of thermo-osmosis on the pitch surface at the Nagaland Cricket Academy Ground, Souvima, Nagaland dated—24.12.2018

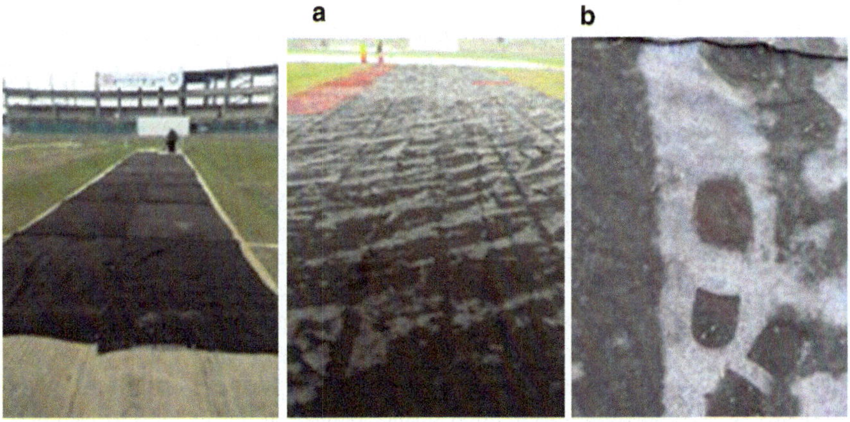

Fig. 5.10 Pitch surface at the time of covering (in the evening) and **a**, **b** condensed (due to thermo-osmosis-sweating) pitch soil moisture on the pitch surface after putting off the pitch covers early in the morning

Though this soil moisture in vapour form is present in a lesser amount within the pitch profile, it affects the swelling and shrinking capacities of the pitch soil to a large extent, thereby affecting and regulating the pitch behaviour as the game progresses.

5.11 Role of Available Soil Moisture Within the Pitch Profile During Pitch Preparation

1. The soil moisture available within the pitch profile in the form of liquid (held in micropores as capillary water and adsorbed hygroscopic water) and vapour phases (at the time of initial or final pitch preparation) affects the nature and behaviour in the following ways:

Soil water directly affects the pitch's behaviour during the course of the game. Even in the fully compacted stage, the pitch profile has sufficient moisture entrapped in the micropores as capillary water and adsorbed hygroscopic water (available in the non-liquid vapour phase), which affects the behaviour of the match.

Clays that swell and shrink act as reservoirs for water in the very large specific internal surface area and tightly adsorbed water in the intralayer lattice structure and double diffused layer. Gradual loss of soil moisture (generating variations in pitch surface density) within the pitch soil profile due to diurnal changes in weather variables (temperature, relative humidity, and radiation), E.T. thermo-osmosis, and vapour pressure regulate the nature and behaviour of the pitch as the game progresses day by day or session by session. This differential soil moisture, present as capillary water and adsorbed moisture within the lattice structures of pitch soil rich in smectite clay, provides equal opportunities for the pacers, batsmen, and spinners by providing variable pitch surface density during the course of the match. During the first and second days of the match, having high soil moisture within the pitch soil profile along with uniform live grass favours the pacers and seamers, but assists the spinners and batsmen later on as the pitch profile gets dried out and desiccates due to the evapotranspirational loss of moisture from the pitch profile. The adsorbed hygroscopic water within the smectite clay lattice acts as a permanent source of the optimum moisture content desired for achieving maximum or variable compaction of the pitch profile during the course of the match (especially in Test matches). Fine clay soils retain at least 15–20% of the hygroscopic moisture as compared to matches). Coarse-grained sandy soils (only 1–2%). The retention and movement of this capillary and adsorbed hygroscopic water in vapour form within the pitch profile produces the desired variations in pitch surface behaviour during the various phases of the match.

Soil moisture available in the pitch soil complex directly controls the behaviour of the pitch by controlling other important components of the pitch soil complex, such as type and quantity of mineral content, organic matter content, proliferation of grass roots, and air voids.

Among the many factors that determine the behaviour of a cricket pitch, other factors affecting the availability of soil moisture include type (structure, texture, and mineralogy), compactive effort, method of compaction of soil, etc. The soil-water content in the pitch profile is one of the major factors affecting and controlling the pitch surface density levels during various phases of the match.

It acts as a lubricant (in the form of capillary and hygroscopic water available in the micropores and adsorbed on the surface of clay and humus micelles) in between mineral particles of pitch soils, which are desirable to have the best achievable compaction level at a predetermined OMC. No dry pitch soil can be compacted to its maximum (at optimum water content, OMC) without the presence of soil water as a lubricant in between the pores.

These adsorbed soil moistures, both available and unavailable, act as an elixir, providing the desired variable pitch behaviour as the Test match progresses.

Proctor (year) explained the typical moisture content-dry unit weight relationship with the help of lubrication theory. At low water contents, the soil is stiff, and the soil grains offer resistance to compaction. As the water content increases, the particles develop larger films around them (forming DDL), which lead to lubrication of the particles and make them easier to be worked around, allowing them to move closer into a denser configuration, resulting in a higher dry unit weight and fewer air voids. The dry unit weight continues to increase until the OMC is reached, a stage when the lubrication effect is at its maximum. However, as moisture content increases, water begins to replace soil particles, and the dry unit weight decreases. Lambe (1958) uses the concepts of soil structure and the electrical double diffused layer theory to explain the effect of water content on dry unit weight (Fig. 5.11).

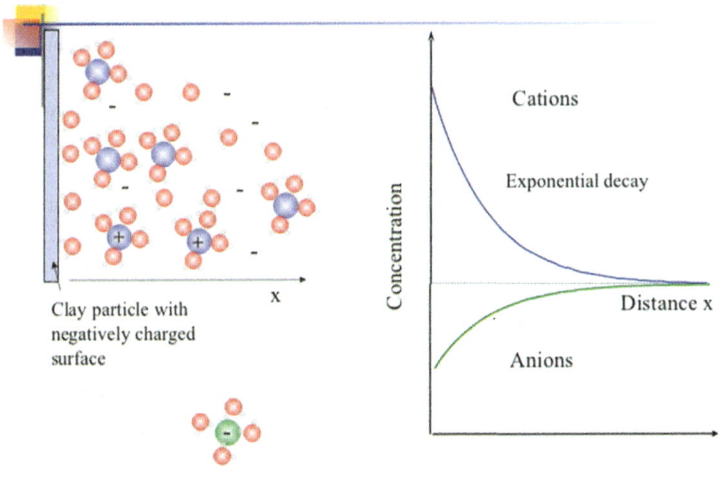

Fig. 5.11 Schematic diagram of an electrical diffused double layer in soil

In soils where cohesive forces are higher, there is an attractive force, namely the van der Waals force, which acts between the soil particles, and a repulsive force that is due to the double layer of adsorbed water tending to come into contact with each other. While the attractive force essentially remains the same in magnitude, the repulsive force is directly related to the size of the DDL. If the net force between the particles is attractive, a flocculated structure is formed, whereas if it is repulsive, the particles tend to move away or disperse. At low water contents, the DDL is not fully developed, the repulsive force is small, and the net force between the particles is a strong attractive force. This makes it difficult for the particles to move about when compactive effort is applied and results in a low dry unit weight. This is the reason why dry pitches should not be rolled. As the moisture content is increased, the DDL expands and the interparticle repulsive forces increase. The particles easily slide over one another and get packed more closely, resulting in a higher dry unit weight. The DDL expansion is complete at the OMC, and that is why the dry unit weight is at its maximum at that stage. Water does not add any additional DDL expansion beyond the OMC, but it does tend to occupy space that would otherwise be occupied by soil grains. Hence, there is a decrease in dry unit weight.

2. Provides available moisture for pitch grass vegetative and root growth, as well as the microbial flora and fauna found within the pitch soil complex. Soil water, along with the dissolved minerals and salts in the soil solution, provides the basic essential micronutrients required for good vegetative growth of pitch grass.
3. Aids in decompacting the pitch soil complex immediately following the match and in regenerating the pitch grasses. Cricket pitch soils with swelling and shrinking clays regenerate themselves naturally through their wide cracks when saturated with water.
4. Helps in aerating the decompacted pitch soil profile and helps in releasing the entrapped poisonous gases such as CO_2 and CO.
5. Helps in flushing out deposited salts (generally through irrigation water, fertilisers, insecticides, or pesticides) within the pitch profile complex through the available subsoil drainage layer below the pitch profile. This harmful and toxic salt deposited within the pitch profile gets leached away along with the gravitational water.
6. Helps in maintaining the SOM/clay ratio within the pitch profile. Soil water is a must for the microbial decay of dead grass roots.
7. Assists in keeping the pitch soil complex's temperature just above and below normal. It also regulates the soil air, which is intricately linked with soil water.
8. The moisture level of the pitch soil also regulates the impact and formation of dew on the pitch surface or outfield. The intensity of infestation and the menace of dew depend on the extent of available soil moisture within the pitch profile or in the outfields.

5.12 Summary

The whole art of pitch curatorship lies in achieving the best available consistent compression levels (at a predetermined optimum moisture content) in between the two extremes of soil moisture available in the form of capillary, hygroscopic, and film moisture (both in liquid and vapour form) entrapped within the pitch profile through the different roller weights and rolling schedules, and uniform deep drying of the pitch profile through the pitch grasses. If the pitch is prepared at different levels of entrapped available water within the pitch profile, it will behave differently. Furthermore, the swell-shrink behaviour of clay and humus-dominated pitches is completely dependent on soil moisture levels to provide the pitch profile with a maximum or variable bulk density. The extent of swelling and shrinking is limited to the presence, absence, or movement of water within the pitch profile only.

Generally, pitches on the first and second days of Test matches have some capillary water and more of the entrapped and adsorbed hygroscopic water (mostly in unavailable form) in vapour form, which favours the fast bowlers initially. As the pitch moisture in its less available forms gets depleted, it favours the batsman on the third and fourth days. The pitch surface cracks or crumbles on desiccation in the absence of available soil moisture on the fourth and fifth day, favouring bowlers, particularly spinners.

Keeping all other factors of the pitch preparation materials and process in common, the pitch soil moisture available in the form of available capillary water held in the macropores and generally unavailable soil water as hygroscopic water (in vapour form) is the single most important tool in the hand of a pitch curator to control and regulate the desirable pitch behaviour during the pitch preparation. Different pitch behaviour, either their cracking or crumbling patterns, structure, wear, and tear of the pitch surface and the turf grass during the pitch preparations or during the match, solely depends upon the retention, migration, and loss of pitch soil moisture within and just above the pitch profile.

By creating an artificial stress through the retention and regulation of the pitch soil moisture (during the pitch preparation or the matches), desired pitch soil behaviour can be directly achieved according to the laid-down objectives of a match. As a result, it is reasonable to regard pitch soil moisture as an elixir for the pitch profile complex.

References/Bibilography

Albalasmeh AA, Ghezzehei TA (2013) Interplay between soil drying and root exudation in rhizosheath development. J Plant Soil

Aldous DE (2016) Handbook on international turf management. CRC Press

Baker SW, Binns DJ, Cook A, Mooney SJ (2001a) The performance of cricket pitches in relation to soil type and moisture content. Int Turf Grass Soc Res J 9:469–478

BCCI Curators Certification Course Manual (2013) BCCI, 2nd edn

Brady NC, Weil Ray R (2008) Nature and properties of soil, 14th edn. Prentice Hall

Bowles JE (1979) Physical and geotechnical properties of soils. John Wiley, New York

Burdet L (1986) Turf cricket pitches. South Australian Cricket Association, Adelaide, 28 p

ECB Staff (2007) Recommended guidelines for the construction, preparation and maintenance of cricket pitches and outfields at all levels of the game, England and Wales Cricket Board, Lord's Cricket Ground, London, UK

Gray DH, Barker D (2004) Root soil mechanics and interactions. Water Sci Appl Am Geophys Union, New York, pp 113–123

IOG/ECB (2006) Management and maintenance of the cricket pitch, The Institute of Groundsmanship, Milton Keynes, UK

Loveday J (1974) Methods for analysis of irrigated soils, Technical communication No. 54, Commonwealth Agricultural Bureaux, Buckinghamshire, England

Mrshall TJ, Holmes JW, Rose CW, Soil physics, 3rd edn. Cambridge University Press, New York

McIntyre K, Don (2001) Cricket wickets—science vs fiction. Horticulture Engineering Consultancy, 5-Brimage Palace Kambah, ACT Australia
McIntyre DK, Jackobsen B (1998) Drainage for sportsturf and horticulture, Horticultural Engineering Consultancy, Canberra, Australia
Mitchell K (2010) Cricket wickets, Copyright ©Queensland Cricket, pp 1–37
Mehra RK (2014) A textbook of soil science, Indian Council fir Agricultural Research Pub.
Parsons S, The aeration of clay soils in cricket, Engineering Doctorate (Academic year 2008–2012), School of Applied Sciences, Cranfield University
Pearson J (2010) Basic guide to turf cricket pitch preparation, VCA
Rose CW (1966) Agricultural physics. Pergamon Press, Oxford
Rathinaswamy A, Saliha BB (2014) Fundamental of soil science. Scientific Publication
Sehgal J (1996) Pedology: concepts and application. Kalyani Publishers, Ludhiana, India
Shimpton P, James L (2009) Optimization of cricket pitch rolling, PhD thesis, Cranfield University, School of Applied Science and Natural Resource Department
Singh A (2009) Modern geotechnical engineering, 3rd edn. CBS Publication
Taiton NM, Klug JR, Edmondson D, Campbell RK, Deventer PWV, Beer MJD (1993) Principles and practice of pitch preparation, United Cricket Board of South Africa
TCCB and NCA (1989) Recommendations on the preparation of cricket pitches, Test and county Cricket Board and National Cricket Association, Lords, London
Wood C (2006) Recommended guidelines for the maintenance of cricket Pitches and outfields at all levels of game. England and Wales Cricket Board in association with The Institute of Groundsmanship

Chapter 6
Cricket Pitch Grass: The Science Behind the Art of Turf Pitch Making

6.1 Introduction

All the major formats of cricket matches are played on the same pitching surface, but the results differ. Turf pitches require not only good-quality clay soil with a high percentage of clay contenets but also the type of clay minerals (either 1:1 or 2:1 kaolinite, illite, vermiculite, or smectite) with uniformly grown grass foliage and deeply developed roots. Even with a lower percentage of clay but a swelling and shrinking type of clay (preferably smectitic), having a uniformly spread Bermuda grass turf produces good bouncy and fast pitches when compared to pitches with soils that have no or little expanding lattice structure, such as illitic or kaolinitic clay minerals, and with less or no grass turf. The recovery of turf grass after a match on cricket pitches with non-swelling and non-shrinking types of clays dominated by kaolinitic or illitic types of clay minerals is very poor.

Rollers only squeeze the moisture out of the soil in the pitch profile and compact the top 04 inches (100 mm) of the surface of the pitch. The top one to two inches of a wicket pitch profile without grass can be dried out by evaporation into the atmosphere. Nearly 2/3 of the moisture in the soil below the pitch surface is lost through evaporation by the grass roots alone. This helps make the pitch surface uniformly hard. Through the process of evapotranspiration, commonly used Bermuda grasses (Cynodon dactylon) on a turf wicket (a warm-seasoned grass variety as Bermuda grass) generally transpire 6–7 mm of water per day (90 mm during 15 days of Test match wicket preparations, 180–210 mm per month, or 2190–2555 mm per year). Deep-rooted turf grasses help in drying the turf wickets in the subsurface layers of the cricket pitch profile, where even the heaviest rollers have limited or no rolling effect. Without uniform spread of turf grass, water removal will be only through

S. B. Singh et al., *Cricket Pitches*, Springer Transactions in Civil and Environmental Engineering, https://doi.org/10.1007/978-981-99-2913-9_6

surface evaporation, which can cause steep moisture gradients and unregulated bad pitch surface cracking. A good cricket wicket management system must use this process to remove water from deep in the profile in order to dry and harden the soil profile. It is therefore essential to have turf grass roots at least up to the top 100 mm of the pitch profile evenly dried out to achieve a uniformly dry and hard surface. If an area on the pitch surface has no roots or only shallow or sparse roots compared to the adjacent areas, it will not be possible to prepare a wicket that has uniform hardness. Zones having different root depths will produce surfaces with different responses to the ball, as different moisture levels between 30 and 100 mm below the surface will generate different bounce heights. Higher moisture content in the top 100 mm as compared to the adjacent area will have lower bounce than the adjacent area, which has been dried out completely by a good root system (at least 100 mm).

6.2 Classification of Turf Grass

Based on how they deal with the weather, turf grasses are put into two groups: warm-season grasses (C4), which do well in tropical and subtropical areas, and cool-season grasses (C3), which do well in temperate and subarctic areas (Huang 2006). Warm-season grasses use significantly less water than cool-season species. Cool-season grasses, on the other hand, are generally more susceptible to moisture stress than warm-season grasses (Double 2006). Warm-season grasses are well adapted to warm climates, humid, and wet conditions, or semi-arid areas (as available in the Indian subcontinent, Australia, Africa, and the West Indies). Warm-season grasses thrive in temperatures ranging from 27 to 35 °C, with root growth occurring at temperatures ranging from 10 to 16 °C. These grasses usually become dormant at or below 10°C and discolour brown. With the exception of Bahia grass, the warm-season turf grasses have creeping or spreading growth habits, due to which they bear the stress of rollers well and recover naturally after the match.

Some of the commonly available warm-season turf grasses are St. Augustine grass (*Stenotaphrum secundatum*), Zoysia grass (Zoysia *spp.*), Bahia grass (*Paspalum notatum*), Bermuda grass (*Cynodon dactylon*), Centipede grass (*Eremochloa ophiuroides*), Seashore paspalum (*Paspalum veginatum*), Buffalo grass (Bouteloua *dactyloides*), etc.

Fig. 6.1 Gradual wear and tear and after-match recovery of Bermuda turf grass during the pitch preparation and match days (Test match between India Vs South Africa, 19 to 23 October 2019, at the JSCA Cricket Stadium Complex, Ranchi, Jharkhand)

Among these grass varieties, Bermuda grass is a medium- to fine-textured, warm-season turf grass that spreads by rhizomes and stolons. Also called wiregrass or couchgrass, this is a mostly suitable and acclimated grass variety to be used in cricket pitches and outfields throughout the world. It has excellent wear, drought, and salt tolerance, is extremely heat tolerant, and is competitive against weeds. It recovers well from close mowing, and after mowing, recovery is good (as shown in Fig. 6.1). However, its disadvantages are that it has a number of cultural and pest problems and, therefore, needs a higher level of maintenance input than other grasses. Generally, it becomes dormant during the cold weather and has very poor shade tolerance.

(2): Cool-season grasses (C3)

Climate limits cool-season grasses to temperate regions (in northern or southern latitudes). They thrive in temperatures ranging from 16 to 24°C. Generally, root growth stops as the temperature approaches 27°C, and shoot growth stops above 32°C. Some of the commonly available cool-season grasses are Kentucky bluegrass (*Poa pratensis*), perennial Rye grass (*Lolium perenne*), tall fescue (*Festuca* arundinacea), creeping bent grass (*Agrostis stolonifera*), etc.

In England and New Zealand, Rye grass is the most common grass used on cricket wickets. However, Bermuda is thought to be better for cricket wickets. Rye grass, devoid of stolons and rhizomes, has a bunchy and shallow root system (generating a moist subsurface), which is one of the major reasons for the low pace and bounce of UK and New Zealand pitches. Rye grass germinates quickly from seeds. It has a high wear tolerance and generally withstands low mowing. Rye grass, being a cool-season grass, requires nearly 20% more water than Bermuda and has a very low drought tolerance. Drought causes cool-season grasses to go dormant and eventually die. Being a bunch of grass Rye grass grows from a crown (apical meristem) in its centre, close to the ground. When this crown is damaged by rolling impact or low mowing, the plant will not regenerate and die. Under these circumstances, the roots can no longer pump water out of the bottom part of the pitch profile, leading to high moisture levels and low pitch surface bounce. In seasonal subtropical countries, cool-season grasses perform well during the winter season.

Considering the desirable low temperature and high moisture availability in Shillong, Meghalaya, Rye grass has been introduced at Polo Ground (from the UK). Though the initial establishment of Rye grass on the cricket pitch at Shillong's Polo Ground was discovered to be very quick, the after-match recovery due to rolling impact and wear and tear was discovered to be very poor (as illustrated in Fig. 6.2 a). Because of the poor after-match recovery of Rye grass and the available inactive fine kaolinite-dominant clay soil at very low pH levels, the only alternate method with these types of wicket conditions is reseeding of annual Rye grass immediately after the match.

Fig. 6.2 Newly introduced Rye grass (from UK) at Polo Ground, Shillong, Meghalaya. **a** Showing very high mortality of pitch surface turf grass and very poor after-match recovery of the turf grass at Polo Ground, Shillong, Meghalaya

6.3 Functions of Grass in Turf Cricket Pitches

The intrinsic properties of cricket pitch soils, such as aggregate stability, infiltration water holding capacity (WHC), bulk density, soil structure and texture, organic, chemical, and colloidal content, shear and tensile strength, pitch soil compaction and consolidation by controlling the soil moisture content (SMC) *vis-à-vis* evapotranspiration, are the most important determinants of pitch behaviour, directly or indirectly controlled by the uniform deep-rooted pitch turf grass cover. Turf wickets are generally defined as the upper surface of the cricket pitch, which is soil bound by grass roots that form a thick mat. Turf grasses have fibrous, shallow roots ranging from 100 to 200 mm, and nearly 85 per cent of the roots are present in the top 120 mm of the pitch profile (Figs. 6.3 and 6.4).

Fig. 6.3 Laying of a turf wicket in sequence from 1 to 3

6.3.1 *Grass as a Pumping System for Drying Pitches*

Pitch grasses extract water through their root systems, and part of the water is lost through their leaves by a process called transpiration. Water is also lost through evaporation from the pitch soil, leading to the loss of water from the pitch through evapotranspiration (ET) (the combination of evaporation by the soil surface and transpiration through the plants). The ET process is the only way to dry a clayey pitch profile below its field capacity. Plant transpiration and soil structural changes that increase initial drainage may cause a faster decline in pitch soil moisture through grasses. Evaporation drives soil moisture upward at the soil surface through capillary action, resulting in the removal of moisture from the lower depths of the soil

(Adams et al. 2001; Baker et al. 1998). Drying the pitch soil profile evenly requires evapotranspiration to remove moisture from the pitch profile. This makes the surface as hard as possible and increases the dry bulk density of the soil as the soil shrinks.

When it comes to removing water from the cricket pitch profile, the size of the roots and the amount of water the grass needs are the most important factors. The depth of the roots will determine how deep into the soil the water is likely to be removed. The rate of water supply from the soil to a root depends on the difference in water potential between the root and the soil, the conductivity of water in the soil around the root, and the soil-root contact. Through evapotranspiration, cricket pitch grass allows for deep and even drying of the pitch. Without a reasonable turf cover, water removal will be mainly by surface evaporation, which can create a steep moisture gradient and an undesired surface crack (Kirkman et al. 1989). Roots, rhizomes, and even rolled surface vegetation together assist in a pitching behaviour. The turf cover affects the performance of the pitch, including the pace, minimising damage (wear) on the wall, and providing the seam.

The ability of the couch to produce rhizomes, which add tensile strength to soil and help hold soil aggregates together, is a significant advantage. Turf grass water use is influenced by environmental factors such as weather (temperature, wind, solar radiation, and relative humidity), soil type, and soil moisture. It is also affected by turf species, genotype, and plant morphological characteristics, since all these factors affect both plant transpiration and soil evaporation. Many investigators have shown that turf grass water availability may vary by species, genotypes, climatic conditions, plant density, depth of the water table, water availability, plant morphological characteristics, etc. (Ekern 1996; Steart et al. 1967; Brown 2003). Kim and Beard (1988) conducted research on turf grass evapotranspiration rates using 12 turf grass species, including warm- and cool-season grasses, growing under well-watered conditions. ET rates were determined by the water-balanced method. Black plastic mini-lysimeters were used.

Table 6.1 Evapotranspiration rates of warm- and cool-season grasses

Relative ranking	ET rate	Grass types	
	mm/day	Cool-season Grasses	Warm-season Grasses
Very low	< 6		Buffalo grass
Low	6–7		Bermuda hybrids, Zoysia, Centipede
Medium	7–8.5	Hard, chewing,	Bahia, Paspalum, St. Augustine, Zoysia
		Red fescue grass	
High	8.5–10	Perennial Rye grass	
Very High	> 10	Creeping bent grass, Kentucky bluegrass, Annual bluegrass	

Source Beard and Kim (1989)

The differences in ET rates between species were linked to morphological traits like the number of shoots, the number of leaves per unit area, how the leaves were oriented, how wide they were, and how fast they grew vertically. Due to having very few shoots (low canopy resistance), St. Augustine grass had a medium–low ET rate of 0.23 in/day. Bahia grass showed a medium ET rate of 0.25 in/day when growing without limiting soil moisture due to its high leaf area. Adelaide seashore paspalum, on the other hand, had a low ET rate of 0.21 in/day. It also had a very fast vertical leaf extension rate, but its leaf width was average. Zoysia grass exhibited significant differences in ET rates due to vertical leaf orientation.

The three Bermudas (Arizona common, Tiffany green, and Tiffany) were in the low range due to low leaf area. Warm-season grasses use significantly less water than cool-season species. Cool-season grasses are generally more susceptible to moisture stress than warm-season grasses (Duble 2006).

This difference in water use comes from changes in the way grasses make food that happened as they grew in hot, dry places. These changes, which include modifications to biochemical reactions and internal leaf anatomy, greatly enhance photosynthetic efficiency, which means that plants can maintain a high level of carbohydrate production and continue to grow even if their stomata are partially closed. The partial closure of the stomata slows water consumption. Cool-season grasses cannot maintain enough carbohydrate production to maintain growth unless their stomata are nearly wide open. When water is limited, transpiration rates are generally higher than those of warm-season grasses (Gilbert et al. 1989). The low ET rates of centipede grass are related to a very slow vertical leaf extension rate. Warm-season turf grasses are characterised by their lower ET rates compared to cool-season grasses. Cool-season turf grasses use more water than warm-season turf grass when water is unlimited. Turf grass ET rates vary among species and cultivars within species, which can be explained by differences in stomatal characteristics, canopy configuration, growth rate, and characteristics of the roots. Increased mowing height and top growth can be expected to increase evapotranspiration by roughening the plant canopy surface, increasing the capacity for absorbing advective heat, and increasing root growth, resulting in greater exploitation of the soil-water source (Kneebone et al. 1992).

6.3.2 Benefits of Turf Grass in Cricket Pitches

1. Soil aggregate stability:

Amezketa (1999) says that the good effects of grass root hairs on soil aggregation are caused by root secretions that bind fine particles together into stable macroaggregates, even when the roots are dead. Several plants' roots release substances that stick together and form a coating of soil particles called rhizosheaths. The rhizosheaths are strongly attached to the roots and to each other. Czarnes et al. (2000) say that rhizosheaths are important for taking in nutrients and water and reducing the effects of environmental stress. Plant root mucilage and microbial mucilage are known to form and stabilise soil aggregates (Chesire 1979).

Mucilages are produced by the root cap cells and mucilage-providing bacteria, such as cytophaga. Soil-binding properties of the root cap and bacterial mucilage of the rhizosphere indicate that mucilage can bind soils (Watt et al. 1943). Thus, the adhesive agent of the rhizosheath includes both root and bacterial mucilage. Soil adhesion by the mucilage is primarily via hydrogen bonds between the hydroxyl groups of these neutral sugars and soil particles (Watt et al. 1993).

The roots and rhizomes of the grasses interact with the soil to produce a composite material in which the root fibres of high tensile strength and adhesion are embedded in a matrix of lower tensile strength. The shear strength of the soil is therefore enhanced by the cumulative effect of the root–soil matrix. Root exudates are generally composed of polyglactouronic acid (PGA). These polysaccharides are effective at binding or aggregating the soil. Polysaccharide persistence in soil has been attributed to the interaction of soil clays with other organic constituents or to the polysaccharide's other resistant nature. Organic exudates cause soil particles to adhere firmly to the root surfaces and thus increase the strength of the soil–root matrix (Tengbeh 1993). By mobilising their tensile strength, roots increase the soil's resistance to shearing. Roots act as a sticky agent, binding soil particles at the soil surface.

Physical entanglement and the exudation of higher-molecular weight polysaccharide material capable of binding to mineral surfaces and bridging gaps between mineral particles are two direct effects of roots on the stabilisation of soil structural form. Furthermore, Sprent (1975) noted that the root hairs in the rhizosheaths were far too short and did not enmesh the soil particles, indicating that the attachment of soil particles to roots is primarily a result of the glueing action of root mucilage and not the engulfing effects of root hairs. Soil aggregation in the rhizosheath is a product of the intricate interplay between the release of mucilage by plant roots and the moisture regime of the rhizosphere.

Specifically, soil drying has been shown to be an essential requirement for rhizosheath development. The role of drying as a trigger for the release of mucilage in larger quantities to cope with environmental stresses is a widely recognised mechanism (Hale et al. 1971). Rhizosheaths of grasses formed under dry conditions are larger, more coherent, and more strongly bound to the roots than those formed in wet soils. Rhizosheaths formed in dry layers (9% SWC) were 5 times the volume of the subtending roots. Whereas, in the wet layer (20% SWC), the rhizosheaths formed were only 1.5 times the root volume. Drier soils enhanced the adhesiveness of rhizosheath mucilages and stimulated the formation of root hairs; both effects stabilised the rhizosheaths. Rhizosheath or the mucilage in dry pitch soil conditions acts as a gel and improves the cohesiveness, adhesiveness, and structure desired to have a compacted pitch with good pace and bounce. This coating of rhizosheath and mucilage also provides the desired optimum moisture content for pitch soil compaction. In particular, the concentration of exudates increases with the degree of dryness, which binds to gel. This gelling is assumed to be the major cause of soil aggregation.

A higher root concentration results in the release of more root exudate. These exudates in turn dissolve $CaCo_3$ present in the soil, which helps improve Ca nutrition in plants as well as form a better soil structure, enhancing aeration and hydraulic conductivity. A closer look at the root distribution indicates a relatively low $CaCo_3$ content (3.5–5.1%) in the surface soil, where most of the roots are concentrated.

2. Increases the water infiltration capacity

The ability of soil to absorb water allows the highly compacted match wicket to decompose and recover quickly. Grass roots penetrating the soil create macropores that improve water movement. This also results in the removal of the undesirable gases, which helps to recover the nearly dead or desiccated grass due to the over-consolidation of the match pitch. They contribute to the system of continuous pores in the soil and enhance its infiltration capacity. Soil infiltration increases because the grass roots improve the non-capillary porosity of the soil and promote the formation of water-stable aggregates of 2–5 mm in diameter as well as aggregates larger than 5 mm in diameter.

3. Decrease in soil bulk density:

When grass roots grow in pitch soil, they take up space that was once filled with soil pores and soil particles. Since root diameter is usually larger than soil pores, soil particles are pushed aside up to 8 mm near the root, and the bulk density of the soil increases (Glinski and Lipiec 1990). The size of the macropores is about 0.08–5 mm, and that of the micropores is less than 0.3 mm in soil. Plant root hairs can penetrate up to 0.005 mm in the soil profile (Brady 2002), which provides scope to increase the size of macropores. When there are more roots, the bulk density goes down because the weight of roots per unit volume is less than the weight of soil minerals.

4. Mechanical strength:

One of the mechanical properties of roots is their strength and high tensile strength. The inorganic constituents of soil, on the other hand, are strong in compression and weak in tension. A combined effect of soil inorganic constituents and roots results in reinforced soil. When shearing the soil (through rolling), roots mobilise their tensile strength, whereby shear stress that develops in the soil matrix is transferred to the root fibres through interface friction along the root length (Gray and Barker 2004) or through the tensile resistance of the roots (Ennos 1990). Therefore, when soil and roots in the pitch profile are combined, the resultant soil-root matrix produces a mass that is much stronger than either the soil or the roots on their own. The roots act by transferring the shear stresses developing in the soil to the tensile resistance in the roots and also distributing stresses through the soil, avoiding local stress build-ups and progressive failures. Grass roots change the shear strength of the soil matrix. Grass roots increase the soil shear strength both directly by mechanically reinforcing it and indirectly through water removal by transpiration. The increased shear strengths of *Imperata cylindrica* and *Cynadon dactylon* grasses are 36 kPa and 11 kPa, respectively. Relative shear strength is also defined and used to quantify the reinforcement capacity of grasses. The relative shear strengths of *Impertia cylindrica* and Cynadon *dactylon* are 03 and 02, respectively.

Fig. 6.4 Root area index (RAI) for a cross section of pitch profile

Knapen et al. (2007) say that the soil's resistance comes from the forces that hold the soil particles and other things in the soil matrix together, such as organic matter, plant roots, and broken pieces of rock. Roots reinforce the soil due to their tensile strength and frictional or adhesion properties. Roots extending perpendicular to the soil surface reinforce the soil by increasing the shear strength of the rooted mass on the sheared surface (Ekanayake et al. 1979; Greenway 1987). Roots growing parallel reinforce the soil by increasing the in-plane tensile strength of the rooted soil zone (Zhou et al. 1998). Shear strength in soils with switchgrass roots can be about five times greater than in soils without grass roots (Goldsmith 2006). A pitch's shear strength is defined as its resistance to shearing stresses (the rolling of pitch through heavy rollers and the impact of a ball on pitch) and the resulting tendency for shear deformation. The shear strength of a soil is the most difficult to comprehend in view of the magnitude of factors known to affect it.

Most of the time, the best soil for a cricket pitch is one with a lot of clay, preferably smectites, which swell and shrink. Also, grass roots with deeply rooted rhizomes and rhizosheaths that can stick together make clay soils stickier and more cohesive, which increases their shear strength as a whole. Greenway (1987) says that the size of root reinforcement depends on the root system's shape (e.g. how the roots are spread out with depth and how they are spread out among different root diameter classes), root tensile strength and root tensile modulus values, root tortuosity, the friction between the roots and the soil, and how the roots are oriented to the main direction of strain. Nilaweera and Nutalong (1999) say that the overall mechanical effect of roots can be thought of in terms of both how strong they are and how they are spread out in the soil. A large number of fine roots would fix the soil more effectively than a small number of coarse roots. (Bischetti et al. 2005; Genat et al. 2005) found that tensile strength, which is usually measured in mass per unit area (MPA), goes up as the root diameter goes down. The finer the roots, the higher the tensile strength (Operstein and Frydman 2000; Gray and Barker 2004).

It can be hypothesised that a larger number of small roots will contribute more to soil reinforcement as compared to a small number of thick roots. Figure adapted from Coopin and Richards (1990) showing the change in shear strength with the addition of roots.

Numerically, it can be expressed as follows:

$$\sum \mathrm{SR} = 1.15\,\mathrm{TR} \;@\; \mathrm{RAR}$$

where summation of SR = increase in shear resistance by roots,

TR = mean tensile strength of individual roots,

RAR = Root Area Ratio, the root cross-sectional area per unit soil surface.

RAR is defined as the fraction of the soil cross-sectional area occupied by roots per unit area (Gray and Leiser 1982).

RD = root density, the dry mass of living roots per unit of soil volume.

RLD = root length density, the total length of the living roots per unit of soil volume.

For grasses, RAR at different soil depth (every 0.10 m) is calculated using the equation:

$$\mathrm{RAR}_{\mathrm{grass}} = RL/0.10.\mathrm{ai}/\mathrm{A}$$

where RL = is total root length per soil depth class (m),

ai = is mean root cross-sectional area of a representative root of the studied grass species (m^2) and A = reference area (m^2), determined by the vertical projection of the above ground biomass of the plant.

5. Soil Shear Strength

The model of Wu et al. (1979) can be used to estimate the increase in shear strength due to the presence of grass. This model assumes that all roots grow up and act like a loaded pile. When the soil is sheared, tension is transferred to the roots. The assumption that fibres are oriented perpendicular to the shear plane is a useful simulation because it yields an average estimate of all possible orientations (Gray and Stoir 1996).

Abernethy and Rutherford (2001) say that plant roots tend to bind the soil together into a solid mass and add to its strength by making it seem to stick together even more.

If there are roots in the soil, the increased soil shear strength can be described as having more cohesion.

$$\mathrm{Sr} = \mathrm{S} + \mathrm{Cr}.$$

where S = soil shear strength (kPa),

Sr is the shear strength of the soil reinforced by roots (kPa),

Cr is the increase in shear strength due to the presence of roots (kPa).

When the shear forces occur, the root fibres deform. This deformation causes the fibre to stretch if the interface friction, confirming stress, and anchorage length are sufficient to lock the fibre in place and prevent slippage or pull-out.

The fibre elongation mobilises the tensile resistance in the fibre (Gray and Stoir 1996). The tension developed in the roots is resolved with a tangential component resisting shear and a normal component increasing the confining pressure on the shear plane.

6.4 Summary

Even if there is only a small amount of clay that expands and contracts, grass turf that is tightly bound together makes good pitches that are fast and bouncy. This is in contrast to pitches made of illite or kaolinite clay minerals, which have less grass turf and do not expand or contract as much. One of the main reasons for a fast and bouncy wicket is that the grass grows in a uniform way with deep roots. This helps the soil dry out deeply through transpiration and gives the soil a firm structure. It acts as the skeleton of a pitch, which gives it a clear shape, just as in people, the skeleton and muscles give the body a clear shape. The same is true for pitches, where the pitch soil is like the muscles and the deep-rooted grass is like the skeleton, giving the pitch a clear and firm shape and structure.

Deeply grown pitch grass roots in turf wickets with a high root area ratio, root length diameter, and rhizosheaths that act as glue improve the soil structure, cohesion, adhesion, shear, and tensile strength of the pitch soil matrix. This helps make a pitch with a good, consistent pace, and bounce.

Grass, with its morphological features, regulates the pitch soil temperature and soil matrix moisture level through the process of evapotranspiration, which is the prime requisite to attaining the best level of uniform pitch behaviour.

Deep-rooted turf grasses help in drying the turf wickets within the subsurface. A roller only squeezes the available soil moisture and consolidates and compacts the pitch up to one-fourth of the pitch surface (100 mm). Nearly 2/3 of the soil moisture below the pitch surface is transpired out only by grass roots, which helps in attaining a uniformly compacted pitch surface.

The goals of a good turf wicket can only be met through an integrated process of pitch management, of which pitch grass is the most important part. This is the result of properly layered, three-layer wickets with laboratory-tested good mineral pitch soils, uniformly grown deep-rooted grass, and good rolling schedules and techniques at a predetermined optimum moisture content.

The desirable pitch variations in the pitch behaviour during the matches can only be achieved by having proportionate combinations and contributions of the above-mentioned variable parameters, of which the role of the vegetative growth of grass leaves, stems, stolons, rhizomes, rhizosheaths, etc., in the turf wickets is second to none in defining and regulating the pitch behaviour.

References

Albalasmeh AA, Ghezzehei TA (2013) Interplay between soil drying and root exudation in rhizosheath development. Plant and Soil

Albrecht BA (1996) Effect of desiccation on compacted clays. Msc Thesis, Department of Civil Engineering, University of Wisconsin, Madison, WI, USA

Aldous DE (2016) Handbook on international turf management. CRC Press

Ali FH, Osman N (2008) Shear strength of a soil containing vegetative roots. J Jpn Geotech Soc Soil Fou 48:587–596

Allen RG, Pereira LS, Raes D, Smith M (1998). Crop evapotranspiration. Guidelines for computing crop water requirements, FAO Irrigation and Drainage Paper 56. In: Food and Agriculture Organization of the United Nations, Rome, Italy, p 300

Aronson LJ, Gold AJ, Hull RJ, Cisar JL (1987) Evapotranspiration of cool-season turf grasses in the humid North East. Agron J 79:901–905

Atkins CE, Green RL, Sifers SI, Beard JB (1991) Evapo-transpiration rates and growth characteristics of ten St. Augustine Grass Genotypes. Hortscience 26(12):1488–1491

Badlock JA (2002) Interaction of organic matter and microorganisms with minerals in the stabilization of soil structure. In: Huang PM, Bollg JM, Sensei N (eds) Interactions between soil particles and microorganisms: impact on the terrestrial ecosystem. Wiley

Baldock JA, Nelson PN (2000) Soil organic matter. In: Summer M (ed) Handbook of soil science. CRC Press, Boca Ralton, FL, p B25

BCCI Curators Certification Course Manual (2013) BCCI, 2nd edn.

Beehag GW Collins and Sons (1995) Turf grass identification manual: an introduction about basic identification of turf grass species. Australian Turf Grass Research institute, Concord West, N.S.W

Brady NC, Weil RR (2008) Nature and properties of soil, 14th edn. Prentice Hall

Chenu C, Guerif J (1991) Mechanical strength of clay minerals as influenced by an adsorbed polysaccharides. Soil Sci Soc Am J 55:1076–1080

Consuelo CR, Michael DD (2009) Turfgrass and ornamental plant evapotranspiration and crop coefficient. University of Florida, Agricultural and Biological Engineering Department

De Baets S, Torri D, Poesen J, Salvador MP, Meersmans J (2008) Modelling increased soil cohesion due to roots with EUROSEM. Wiley, pp 1948–1952

Dury PLK (1978) Practical research on grass cricket pitches. J Sports Turf Res 54

Gobat JM, Aragan M (2004) The living soil fundamental of soil science and soil biology. Wiley

Gray DH, Barker D (2004) Root soil mechanics and interactions. Water science and application American Geophysical Union, New York, pp 113–123

Guidi G, Poggi G, Petruzelli G (1995) The porosity of soil aggregates from bulk soil and soil adhering to roots. Plant Soil 87:311

Kim KS, Beard JB (1988) Comparative turf- grass evapo- transpiration rates and associated plant morphological characteristics. Crop Sci 28:328–331

Muntohar AS (2012) Effect of the roots densities on the shear strength of root- reinforced soil. In: National Seminar viii, Teknik Sipil ITS Surabaya, Department of Civil Engineering, Universitas Muhammadiyah Yogyakarta, Indonesia, ISBN 978–979–99327–7–8

Osano SN, Sixtus KM (2011) Root tensile strength of 3 plant species and their contribution to soil shear strength. J Civ Eng Res Pract 8:57–73

Pearson J (2010) Basic guide to turf cricket pitch preparation, VCA

Rai R, Shrivastava BK (2012) Effect of grass on soil reinforcement and shear strength. Proceedings of the Institute of Civil Engineers Ground Improvement 165(3):127–130

Romero CC, Dukes MD (2009) Turf grass and ornamental plant evapotranspiration and crop coefficient. University of Florida, Agricultural and Biological Engineering Department

Russell RS (1977) Plant root systems: their function and interaction with the soil. McGraw- Hill Book, London

Shimpton P, James L (2009) Optimization of cricket pitch rolling. Ph.D. Thesis, Cranfield University, School of Applied Science and Natural Resource Department

Taiton NM, Klug JR, Edmondson D, Campbell RK, Deventer PWV, Beer MJD (1993) Principles and practice of pitch preparation. United Cricket Board of South Africa, May

Tovey R, Spencer JS, Muckel DC (1969) Turf grass evapo- transpiration. Agron J 61:863–867

Chapter 7
Cricket Pitch Soil Organic Matter: "Importance and Role of Pitch Soil Organic Matter (SOM) in Controlling the Nature and Behaviour of Turf Wickets"

7.1 Introduction

Soil organic matter (SOM) consists of an accumulation of partially disintegrated and decomposed plant and animal residues and other organic compounds synthesised by the soil microbes as the decay occurs. Such materials are continuously broken down and resynthesised by soil micro-organisms. Over time, SO_2 is lost from the soil as carbon dioxide produced by microbial respiration. Consequently, SOM is a rather transitory soil constituent, lasting from a few hours to several hundred years. The SOM content of a typically well-drained mineral soil is small, varying from about 0.5–2% under Indian conditions and about 1–6% by weight under temperate conditions in the top soil and even less in the subsoils. The influence of SOM on soil behaviour and properties is far greater than the low percentage would indicate.

The main four parts of soil—minerals, SOM, soil moisture, and air—work together to make the soil behave in a certain way. However, the impact of one component is rarely independent of the impact of another. Soil moisture, which supplies plants with water, simultaneously controls much of the air supply to the plant roots desired for their aeration and regeneration under comparative stress. Likewise, SOM, by binding mineral particles into clusters, increases the porosity of soil and its water holding capacity. These complex materials, along with some that are synthesised by the soil micro-organisms, are collectively known as "humus." This material, usually black in colour, is very fine (colloidal) in nature. For a given mass, its capacity to hold water and nutrient ions greatly exceeds that of clay, its inorganic counterpart. Humus is a collection of very complex organic compounds that accumulate in soil because they are relatively resistant to decay. Just as clay is the colloidal fraction of soil mineral matter, so humus is the colloidal fraction of SOM. Because of their charged surfaces, both humus and clay act as contact bridges between larger soil particles; thus, both play an important role in the formation of soil structure. The surface charges of humus and clay attract and hold both nutrients and water molecules. However, gramme for gram, the capacity of humus to hold nutrients and water is far greater than that of clay. Unlike clay, humus contains certain components that can have a hormone-like stimulatory effect on plants (Fig. 7.1).

S. B. Singh et al., *Cricket Pitches*, Springer Transactions in Civil and Environmental Engineering, https://doi.org/10.1007/978-981-99-2913-9_7

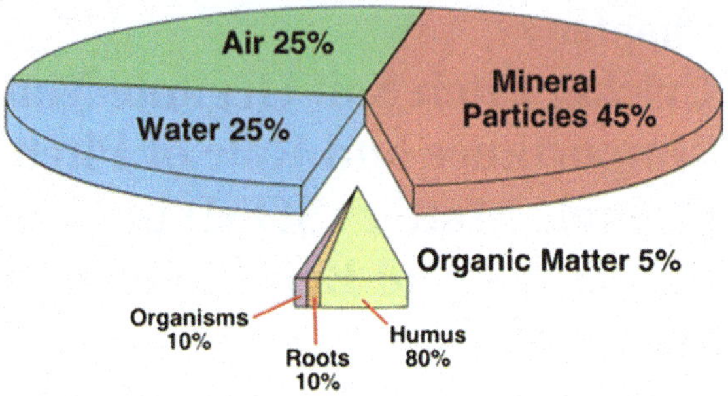

Fig. 7.1 Composition of a well-drained loamy soil

7.2 Importance of Soil Organic Matter (SOM) in Controlling the Pitch Behaviour

1. It usually makes up one-third or more of the cation exchange capacity (CEC) of surface soils and is a big reason why soil aggregates are stable. SOM binds mineral particles together into a granular soil structure. This structure is largely responsible for making conditions for plant growth that are loose and easy to control. Because of their extremely small size, clay and humus have a larger surface area per unit weight. Most physical properties of soils are also controlled by clay and humus. Their charged surfaces permit them to act as "contact bridges" between larger particles, thus helping to create and maintain stable aggregates of soil particles (Fig. 7.2).

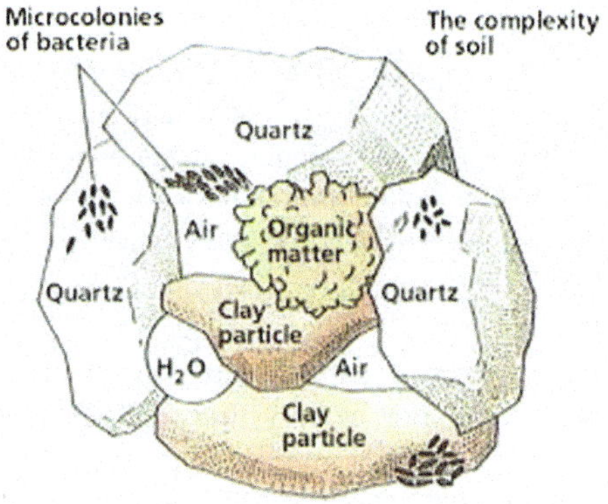

Fig. 7.2 Clay—humus interaction in soil

A cation, which is usually Ca^{+2} or Fe^{+2}, forms a bridge between clay minerals and organic polymers to hold them together. Ca^{+2} gives very strong bonds that are very stable, which impede very rapid mineralisation of humic SOM and oppose dispersion of the clays (Figs. 7.3 and 7.4).

2. The humus-Ca^{2+} clay complex gives the soil its dark colour, which is often seen in soils with a lot of carbonates. Fe^{+2}replaces a little or much Ca^{+2} in decalcified

soils or in Fe^{+2}-rich calcic soils. The bonding in them is weaker because of the water from hydration around the cations. The humus-iron-clay mineral complex is responsible for the brown colour of the soil. When finely divided humus is placed in intimate contact with moist clays, the two substances adhere to one another so tenaciously that a small quantity of humus is combined with each clay mineral particle, resulting in a dark mixture that looks like a fine-grained earthly substance when dried slowly (Muller 1989).

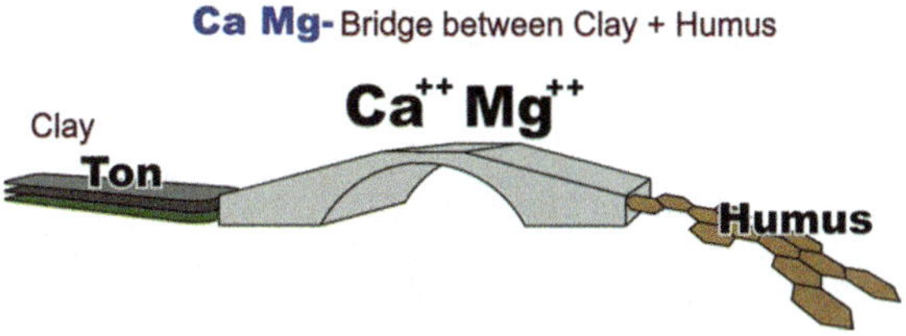

Fig. 7.3 Ca^{2+}/Mg^{2+} bridge between the clay and humus to form a clay-humus complex

A cation, which is usually Ca^{+2} or Fe^{+2}, forms a bridge between clay minerals and organic polymers to hold them together. Ca^{+2} makes very strong, stable bonds that slow down the mineralisation of humic SOM and stop clays from spreading out.

The main four parts of soil—minerals, SOM, soil moisture, and air—work together to make the soil behave in a certain way. However, the impact of one component is rarely independent of the impact of another.

Soil moisture, which gives plants water, also controls a lot of how much air gets to plant roots, which is important for aeration and growth when plants are under stress. In the same way, SOM makes soil more porous and better at holding water by holding mineral particles together in clusters.

3. Interaction of organic matter and micro-organisms with minerals in the stabilisation of soil structure in a cricket pitch profile

It has been suggested that as soils dry, the strength of the bond between the inorganic particles and the organic binding increases. This makes it possible for the mucilaginous materials from roots and the activity of microbes to strengthen the soil's structure.

The grass root system in the pitch distributes itself within the surface layer of mineral soil in the form of a dense network. Physical entanglement and the exudation of high-molecular weight polysaccharide materials capable of binding to mineral surfaces and bridging gaps between adjacent particles are two direct effects of roots on the stabilisation of soil structural form.

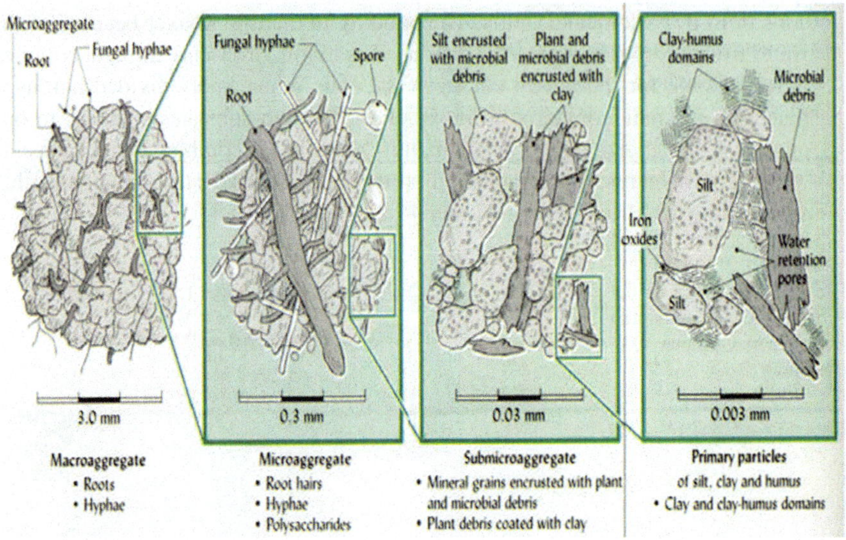

Fig. 7.4 Macro-aggregate model and hierarchy (*Source* Tisdell and Oades 1982)

The direct and indirect processes involved in the stabilisation of soil aggregates by the growth and presence of living roots are as follows:

Direct effects:

1. Physical entanglement by roots and root hairs
2. Exudation of mucilaginous materials capable of binding soil mineral particles together.

Indirect effects:

1. Physical entanglement by root-associated mycorrhizal fungi (as V.A.M.—vesicular arbuscular mycorrhizal fungi).
2. Organic substrates for rhizosphere micro-organisms that produce metabolic products are capable of binding mineral particles together.

Pojasok and Kay (1998) found that the organic materials that roots release also make soil aggregates more stable. This is in addition to the fact that roots physically bind aggregates together. The direct effects of plant roots on structural stabilisation include the provision of an organic substrate for soil micro-organisms and the influence that roots have on soil-water content.

Some people think that as soils dry, the bond between the inorganic particles and the organic bindings gets stronger. This makes it possible for the mucilaginous materials from roots and the activity of microbes to strengthen the soil's structure.

3. It is also a major source of phosphorous and sulphur and the primary source of nitrogen. SOM is the main source of energy for soil organisms. In addition to the original plant and animal residues and their partial breakdown products,

SOM includes complex compounds that are resistant to decay. These complex materials, along with some that are synthesised by the soil micro-organisms, are collectively known as "humus."

4. This material, usually black in colour, is very fine (colloidal) in nature. For a given mass, its capacity to hold water and nutrient ions greatly exceeds that of clay, its inorganic counterpart. Humus is a collection of very complex organic compounds that accumulate in soil because they are relatively resistant to decay. Just as clay is the colloidal fraction of soil mineral matter, so humus is the colloidal fraction of S.O. Because of their charged surfaces, both humus and clay act as contact bridges between larger soil particles; thus, both play an important role in the formation of soil structure.

The surface charges of humus and clay attract and hold both nutrients and water molecules. However, gramme for gram, the capacity of humus to hold nutrients and water is far greater than that of clay. Unlike clay, humus contains certain components that can have a hormone-like stimulatory effect on plants.

7.3 Important Distinct Characteristics of Soil Organic Matter (SOM)

1. The clay and humus particles in soils are referred to collectively as the colloidal fraction because of their small size and colloidal-like behaviour (showing adsorption, electronegative charge, Tyndall effect, Brownian movement, cataphoresis, electro-osmosis, coagulation, etc.). Organic colloids (humus) are not minerals, nor are they crystalline. Instead, they consist of convoluted chains and rings of carbon atoms bonded to H, O, and N. Humus particles are frequently among the smallest soil colloids, with extremely high water adsorption capacities but almost no plasticity or stickiness. Because humus is non-cohesive, soils composed mainly of humus (histosols) have very little bearing strength. The breakdown of humus can be approximately split into two categories: the production of humic and non-humic substances. Non-humic substances, such as polysaccharides, act as glues to help bind the soil particles together, encouraging an aggregate structure.
2. Humus has high amounts of both negative and positive charge per unit mass, but the net charge is always negative and varies with soil pH. The negative charge on humus is extremely high in neutral to alkaline soils.
3. No figures can be given for the humus content of mineral soils because there is no satisfactory method for its determination. Likewise, there is no satisfactory method of measuring the exact quantity of SOM. The procedure is to measure the organic carbon content, which is multiplied by the factor 1.72 to derive the SOM content. A conversion factor of 1.72 is generally used to convert organic carbon to organic matter.

Table 7.1 Some properties of inorganic and organic soil colloids

Colloid type	Type	Size (um)	Shape	Surface area, (m^2/g)		Interlayer	Net charge
				External	Internal	spacing	(cmol/kg)
Humus	Organic	0.1–1	Amorphous	Variable			(-)100 to 500
Smectite	2:1 layer silicate	0.01–1.0	Flakes	80–150	550–650	1.0- 2.0	(-) 80–150
Vermiculite	2:1 layer silicate	0.1–0.5	Plates, flakes	70–120	600–700	1.0–1.5	(-) 100–200
Kaolinite	2:1 layer silicate	0.1–5.0	Hexagonal	May-30		0.72	(-) 1–5
Fine micas	2:1 layer silicate	0.2–2.0	Flakes	70–150		1	(-)10–40

4. Humus colloids have a very high surface area per unit mass, usually more than silicate clays. The surface area of SOM, which ranges between 20 and 800 m^2/g (Table 7.1), is extremely difficult to calculate. The colloidal surfaces of humus are negatively charged, the surface charge being hydroxyl (–OH), carboxylic (–COOH), or phenolic (C_6H_5–OH) groups. The extent of the negative charge is pH dependent (high at high pH values). At high pH values, the C.E.C. of humus is 150–300 cmol/kg, which far exceeds that of most silicate clays (Table 7.2).

Table 7.2 CEC of different soils (from various sources)

Constituents or soil horizon type	CEC (cmol / kg)
SOM	
Scarcely humified SOM, peat	100
All SOM, in situ	6–280
Pure humified SOM	200–500
Pure humic acids	485–870
Pure fulvic acids	1400
Minerals	
Kaolinite	2–15
Illite	10–50
Montmorillonite	80–150
Vermiculite	100–150

5. The water holding capacity (WHC) of humus on a mass basis is 4–5 times that of the silicate clays. This high W.H.C. (as optimum moisture content) is desirable to attain the maximum possible bulk density of the pitch surface through compaction.

6. Humus has a very favourable effect on aggregate formation and stability.
7. The black colour of humus tends to distinguish it from most of the other colloidal constituents.
8. C.E.C reactions with humus are quantitatively similar to those occurring with silicate clays. The humic micelles, like clay particles, carry a swarm of adsorbed cations (Ca^{+2}, Mg^{+2}, K^{+}, Na^{+}, etc.). Thus, colloidal humus may be represented by the same illustrative formula used for clays.

7.4 Effect of SOM on Soil Properties

1. Effect on soil colour: Soil organic matter (SOM) tends to be dark brown to black in colour. Soils with a high SOM content are darker in colour than those with a low SOM content. Organic coatings tend to darken and mask the brighter colours from compounds such as iron oxides. The black colour of black soils is not due to a high content of SOM but rather to the high base saturation of the SOM, its intimate mixing with clay, and less dehydration of iron oxides than in the red soils.
2. Effect on soil physical properties

   a. Helps to form granular structure
   b. Reduces plasticity, cohesion (reason for maintaining total SOM % in the pitch profile below 5%) as the reduced plasticity and cohesiveness (with amorphous structure) reduce the mechanical strength (bulk density) desired to have a uniformly compacted pitch soil profile.
   c. Effect on water holding capacity of soil.

3. Effect on cation exchange capacity:

   a. CEC of SOM is about 2 to 3 times as great as mineral colloids which affects the swelling and shrinking capacity of the pitch soil matrix thereby directly affecting the pitch behaviour.
   b. CEC accounts for 20–90% of the adsorptive power of mineral soils.

4. Effect on availability of nutrients:

   a. Easily replaceable cations present on humus colloids.
   b. N, P, S, and micronutrients held in organic form indicating their ready availability.
   c. Release of nutrients from minerals by acid humus becomes more easy.

Humic substances reduce the plasticity, cohesion, and stickiness of clay soils aiding the formation of granular loose soils (Brady and Weil 2001). High SOM, especially large, non-decomposed material, has been negatively correlated with soil binding strength (Adams et al. 1994).

5. Effect on other soil properties:

The fibrous nature of SOM forms soft spongy layers which resist compaction from the roller (Shipton, 2008) (as they are easily compressed but highly elastic), negatively affecting the ball bounce (Adams et al., 1994; Baker et al., 2003).

The increased SOM makes the surface absorb more energy, which makes the ball lose energy after hitting the surface. This makes the pitch less flexible (Baker et al., 1998; Baker and Adams, 2001).

A reverse correlation exists between SOM and bulk density: as the SOM increases, the bulk density decreases (Sakin et al., 2000) (Figs. 7.4 and 7.5). This is why cricket pitch soils with SOM below 5% are preferred, as more of it reduces the bulk density of the profile, resulting in slow and dead pitches. A proportionate ratio is to be maintained between the organic and inorganic portions of the playing surface so as to have a balanced synergistic effect of the above-mentioned two basic ingredients of the pitch profile.

Table 7.3 Targeted dry bulk density for the top 0–80 mm of prepared pitches

SOM (%)	4	6	8	10	12
Dry BD (gm/ cm^3)	1.75	1.65	1.55	1.45	1.35

Source: (Adam et al.; 2004)

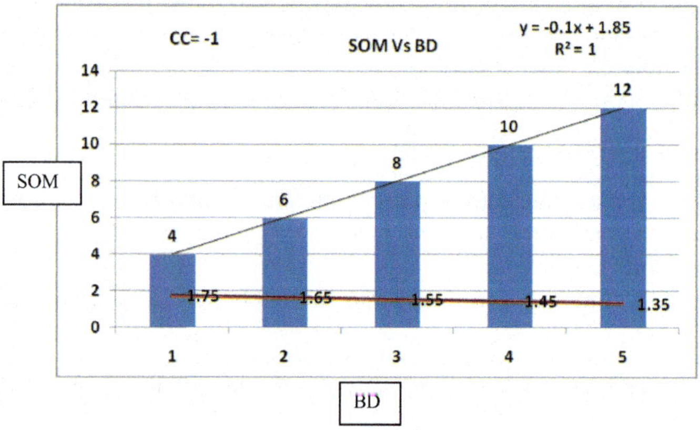

Fig. 7.5 Correlation between soil organic matter and bulk density for the top 0–80 mm of prepared pitch (*Source* Adam et al., 2004)

This is why cricket pitch soils with SOM below 5% are preferred, as more of it reduces the bulk density of the profile, resulting in slow and dead pitches. A proportionate ratio is to be maintained between the organic and inorganic portions of the playing surface so as to have a balanced synergistic effect of the above-mentioned two basic ingredients of the pitch profile (Figs. 7.6, 7.7 and 7.8).

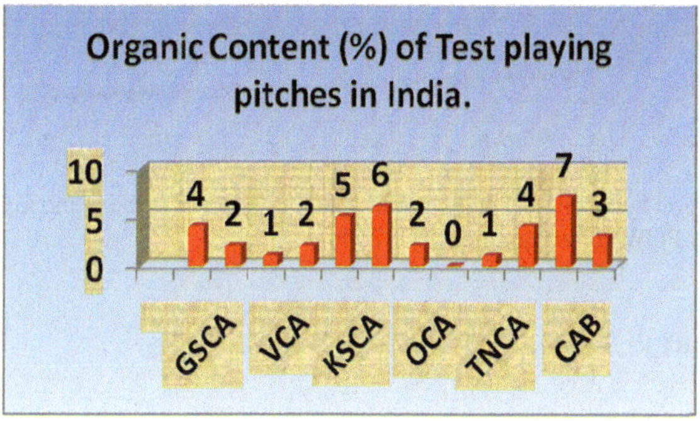

Fig. 7.6 Proportion of clay content to SOM for Test match playing pitches of India (*Source* IIT POWAI through the BCCI 2008)

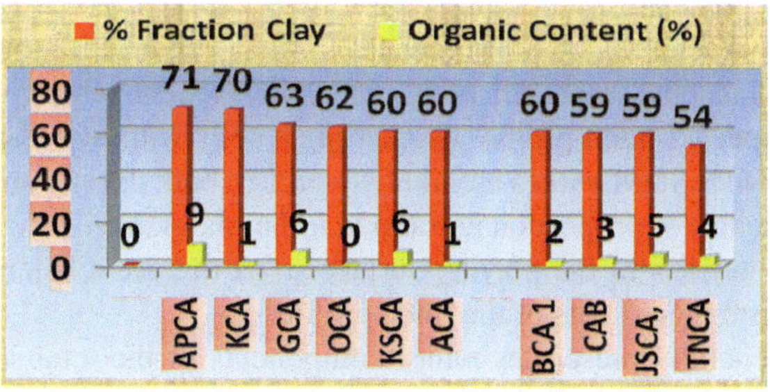

Fig. 7.7 Proportion of clay content to SOM for some of the important cricket pitches of India (*Source* IIT POWAI through the BCCI 2008)

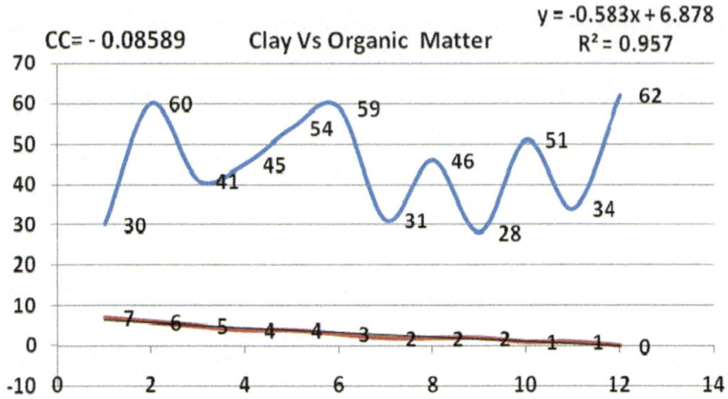

Fig. 7.8 Correlation between clay and organic matter content of some of the major Test pitches of India. (*Source* IIT POWAI through the BCCI 2008)

7.5 Summary

Pitch soil organic matter, as an organic colloid, provides essential basic nutrition to pitch soil biomass, increases and retains the structure of pitch soil aggregates (providing a gluing hormonal like effect through microbial secretions and thus increasing the clay-humus interactions), and controls nearly one-third of the CEC of the surface layer, thereby affecting and controlling the swell-shrink behaviour of the pitch soil complex (CEC of organic colloids are 2–3).To provide the best pitch condition, the proportion of SOM and clay content should be optimum to maintain the desired soil properties.

Amorphous structure pitch SOM reduces mechanical strength (bulk density) due to its reduced plasticity, very high WHC, and cohesiveness, which is desirable for a uniformly compacted pitch profile. Cricket pitch soils with SOM less than 5% are preferred, as higher SOM values are likely to reduce bulk density and result in slow and spongy pitches. A proportionate ratio is to be maintained between the organic and inorganic portions of the playing surface so as to have a balanced synergistic effect of the basic ingredients of the pitch profile.

Although generally present as minor components of the total soil mass, living (e.g. plant roots and micro-organisms) and non-living (e.g. particulate debris or molecules) organic materials exert strong controls over many soil physical, chemical, and biological properties, which may control the behaviour of cricket pitches.

References

Aldous DE (2016) Handbook on international turf management. CRC Press

Ali FH, Osman N (2008) Shear strength of a soil containing vegetative roots. J Jpn Geotech Soc Soil Fou 48:587–596

Ammar AA, Teamrat AG (2013) Interplay between soil drying and root exudation in rhizosheath development. J Plant Soil
Badlock JA (2002) Interaction of organic matter and microorganisms with minerals in the stabilization of soil structure. In: Huang PM, Bollg JM, Sensei N (eds) Interactions between soil particles and microorganisms: impact on the terrestrial ecosystem. Wiley
Baldock JA, Nelson PN (2000) Soil organic matter. In: Summer M (ed) Handbook of soil science. CRC Press, Boca Ralton, FL, p B25
BCCI Curators Certification Course Manual (2013) BCCI, 2nd edn.
Brady NC, Weil RR (2008) Nature and properties of soil, 14th edn. Prentice Hall
Chenu C, Guerif J (1991) Mechanical strength of clay minerals as influenced by an adsorbed polysaccharides. Soil Sci Soc Am J 55:1076–1080
De Baets S, Torri D, Poesen J, Salvador MP, Meersmans J (2008) Modelling increased soil cohesion due to roots with EUROSEM. Wiley, pp 1948–1952
Gobat JM, Aragan M (2004) The living soil fundamental of soil science and soil biology. Wiley
Gray DH, Barker D (2004) Root soil mechanics and interactions. Water science and application American Geophysical Union, New York, pp 113–123
Grim RE (1953) Clay mineralogy. Mc Graw-Hill, New York
Huang PM, Bollang JM, Senesi N (2002) Interactions between soil particles and micro- organisms. Wiley
Russell RS (1977) Plant root systems: their function and interaction with the soil. McGraw- Hill, London
Shimpton P, James L (2009) Optimization of cricket pitch rolling. Ph.D. Thesis, Cranfield University, School of Applied Science and Natural Resource Department
Taiton NM, Klug JR, Edmondson D, Campbell RK, Deventer PWV, Beer MJD (1993) Principles and practice of pitch preparation. United Cricket Board of South Africa, May
Wood, C (2006) Recommended guidelines for the maintenance of cricket Pitches and outfields at all levels of game. England and Wales Cricket Board in association with the Institute of Groundsmanship

Chapter 8
Cricket Pitch Soil Rolling Management—Integrated Rolling Management (IRM)

8.1 Introduction

Cricket pitches are not laid on barren soils but on turf, which is a fine clay-dominant soil surface that is uniformly embedded with turf grasses that have moderately dense leaf density and maximum possible root penetration (150–200 mm), through which they breathe (transpire) and provide tensile strength to the pitch profile, generating entirely different levels of pitch surface desirable compression levels (compaction + drying-induced consolidation or desiccation) (Fig. 8.1).

Fig. 8.1 **A** Pitch surface devoid of grass. **B** Pitch surface with uniformly developed turf grass just before the start of pitch preparation. **C** Final turf match pitch just before the toss

The whole impact of the natural forces (gravity, weather, and climatic variations in the form of temperature, relative humidity, moisture (in the forms of rain, fog, mist, dew, etc.), wind, solar radiations, etc.), and the phenomena (such as evaporation, condensation, transpiration, capillary action, retention, movement, and exchange of moisture and air, etc.) on the turf pitches gets altered once we conceive the concept

S. B. Singh et al., *Cricket Pitches*, Springer Transactions in Civil and Environmental Engineering, https://doi.org/10.1007/978-981-99-2913-9_8

of the turf pitch surface as compared to the soil itself. The whole basic, fundamental concept (related to engineering and agronomic properties, both theoretically and in the laboratory) of soil compaction, consolidation, cracking or crumbling patterns, swelling and shrinking patterns, and the drying behaviour of turf pitches are entirely different from those we conceive in barren soil.

Cricket match or turf wicket preparation is done under the sun in which the available dynamic diurnal climatic and weather variables along with the natural phenomena control the resultant pitch surface density levels by controlling the retention, movement, and migration of the available pitch surface moisture levels (within and just above the pitch profile). The total compression levels achieved (through rolling schedules and dessicative shrinkage) on the finished match pitch surface under the direct influence of the above-mentioned natural forces and phenomena are entirely different from those we achieve and analyse under limited, simulated, or controlled laboratory conditions. The individual effects of the above-mentioned weather and climatic conditions and the impact of turf grasses are not considered while analysing and comprehending the cohesive behaviour of the pitch soils and surface under laboratory conditions.

The pitch is not only a rectangular area of 22 yards (20.12 m) in length and 10 feet (3.05 m) in width, but its depth matters (as shown in Fig. 8.2) in defining and regulating the pitch's behaviour by controlling the retention and movement of the pitch's moisture (held, capillary, and hygroscopic moisture within the pitch profile) and provides the variable pitch surface density levels during the different phases of the match. Fine clay-dominant pitch soil acts as a storehouse of moisture (hygroscopic, held, and capillary moisture), which regulates the pitch behaviour during the course of the match due to their migration in vapour form within the pitch profile.

Fig. 8.2 Turf pitch profile with turf pitch surface and drainage layer

Depending upon pitch soil texture, structure, physical, chemical, and mineralogical characteristics (cracking clays or non-cracking clays), and the predominant climatic/weather variables (temperature, relative humidity, evapotranspiration rates, dew, rainfall patterns, etc.), the depth of the pitch soil layer is around 8 to 12 inches.

James, Carre, and Haake (2004) explain how important the interaction between the ball and the surface is in cricket. This interaction is controlled by the mechanical and physio-chemical–mineralogical properties of the dominant swelling and shrinking fine clay pitch surface, which is changed using smooth-wheeled rollers with different compactive potential at optimum moisture content (Shimpton et al. 2006). The aim is to achieve a flat, uniform surface that maximises soil strength by increasing soil particle packing and minimising moisture content but still permits grass growth to aid the playing performance of the pitch, reinforces the soil with the grass root network, and facilitates drying of the soil through evapotranspiration. A soil profile's critical soil depth for playability ranges from 75 to 100 mm (Adams and Gibbs 1994; Baker et al. 2003).

Based on the format of the match (ODI/T-20, Test match), the available moisture content of the pitch profile, existing climatic and weather variables above and within the pitch profile, the water holding and retention capacity of the pitch soil, etc., the aim of rolling is to achieve the desired match target density. This is achieved in conjunction with the shrinkage of the pitch surface during the pitch preparation. As the soil moisture content required for efficient compaction is higher for optimum playability (Shimpton et al. 2006), a drying period after rolling and before the play is necessary.

The cricket pitch surface is packed down with smooth-wheeled rollers of different weights and sizes (compaction potential) to get the right density. Rolling compaction increases the density and shear strength of the pitch surface and produces a level, smooth surface that generates predictable, uniform pace, bounce, and turn. While clay soil has high mechanical stiffness and shear strength (a key property for resistance to wear) when compacted and dry, its shear strength is known to be highly sensitive to moisture content (Henkel 1959).

Cricket pitches are purposely packed down with a smooth-wheeled roller to make a hard, solid surface that absorbs less of the ball's energy when it hits the ground (Shipton 2004). Soil compaction in clay soils is closely related to increased soil strength, and this increased soil strength has been implicated as a factor in mechanically limiting plant growth (Stephaniewshi and Gkinski 1985). Recent research has shown that mechanical impedance may not be the primary cause of poor root growth in turf grass (Shipton 2004), but that total root mass was not decreased due to compaction but that root density became more concentrated towards the shallower region of the profile.

The goal of rolling is to get the pitch surface to the desired density, which is harder and more even. The harder the pitch, the less the ball will change the shape of the field when it hits it. This means that the ball will move and bounce in the same way everywhere on the field. Rolling is part of a good pitch preparation method but not the single most important factor in generating a hard, compacted pitch surface. The important natural phenomena that impart the total desirable pitch surface behaviour during different sessions of the match are deep drying and desiccation of the pitch surface (nearly 4 inches of top surface) achieved through evapotranspiration of turf grass and hardening of the pitch surface due to shrinkage of the predominant swelling and shrinking fine smectite-dominant clay in the pitch profile.

So, during the game, the soil moisture should be controlled and kept at a good level, taking into account things like temperature, relative humidity, dew point, evapotranspiration, the movement of water through the pitch profile, etc. For the pitch to dry evenly (at least 4 inches from the top), the grass must grow evenly and have deep roots. This is because evaporation and transpiration are the only ways to get rid of excess moisture in the pitch soil. Therefore, the bulk density and shear strength must not exceed critical values, which prevent grass growth after the match on an over consolidated pitch surface. With non-swelling and shrinking clays, the pitch surface is bound to have lower limits of shear strength for desirable pitch surface hardness that generates pace, bounce, and wear resistance and an upper limit of shear strength for the grass to recover and regenerate well after the match. Generally, pitch soils with swelling and shrinking fine smectite clays have their natural ability to regenerate the match pitch and turf grass after the match through cracks of various shapes, sizes, and depths.

8.2 Science Behind the Art of Cricket Pitches Profile Compression

The soil is a particulate material consisting of solid grains and void spaces enclosed by the grains. The voids may be filled with air or other gases, with water or other liquids, or with a combination of both. The property of a soil by virtue of which volume decrease occurs under applied pressure is termed compressibility. According to the permeability characteristics of the soil, the compression of the soil mass caused by the imposed stress may be almost immediate or time dependent. Cohesion-less soils, which are highly permeable, are compressed in a relatively short period of time as compared to cohesive soils, which are less permeable.

The compressibility characteristics of a soil mass might be due to any or a combination of the following factors:

A. Compression of the solid grains.
B. Compression of the pore water or the pore air.
C. Expulsion of pore water or pore air from the voids, thus decreasing the void ratio and porosity due to applied stress (rolling and through the process of evapotranspiration by the turf grass deeply embedded in the turf pitch profile).

It is quite reasonable and rational to assume that the soil matter and the pore water are relatively incompressible under the loads usually encountered in soil masses. The change in volume of a mass under imposed stresses must be due to the escape of water if the soil is saturated. But if the soil is partially saturated (as in a cricket pitch surface just before the start of first rolling), the change in volume of the mass is partly due to the compression and escape of air from the voids and partly due to the dissolution of air in the pore water.

Thus, compression of pore air (compaction), expulsion of pore water by the rollers, and evapotranspiration by the turf grass (drying-induced consolidation) are the primary sources of volume decrease in the cricket pitch profile. A partially saturated soil may experience an appreciable volume decrease through the compression of pore air before any expulsion of pore water takes place.

For partially saturated soils (as in the pitch profile during the initial phase of rolling), air and water vapour are considered highly compressible, with the presence of air reducing pore pressure (Bishop 1955). Cricket pitch soil compaction is the densification of soil by the application of mechanical energy (smooth, wheeled, cylindrical rollers). It is a process by which the soil grains get rearranged more closely, the volume of air voids gets reduced, and the density of the soil is increased. It is the rearrangement of soil aggregates (both inorganic and organic) and/or particles in a denser way when the voids and pores mainly between the aggregates and particles become smaller or disappear in comparison with the arrangement of similar non-compacted soils. The orientation, size, and shape of soil aggregates are evidence of soil compaction.

The degree of compaction of a soil is measured in terms of dry unit weight, either, the amount of soil solids that can be packed in a unit volume of the soil. The compaction characteristics and the degree of compaction can be obtained from the laboratory test known as standard/modified Proctor test. In these tests, a specified amount of compactive effort is applied to a constant volume of soil mass. R. R. Proctor (1933) showed that there exists a definite relationship between the soil moisture content and the dry density on compaction and that, for a specific amount of compaction energy used, there is a particular moisture content at which a particular soil attains its Maximum Dry Density (MDD) (Fig. 8.3).

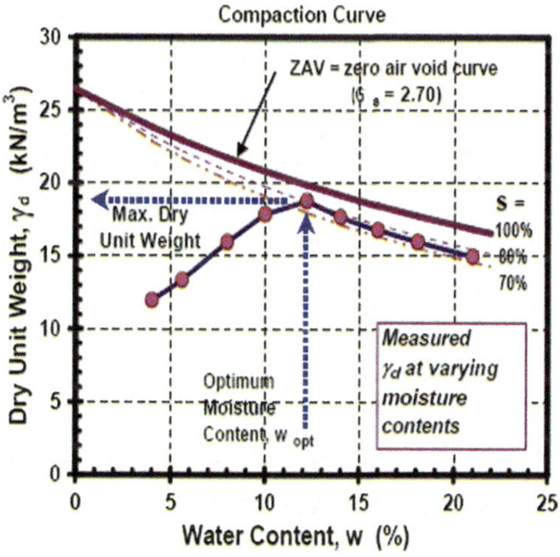

Fig. 8.3 Compaction curve

Each data of the compaction curve represents a single compaction test. Usually, 4 to 5 points are required to obtain the compaction curve. This curve is unique for a given type of soil, method of compaction and compactive effort. The peak point of compaction curve corresponds to the maximum dry unit weight. The water content corresponding to the maximum dry unit weight is known as optimum moisture content (OMC).

Typical values of maximum dry unit weight range from 16 to 20 KN/M^3 with the widest range being to 24 KN/M^3. Typical OMC value ranges from 10 to 20% with a maximum range of 5–30%. In turf cricket pitches, the densification of soil is achieved by rolling compaction using smooth-wheeled rollers having different compactive potential and also through different rolling techniques and schedules. Further variations in the pitch surface density are due to moisture changes taking place due to atmospheric variations within and just above the pitch profile. The expelled pore water from inter- and intralayer clay lattice and the available water in the macro- and micropores as capillary water is evapotranspired uniformly spread grasses from the pitch surface. Slow and gradual evapotranspirative loss of soil moisture within the pitch profile desiccates the pitch from the bottom and helps in achieving the desired compaction level of the pitch.

The overall increase in pitch surface density is due to (1) the compactive effort of the roller, (2) rest hardening and shrinkage effect on drying, and (3) shrinkage effect on compaction-induced drying.

8.2.1 *Consolidation of a Cricket Pitch Soil*

When a saturated clay water system is subjected to an external pressure, the pressure applied is initially taken by the water in the pores resulting thereby in an excess pore pressure. If drainage is permitted, the resulting hydraulic gradients initiate a flow of water out of the clay mass and the mass begins to compress. A portion of the applied stress is transferred to the soil skeleton, which in turn causes a reduction in excess pore pressure. This process involving a gradual compression occurring simultaneously with a flow of water out of the mass and with a gradual transfer of the applied pressure from the pore water to the mineral skeleton is called consolidation. Consolidation may be due to one or more of the following factors:

1. External static loads from structures.
2. Self-weight of the soil-geostatic load.
3. Lowering of the groundwater table.
4. Desiccation/drying induced.

Drying-induced consolidation (DIC) in soil is defined as volumetric shrinkage by interparticle stresses (suction stress) during drying under conditions of zero external or total stress. Drying-induced consolidation for unsaturated soils exhibits a similar behaviour to traditional consolidation behaviour of saturated soils.

At the first instant of loading for saturated soils, the applied stress is almost entirely by the pore water due to the contrast between the essentially incompressible liquid phase and the more compressible soil skeleton. In time, drainage of the pore water produces a gradual transfer of the applied stress to the soil skeleton and consequent increase in effective stress. The dissipation of excess pore water pressure, also referred as "hydrodynamic lag" by Terzaghi (1913), is key to understanding the consolidation process.

Soils under unsaturated conditions involve three phases and introduce substantial complexity to the consideration of volume change behaviour. An instantaneously applied stress is not fully carried by incompressible water and will transfer to the soil skeleton and the highly compressible air phase, causing instantaneous settlement and increase in effective stress. On the other hand, deformation of the soil skeleton may be resisted by soil hardening due to the effects of capillarity and adsorption through suction stress (Lu and Likos 2004), where the capillary effects are mainly due to the surface tension on the curved air-water interface, and adsorption is induced but physicochemical interactions such as van der Waals and electrical double diffused layer (DDL) forces. In this case, all the physicochemical and capillary forces at or near soil particle contacts in unsaturated soils have been conceptualised and defined as a stress called suction stress. Suction stress, as fully defined by Lu and Likos (2006), is the effective stress due to changes in soil-water content.

Volume change due to environmental loading (i.e. drying or wetting) for expansive soils has received much attention. For instance, the effect of interparticle stress on volume change due to soil desiccation, without compounding the effects of external stress change, has not been exclusively investigated.

The shrinkage behaviour of the soil by suction stress under drying conditions is defined as drying-induced consolidation (DIC).

8.2.2 Consolidation and Soil Shrinkage

Consolidation behaviour for saturated soils is characterised by nonlinear volumetric response to external loading, including the preconsolidation effect of stress history and time-dependent settlement due to excess pore water pressure dissipation. The transfer of total applied stress from excess pore water pore water to effective stress acting on the soil skeleton is the cause of soil deformation in the case of saturated soils.

However, under unsaturated conditions, soil can develop another type of effective stress depending on water content and soil type.

1. Passive effective stress occurs due to applied external stress and it is a transferable, boundary- and condition-dependent global stress in soil, which can be explicitly calculated using stress and pore water pressure if soils are fully dry or saturated (e.g. Terzaghi 1943).

2. Suction stresses (Lu et al. 2010) or interparticle stresses (Wei 2014), on the other hand, occur in soil under unsaturated conditions due to soil-water interaction or environmental loading, i.e. soil water in equilibrium with ambient pore air humidity. As such, suction stress are non-transferable and water potential-dependent local stress around particle contacts (Lu and Likos 2006). Suction stress can be indirectly quantified through shear strength (Lu and Likos 2006), tensile strength (Lu et al. 2009), deformation (Lu and Kaya 2013), or soil-water retention curve (Lu et al. 2010).

If a zero external loading or total stress is applied to soil under un-saturated conditions like soil shrinkage testing or many in situ conditions, suction stress is the only effective stress causing changes in soil stiffness and deformation. In such drying-induced consolidation case, stress and displacement will take place concurrently upon a new suction stress condition. Similar to the diffusion process in saturated consolidation where external stress is initially sustained initially by incompressible pore water followed by gradual transformation of pore water pressure to deformable soil skeleton, the drying-induced consolidation process lowers the water potential as the soil water diffuses with evaporation, transforming the soil-water potential into the work done by suction stress and soil shrinkage deformation (Lu and Likos 2006).

Apparently, the mechanisms of stress transfer in saturated consolidation differ from each other but are similar in water diffusion process. The time dependency of drying-induced consolidation relies on the phase equilibrium between pore water and water vapour through evaporation/condensation processes, which are governed by soil-water diffusibility and environmental boundary conditions.

When a soil is unsaturated, an active and non-transferable local stress is developed at particle contacts termed suction stress. Suction stress is independent of external loading but highly controlled by the water content through physio-chemical interactions of soil particles such as water adsorption and capillary mechanisms (Lu and Likos 2004, 2006; Lu 2016). When a soil is under environmental loading solely (i.e. drying or wetting processes), shrinkage and swelling occur from changes in suction stress similar to volume change in saturated due to external loading.

8.2.3 Consolidation Process

Consolidation (in turf cricket pitches, it is generally achieved through scheduled rolling and desiccative drying of the pitch surface by evapotranspiration through the pitch turf grass surface) is the expulsion of water from soil due to excess pore pressure with a gradual reduction in soil volume accompanied by a transfer of pressure from pore water to the soil skeleton. When an external pressure (referred to as "total pressure") is applied to a saturated, cohesive soil (either partially saturated soil or the soil at the field capacity), the entire pressure is at first carried by the pore water because the soil structure must compress to take any additional load. The pore water pressure that is induced is called the "excess pore water pressure" since it is greater than the original hydrostatic pressure. The initial excess pore pressure is thus equal to the total external pressure. If the soil is bounded by surfaces from which water can escape (such as an 8-inch drainage layer lying beneath the 8-inch playing surface), the excess pore pressure will cause the water (which will squeeze out from the micro- and macropores) to drain out (especially in the case of the cricket pitch soils high in colloidal clay) into the adjoining layers. This generally occurs very slowly because of the low permeability and hydraulic conductivity, also known as hydrodynamic lag, and the plasticity of clay-dominant cricket pitch soils. As the gravitational water drains out, the soil begins to compress, and a portion of the additional pressure is transferred to the soil grains forming the soil skeleton and the adjoining turf grass roots (providing shear and tensile strength), due to which pore pressure correspondingly falls.

At any given time, the effective pressure on the soil particles is the difference between the total pressure from the outside and the pressure in the pores. At equilibrium, when drainage water (gravitational and capillary water) stops, all of the external pressure is carried by the soil particles as the effective pressure, which is called the consolidation pressure, and the excess pore pressure goes down to zero.

It happens over time, and the time it takes for the compression to happen is called the "hydrodynamic and plastic lag." Even though most of the extra water pressure in the pores has gone away and no more water is draining out, some compression is still going on. This part of compression is called "secondary compression," and it happens because particles keep moving as the structure of the soil adjusts to the increasing effective stress. The double layer of saturated water that surrounds the particles could also be deformed and pushed away.

8.3 Time Lag During Compression of Clay

For a clay soil, it takes a long time for full compression to occur under a given increment of stress. Although it may not take more than 24 h for the full compression to occur in a laboratory sample, it may take a number of years in the case of a field deposit of clay.

Two phenomena are responsible for this time lag. The first is due to the low permeability and hydraulic conductivity of clay and the consequent time required for the escape of pore water. This is called the "hydrodynamic lag." The second action is due to plastic action in adsorbed water near grain-to-grain contacts, which does not allow quick transmission of the applied stress to the grains and the active stress to reach a constant value. This is known as the plastic delay. The time lag (both hydrodynamic and plastic lag) varies depending on the pitch soil, which has variable textures, structures, clay quantity and mineral type, exchangeable cations, available soil moisture within the pitch profile, and macro- and microclimatic variations. This is the reason why considerable time is required to get the maximum bulk density of a cricket pitch through "hydrodynamic lag" and "plastic lag" phenomena. Given the phenomenon of time lag during clay compression due to hydrodynamic lag, plastic lags, or rest hardening due to thixotropic effects, a significant pitch preparation time (PPT) should be used to prepare a Test match wicket. Desiccation is the phenomenon of the drying of soil by ET, accompanied by shrinkage, drying, and possibly cracking.

Soil that is only partially wet (like a cricket pitch on the first day before it is rolled) loses a lot of volume because the pore air is squished together (compaction) before the pore water is pushed out (consolidation). In particular, the compressibility of a soil depends on how the soil particles are arranged or rearranged and, in fine-grained soils like pitch soils or smectitic clays, how well the particles next to each other are stuck together. A structure that is more porous, such as a honeycomb structure, is more compressible than a dense structure. A soil with mostly flat grains can be pushed together more easily than one with mostly round grains.

A soil in a remoulded state may be more compressible than the soil in its natural state, which can be observed through sensitivity and thixotropic analysis. Generally, the cricket pitch soil gets remoulded every time it is uniformly wetted at the time of preparing a new match pitch. Their structural arrangement of a cricket pitch profile gets rearranged through every swell (wetting) and shrink (drying) cycle. The change in consistency or strength of clay upon remoulding is measured as sensitivity. In terms of strength, it is the ratio of the unconfined compressive strength of the undisturbed soil to the remoulded or disturbed strength of the soil reconstituted at the same natural density and moisture content (Table 8.1).

Table 8.1 Sensitivity range for different soils

Clay type	Sensitivity range
Most clays	2–4
Marine clays	1.6–26
Peat	1.5–10

Source Winterkan and Fang (1975)

The decrease in strength on remoulding soil is primarily due to the destruction of soil structure and bonds that exist in natural soil. Also, the structure of the adsorbed water is broken up. With the passage of time, however, the strength increases, though not necessarily to the original value. This phenomenon of strength loss on remoulding and strength gain on standing with no change in volume or moisture content is known as "thixotropy." The strength gain is called thixotropic regain or rest hardening. The loss of strength on remoulding is partly due to the permanent destruction of the structure and partly due to the reorientation of the molecules on the adsorbed layers. The gain in strength is due to the rehabilitation of the molecular structure of the soil (thixotropy = *thixis*, tough, shaking, and *tropo*, to turn, to change).[1]

Thixotropy is the ability of some suspensions of very small particles to harden (form a gel) at rest and soften or become liquid (form a sol) when they are reshaped.

This phenomenon is different from flocculation (coagulation) in that in flocculation, separate flocks form with no bond between them, whereas in thixotropic hardening, all the particles form a continuous framework to develop a thixotropic

[1] Ranjan G, Rao ASR (2002) Basic applied mechanics, p 45.

structure (Wilun and Starzewski 1971). The thixotropic bonding can be destroyed by remoulding, shock, or vibration, resulting in strength loss. Rest hardening reverses the process so as to restore interparticle force equilibrium (or increase the attraction).

The percentage increase in strength will be greater for an initially partly dispersed fabric than for a flocculated fabric. Thixotropic effects are greater for active clay (such as smectites) that absorb large quantities of water than for less active or inactive clay (such as kaolinite and illite clays).[2]

The loss in strength on remoulding and the extent of gain over a period of time are dependent on the type of clay minerals involved; generally, the clay minerals that absorb large quantities of water into their lattice structures, such as montmorillonites, experience greater thixotropic effects than the other, more stable clay minerals. Generally, in cricket pitch soils with a higher amount of cracking clay, thixotropic effects are the main phenomenon responsible for attaining secondary compression (with wide cracks) in Test and Ranji pitches.

8.4 Factors Affecting the Pitch Profile Compression

1. Soil-water content.
2. Soil type (structure, texture, and mineralogy).
3. Compactive effort (based on roller and pitch compactive potential).
4. Methods of compaction (rolling techniques and schedules).
5. Pitch grass density, type, leaf area index (LAI), and root area index (RAI), which generally help in drying and desiccating the pitch profile through its transpiration rate and by increasing the tensile and shear strength of the pitch profile through a network of stolons and rhizomes deeply embedded in the turf wickets.
6. Macro- and microclimatic and weather variations and conditions prevail within and above the pitch profile in the form of temperature, relative humidity, dew point, and sun radiation energy. These weather conditions govern and regulate the movement, retention, and loss of pitch available water content (PAWC), which has a direct impact on the swelling and shrinking (cracking and crumbling) of the cricket pitch profile. Cricket pitch behaviour can be controlled and influenced by controlling and regulating each of the aforementioned factors.

8.4.1 Soil-Water Content

Water has a remarkable effect on soil structure. When soil crumbs are quickly wet, forces develop in the particle that tend to cause aggregate breakdown. If the forces

[2] Singh A. Modern geotechnical engineering. 3rd edn, p 75.

exceed the binding force of the particles, the aggregates will disintegrate. When water is added, care must be taken to avoid differential swelling and shrinkage in cricket pitches, as it results in non-uniform behaviour. This may also result in aggregate breakdown, in which the particles disperse into larger clods rather than desirable crumbs. This explains the reason for heavy crust formation in some soils that are submerged during irrigation, while in other cases the granular structure is retained.

Proctor explained the typical moisture content-dry unit weight (OMC) relationship with the help of lubrication theory. Swelling and shrinking of cohesive soil are due to expansion and contraction of the DDL, which is the function of the minerals present (both quantitatively and qualitatively), the nature of exchangeable cations on the DDL, pH, the available moisture, etc. At low water contents, the soil is stiff, and the soil grains offer resistance to compaction. As the water content increases, the particles develop larger and larger films around them (forming a double diffused layer), which lubricate the particles and make them easier to work with, allowing them to move closer into a denser configuration, resulting in a higher dry unit weight and fewer air voids. The dry unit weight continues to increase until the OMC is reached, a stage when the lubrication effect is at its maximum. However, as moisture content increases, water begins to replace soil particles, and the dry unit weight decreases.

Lambe (1958) used the concepts of soil structure and the electrical double diffused layer (DDL) theory to explain the effect of water content on dry unit weight. In cohesive soils, there is an attractive force, namely the van der Waals force, which acts between the soil particles, and a repulsive force, which is due to the double layer of adsorbed water tending to come into contact with each other. While the attractive force essentially remains the same in magnitude, the repulsive force is directly related to the size of the DDL (Fig. 8.4).

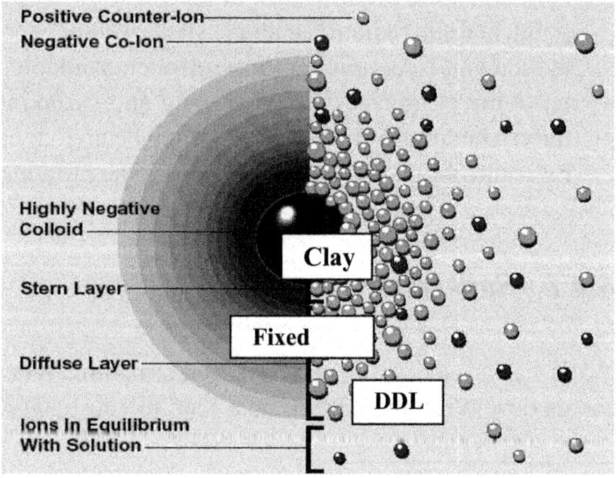

Fig. 8.4 DDL, fixed, and stern layer adsorbed on the clay micelle

If the net force between the particles is attractive, flocculation is the result. If the particles are repulsive, they tend to move away—"disperse." At low water contents, the DDL is not fully developed, the repulsive force is small, and the net force between the particles is a strong attractive force. This makes it difficult for the particles to move about when compactive effort is applied. A low dry unit weight is the consequence.

8.4.1.1 Reason Why Dry Pitches Should Not Be Rolled?

As the moisture content is increased, the DDL expands and the interparticle repulsive forces increase. The particles easily slide over one another and get packed more closely, resulting in a higher dry unit weight. The DDL expansion is complete at the OMC, and that is why the dry unit weight is at its maximum at that stage. Water does not add any additional DDL expansion beyond the OMC, but it does tend to occupy space that would otherwise be occupied by soil grains, hence, the decrease in dry unit weight. The theoretical maximum value of dry unit weight for a compacted soil is obtained for a given water content, corresponding to the situation when no air voids remain and the degree of saturation reaches 100%. Zero air void ratio (ZAVD) is the maximum dry unit weight that one can ideally obtain for a soil at a given water content by applying compaction. It is said to be theoretical because it can never be reached in practice, as it is impossible to expel the entrapped pore air completely by compaction.

8.4.2 Type of Soil (Structure, Texture, Physical, Chemical, Mineralogy, etc.)

The coarse-grained soils do not absorb water as well as the fine-grained soils, and hence, Lambe's DDL is not applicable to them. In fact, the compaction curve is of little importance for sand. Maximum dry unit weight results when the soil is either dry or completely saturated. Therefore, sands are usually compacted either in a dry or saturated state by flooding with water. Cohesion-less soils do not show a marked OMC. The application of vibrations is the most effective method of compacting cohesion-less soils.

Clay and loamy soils respond much more to pressure than sandy soils. The soil density increases almost linearly with the logarithm of the pressure applied, as each doubling of the pressure gives a certain step of increase in the soil density. For example, if an increase in pressure from 1 to 2 kg/cm^2 causes the density of a soil to increase from 1.4 to 1.5 g/cm^3, then the application of 4 kg/cm3 of pressure would result in a density of 1.6 g/cm^3, and 8 kg/cm^3 in a density of 1.7 g/cm^3. Each doubling of the pressure has increased the density by 0.1 g/cm^3.

For the same soil with lower water content, a lower soil density would be achieved at each pressure applied, while the increase for each doubling of pressure would remain similar, although lower, e.g. for 2, 4, and 8 kg/cm^3, the densities may be 1.46, 1.52, and 1.58 g/cm^3, respectively.

The step in density increase for a doubling of pressure for a uniform sand may be only 0.05 g/cm^3, or half of that required for the above clay or loam soil, while some poorly structured loamy soils can be compacted up to 0.50 g/cm^3 for each doubling of pressure.

This is why cohesive soils with very high water holding capacity (WHC), swelling and shrinking capacities, high cation exchange capacity (CEC), high elasticity and plasticity, high entrapped and adsorbed capillaries, film hygroscopic moisture, and high compressive strength, such as clay with smectite or vermiculite type of clay mineralogy, are considered for constructing cricket pitches.

8.4.3 Cricket Pitch Soil Structure and Compressibility of Turf Pitches

"How the cricket pitch soil structure controls and regulates the pitch behaviour at the time of pitch profile construction or at the time of match pitch preparation"

Structural arrangement of soil particles, which depends on various parameters such as particle shape, size distribution, origin and its formation, chemical characteristics, affects engineering behaviour such as compressibility, especially in fine-grained soils. The material used in cricket pitches and field turf is mostly fine-grained clayey soils, which, by nature, exhibit high compressibility. In this regard, the compressibility characteristics of the soil in relation to its soil structure, particularly that used for cricket pitches, are presented in this chapter. The study also summarises the significance of retaining the soil structure during pitch preparation and before laying the field turf. Soil structure means the arrangement of the soil particles relative to each other and the forces acting between particles with respect to each other due to interstitial forces acting between them that hold them together. The concept is further extended to the physical characteristics, the mineralogical composition of grains, the electrical properties of the particle surface, the ionic composition of pore water, the interactions among solid particles (both inorganic and organic parts, namely the humus), pore water, and the adsorption complex.

1. **Classification of Soil Structure**

   Soil structure can be classified as single-grained, honeycombed, or flocculated.

1. Single-grained: This is characteristic of coarse-grained soils with a particle size greater than 0.002 mm, where gravitational force predominates the surface forces and hence grain-to-grain contact results.
2. Honeycomb generally occurs in fine-grained soils, especially silts. Due to the relatively small size of grains, besides gravitational force, interparticle surface forces exist. Miniature arches are formed, which bridge over relatively large void spaces. This results in the formation of a honeycomb structure, with each cell of

the honeycomb being made up of numerous individual soil grains. The structure has a large void space and may carry high loads without a significant volume change.

3. Fine-grained soils, like clay, tend to be flocculated because their physical, chemical, and engineering properties are mostly determined by the forces between the soil particles. Mutual repulsion between the particles may be eliminated by means of appropriate chemicals, which result in the grains coming closer together to form a floc. This is known as flocculation. But the floc, being unstable by nature, tends to settle into a honeycomb structure, in which, in place of each grain, a floc is formed. Thus, flocs form when grains group together around void spaces larger than the grain size, and flocs form when grains group together around void spaces larger than the flocs. Very fine particles of colloidal size (less than 0.0002 mm) may be in a flocculated or dispersed state. In a flocculated structure, the flaky particles are oriented edge-to-edge or face-to-face with respect to one another. Flaky particles of clay minerals tend to form a card-house structure when flocculated (Lambe, 1953).
4. **From a morphological point of view, 10 different kinds of soil structure exist, viz.**

(1) crumb structure of heavy clay soil, (2) granular, (3) subpolyhydric (subangular), (4) polyhydric (angular), (5) blocky polyhydric, (6) massive (compound) polyhydric, (7) prismatic, (8) columnar, (9) complex prismatic, and (10) complex (Figs. 8.5 and 8.6).

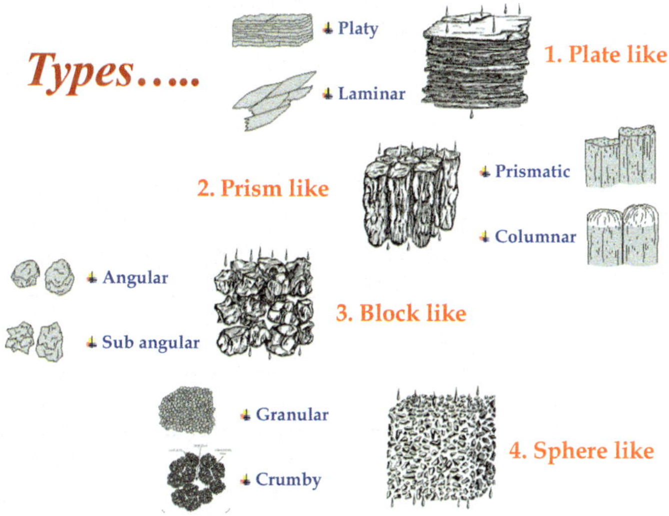

Fig. 8.5 Different types of soil structure (morphological classification)

Fig. 8.6 Different types of soil structures existing at the some of the important cricket pitches located in India

3. **Formation of Soil Structure**

The formation of soil structure is governed by several factors. In coarse-grained soils, the force of gravity is the main factor, while in fine-grained soils, the surface bonding force becomes predominant. Another fact that emerges from the concept of specific surface area is that, for the same void ratio, water content is higher in fine-grained soils than in coarse-grained soils. Also, the quantity of water needed to just wet the surface of smaller particles in a given volume of soil is many times more than what is required in the case of large particles. The presence of water, which is relatively unimportant in coarse soils, plays a decisive role in the compressive behaviour of clay soils.

Grain size distribution and grain shape influence the engineering behaviour of granular soils and hardly affect the behaviour of clay soils. Silts, which can be both granular and fine-grained, are indeterminate. They are neither cohesive nor plastic, but the presence of water affects their behaviour. Although their strength is not greatly affected by water content, dilation is observed in such soils.

4. **Suitability of Soil Structure in Cricket Pitches**:

It is partly true that the engineering behaviour of a soil mass is greatly influenced by the mineral composition of the soil grains forming the soil mass. In the case of coarse-grained soils, the mineralogical composition of the grains hardly affects the engineering behaviour of the soils. Perhaps the grain-to-grain friction is influenced to a greater degree. In such soils, interparticle forces other than those due to gravity are of no consequence. In contrast, the greater the force associated with the surface area of the grains, the finer the particles are. The chemical characteristics of the individual grains assume importance, especially when the surface area is large relative to the size of the grains, a condition that is associated with fine-grained soils. Thus, interparticle attraction holding the grains together becomes increasingly important as the size decreases. A honeycomb structure has a large void space and may carry high loads without a significant volume change. This is one of the major reasons why pitch soils with high silt content are generally flat, slow, and lack pace and bounce.

Other than the inherent mineralogical composition of the pitch soil complexes, all the above-mentioned factors are variable, and their concentration and proportion get changed during the pitch preparation process, thereby altering or changing the basic structure of the pitch profile. Thousands of such combinations exist by altering and manipulating the above-mentioned factors affecting the pitch soil structure and its behaviour.

In a broader sense, consideration of mineralogical composition, electrical properties, orientation and shape of soil grains, nature and properties of soil water, and the interaction of soil water and soil grains may also be included in the study of soil structure. Many engineering properties, such as permeability, compressibility, and shear strength, are influenced by soil structure.

Thus, interparticle attraction holding the grains together becomes increasingly important as the size decreases. A honeycomb structure has a large void space and may carry high loads without a significant volume change. This is the reason why pitch soils with high silt content are generally flat, slow, and lack pace and bounce.

5. **Factors Influencing Soil Structure of Cricket Pitch**:

Particle shape: The shape of the grain affects the compressibility and bearing capacity of the pitch profile. A soil composed primarily of flat grains is more compressible than one composed primarily of spherical grains. Angular particles tend to interlock, form a denser mass, and hence become more stable than rounded particles, which can roll or slide past one another. Such particles' compressibility is low. The majority of clay-rich cricket pitch soils have an angular to subangular blocky structure.

Particle size distribution: When laying a three-layered wicket with 8 inches of paying surface and the first 5 inches laid on top of each other, the clod size should be between 20 and 25 mm. Five to 6 mm of clods are laid upon the 5 inches of laid surface after planting grass upon it. Special care is taken while laying the pitch clods to minimise damage to its natural developed soil structure. Pitch soil should not be applied in powered form by destroying the soil structure fabric and aggregates complex. This is done to retain the maximum possible soil structure of the aggregates when it is transported from its natural source and the pitch is laid upon. Smaller clods of 5–6 mm of 1-inch-thick strip (up to 3 inches) are laid upon the grass so as they can grow and knit the playing surface.

Figures 8.7, 8.8, and 8.9 give a brief idea about laying of cricket pitch and retaining the soil structure while laying.

Fig. 8.7 Excavation of cricket pitch soil from the river basin area

Fig. 8.8 Laying of pitch lying with 05 to 20 mm size clods to retain its structure

Fig. 8.9 Planting of the turf grass (Bermuda, Selection 1 variety) and top dressing of pitch soil having clods of smaller size (5–10 mm)

6. Soil Stabilisation Is the Process of Changing One or More of a Soil's Properties

Soil stabilisation is the process of changing one or more of a soil's properties. Soil stabilisation can be achieved with or without the help of additives. Stabilisation without additives is the mechanical rearrangement of particles through compression, addition, or removal of certain sizes of soil particles. According to Leonard's (1962), mechanical stabilisation means improving the soil particles through rearrangements of particles and densification by compaction, or by changing the gradation through the addition or removal of soil particles.

The properties of soil may be significantly altered by forming textured soil in proper proportions so that the mixtures, when compacted, attain maximum density and strength, or by removing some fraction of the soil. This involves the blending of two or more naturally available soils in suitable proportions to achieve the desired engineering properties for the mixture after necessary compaction. The granular or "aggregate" fraction provides strength by internal friction and hardness or compressibility, while the binder provides cohesion or binding properties, water retention properties, or impermeability, and also acts as filler for the voids in the aggregates.

The pitch soil in Wankhede, Mumbai, is a great example of how mixing two different soils in a 2:1:3 ratio can produce the best results. The relative amounts of aggregate and binder determine the physical properties of the compacted, stabilised soil. The optimum amount of binder is reached when compacted binder fills the voids without destroying all the grain-to-grain contacts of coarse particles. An increase in the binder beyond this limit results in a reduction of internal friction, a slight increase in cohesion, and greater compressibility (Table 8.2, Fig. 8.10).

Fig. 8.10 Wankhede yellow (fine soil) (1) and brown soil (coarse soil) (2) mixed in 2:3 ratios

Table 8.2 Particle size distribution of Wankhede soil

Soil mixed in 3:1 ratio	Clay (%)	Silt (%)	Sand (%)
Coarse murum soil	40	38	20
Fine yellow soil	28	37	26
Red-yellow mix	44	36	15

Source Test report by- GEO Test House, Krishna Industrial Estate, and Vadodra 2010

Cricket pitch soil high in fine calcium, magnesium, or sodium smectites (swelling and shrinking clays) has a greater tendency to rearrange itself, resulting in swelling or shrinkage of the double diffused layer (DDL)) layer, leading to the desirable variation in pitch bulk density during the course of the Test match. Hence, the smectite-dominant clay pitch profile with a high specific surface area, high water retention capacity, and high OMC levels should have a different rolling technique and schedule compared to the pitch soils having medium to fine soil textures.

8.4.4 Compactive Effort

For a given type of compaction, the higher the compactive effort, the higher the maximum dry unit weight and lower the OMC (Fig. 8.11).

Fig. 8.11 Rollers of different weight, diameter, and compactive potential being used at the JSCA Cricket Stadium Complex, Ranchi, Jharkhand. The effect of the diameter and contact angle of the roller with the pitch surface. The direction of stress becomes more vertical as roller diameter increases. This is because of the decreasing contact angle between the roller and pitch surface

8.4.5 Method of Compaction (Rolling Techniques and Schedules)

Different soils attain different maximum dry densities at different moisture contents with specific compactive efforts and rolling schedules. Clay soils have much higher OMC and consequently lower maximum dry densities (Table 8.3).

Table 8.3 The maximum dry density range at an OMC for different soil textures

Soil type	Maximum dry unit weight under light compaction	OMC (%)
Gravel sand-clay mixture	20 kN/m^3	9
Sand	19	11
Sandy clay	18.1	14
Silty clay	16.3	21
Clay	15.2	28

Source (HMSO 1952)

Under natural conditions, the hygroscopic adsorbed water on a clay or peat surface is nearly 15–20%. It is due to their very large external SSA and internal surface area. Sandy soils only contain 1–2% of hygroscopic water (Table 8.4).

Table 8.4 Maximum dry density at different soil textures and OMC

Volume %			
	FC	HC	Capillary water
Soil type	(−0.3 bar)	(−31 bar)	(col. 1–col. 2)
Sandy loam	12	3	9
Silt loam	30	10	20
Clay	35	18	17
FC—Field capacity		HC—Hygroscopic coefficient	

Note that the clay soil retains most water at the field capacity, but much of that water is held tightly in the soil at -31 bars potential by soil colloids (HC).

This is why a cricket pitch with a lot of smectite clay and a high water holding capacity (WHC) needs heavy rolling in the first few days of preparation, either to squeeze out the water that isn't available in the form of capillary water in the pitch profile or both. Clay soils have a high proportion of unavailable hygroscopic water that is tightly adhered to the clay micelle in vapour form and cannot be extracted by the stress developed due to evapotranspiration (Fig. 8.12).

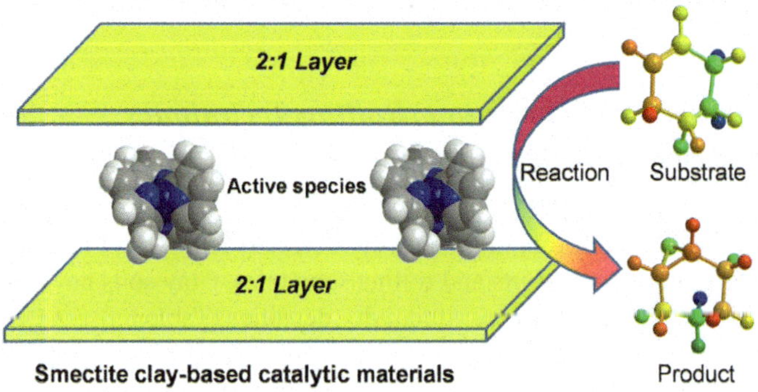

Fig. 8.12 Interlayer structure of smectite clay

Smectite clays with very high specific surface area (SSA), very large internal surface area containing large reservoir of water, and high water holding capacity need heavy rolling to squeeze water from their micropores and 2:1 lattice (Fig. 8.13).

Four inches of uniform wetting   Light showering to regulate the surface cracks

Fig. 8.13 Different patterns of rolling with rollers having different compactive potential

8.4.6 Rollers Do Not Dry the Pitches

Rollers do not dry the pitches. It only extracts available moisture by applying a point load or stress to or within the pitch profile (held as capillary water in the micropores due to surface tension in a partially available form and hygroscopic water adhered tightly to the clay micelle in an unavailable moisture form). The available water is between the capacity of the field and the permanent wilting point. The squeezed-out water from the micropores to the macropores either drains out of the drainage layer lying beneath the playing surface by the force of gravity, is evaporated by the sun's radiation and thermal energy, or is transpired by the adjoining turf root grasses. Moisture lost within the pitch profile increases the air voids and porosity, which have to be removed by the rollers to compact the pitch. Further, the rollers compact the extra air voids generated due to the replacement of water in the pitch profile. As the soil moisture within the pitch profile is incompressible, it can only be squeezed out from the micro- and macropores, which are further evapotranspired by the solar thermal energy and through the turf grass roots.

The top surface (1.25–1.5) inches of the pitch will be dried and hardened by solar thermal energy (in the form of evaporation), while the remaining 2.5 inches of the total 4 inches of pitch profile will only be dried and hardened by transpiration by the turf grass roots.

All of the steps of compaction and consolidation in the profile of the pitch can be compared to how a washing machine cleans clothes. The washing machine does not dry the clothes. Instead, it uses centrifugal force to squeeze the dirt and water molecules out of the tiny pores of the clothes. The actual drying of clothes occurs only when the clothes are hung up to dry in the sun.

8.5 Cricket Pitch Soil Compression Through Live Case Studies (Test/ODI/T-20 Matches Coordinated and Supervised by the Author as a Curator)

The whole scientific phenomenon involved in cricket turf pitch profile compaction, consolidation can be visualised, analysed, and correlated with the existing rolling schedules and techniques involved in attaining a desired pitch surface density through the live case studies (case study 9.5 inserted in Chapter 9, pages 245 to 268), and the following case study:

The purpose of this case study is to understand the basic principles of science and logic behind the art of rolling with the help of group discussions, analysis, and interpretations done on the above-mentioned topic during the Level-1 course organised by the Pitches and Grounds Committee, BCCI, from 25 July to 3 August 2016, at Chinnaswamy Stadium, Bangalore.

a. **It's always a million-dollar question in the mind of a curator as:**

1. Why do we roll, and what does the roller do?
2. When should I start and stop rolling?
3. How much to roll and at what speed?
4. What should be the direction of the roll?
5. With what type of roller to roll, what are the rolling compressive potential of the roller and the maximum compactive potential of a pitch?
6. What are the limitations of a roller?

7. Why is block density generated by the use of a heavy roller?

b. **The whole art of curatorship lies in rolling at the**

1. At the right time: When the moisture range is between FC (field capacity) and PWP (permanent wilting point) or in the plastic range, no pitch profile should be rolled when there is too much or no moisture.
2. Right proportion: Based on the pitch soil type, 6–8 passes will provide the desired density when rolling with the roller for the desired compactive potential. Once the desired compaction has been achieved at OMC, rolling the wicket is pointless.
3. Select the proper rollers: those with the appropriate mass and width factor for either compactive potential or friction.
4. Right rolling patterns: cross, diagonal, and straight rolling.
5. Right moisture levels to achieve the desired compactive potential of the pitch.
6. Maintaining the proper speed: The wetter the surface, the slower the speed, and vice versa.

c. **Why do we roll and what does the roller do?**

Rollers do not dry the pitches. It only extracts available moisture by applying a point load to or within the pitch profile (held as capillary water in the micropores due to surface tension in a partially available form and hygroscopic water adhered tightly to the clay micelle in an unavailable vapour form). The available water is between the capacity of the field and the permanent wilting point. The squeezed-out water from the micropores to the macropores either drains out from the drainage layer lying beneath the playing surface by the force of gravity, is evaporated by the sun's radiation and thermal energy, or is transpired by the adjoining turf root grasses. Further, the rollers compact the extra air voids generated due to the replacement of water in the pitch profile, making it denser. As the soil moisture within the pitch profile is incompressible, it can only be squeezed out from the micro- and macropores, which are further evapotranspired by the solar thermal energy and through the turf grass roots. The top surface (1.25 to 1.5) inches of the pitch will be dried and hardened by sun radiation thermal energy (in the form of evaporation), while the remaining 2.5 inches of the total 4 inches of pitch profile will only be dried and hardened by transpiration by the turf grass roots.

d. **When to start and stop rolling?**

Start rolling after the excess gravitational water is drained off and the pitch profile is at the field capacity, which depends on the WHC and hydraulic conductivity of the pitch profile. Depending upon the WHC and the HC of the pitch profile, it takes a few hours to 24 h to drain off the excess gravitational water from the pitch profile. This is also done to ensure uniform wetting of the whole pitch profile (at least 4 inches from the top of the pitch surface) and to help attain a uniformly desirable bulk density of the playing surface. Rolling at the FC is the best time to start, which can be judged by the feel of the thumb making slight inundation, by spikes, or by taking a core sample.

For a cohesive soil, 100 per cent maximum dry density (MDD) is achieved on a soil surface after 8 passes of rolling with a roller having 2100 kg/m compactive potential and up to a depth of 125 mm (having 5 per cent air voids).*

Table 8.5 Number of passes and the per cent compaction level achieved after subsequent number of passes (From agricultural point of view), data source: National Soil Resource Institute (NSRI), UK, 2001

No. of passes	% Compaction
1	50
2	10
3	6
4	3

* According to the Department of Science and Industrial Research (DSIR), the United States Department of Transportation (DOT), the Transport Research Laboratory (TRL), Berkshire, UK, and Parsons 1992.

The rolling schedules should be stopped once the pitch profile attains the desired bulk density at a particular optimum moisture content (OMC). This can be checked by the use of a Clegg hammer readings, or with the help of spikes, core sampling, ball rebound, feeling the shape, size, and intensity of cracks developed on the pitch surface, or by analysing the colour of the grass or the finished surface (Fig. 8.14, Table 8.5).

Conceptualized by Dr. S. B. Singh, Curator, BCCI

Fig. 8.14 Concept of integrated rolling (IRM) and water management concepts

e. **Why do we continue rolling holidays after the primary compaction (maximum compaction at OMC) is achieved after 8–10 passes of initial rolling only?**

Rolling holidays or a gap of one or two days are followed in between the two rolling schedules after the pitch has achieved the maximum possible compression levels. This is due to the phenomenon called secondary compression and "thixotropic" effects, or rest hardening, which come into effect after the primary compression has taken place. The loss of strength during remoulding (repeated wetting and drying) of soil is primarily due to the destruction of natural soil structure and bonds. Also, the structure of the adsorbed water is broken up. With the passage of time, however, the strength increases, though not necessarily to the original value.

Once the desired maximum compression at OMC is achieved due to primary compression, further increase in density or hardening of the pitch profile is accomplished due to secondary compression or due to thixotropic or rest hardening, though at that stage there is no change in moisture or volume levels of the pitch profile.

Primary compaction takes place during this period due to the loss of soil moisture within the pitch profile (due to the expulsion of moisture from the clay micelles and the micropores and subsequently evapotranspired by the turf grass), which is a slow process (due to the very low HC of the pitch profile, which is aggravated by the higher BD of the pitch profile also).

Secondary compression takes place due to the plastic rearrangement of clay particles (due to further shrinkage of the DDL). This phenomenon can be observed through the cracks developed on the pitch surface, which increase in size and shape with further loss of moisture from the surface.

Usually, a total of 6–8 passes (depending upon the weather variables, available moisture within the pitch profile, and roller and pitch compactive potentials) may be adequate to achieve the equivalent level of compaction achieved in the Standard Proctor Test.

BD/primary or secondary compaction levels and consolidation due to elastic deformation of clay particles (either expulsion of pore water through the rolling schedules) are generally achieved by the first 5–6 passes only, depending on the pitch soil structure, texture, available PAWC within the playing surface, and other dependent and independent variables. The maximum possible or desired BD at a particular OMC is achieved through the initial 5–6 passes only. Further rolling after achieving the desired compaction levels (either primary or secondary compression) is usually a waste of time and money, and it damages the turf grass due to root break. Due to overcompression of the pitch profile, pitch and grass recovery becomes impossible or extremely difficult after the match.

During this stage, plastic rearrangement of clay particles takes place, and other effects as well as colloid-chemical processing and surface phenomena such as induced electro-kinetic potential occur. These are, by their very nature, very slow. At least 7–14 days are required to complete one full wetting and drying (swell and shrink cycle) cycle to attain the primary or secondary compression levels. This is the reason why at least 10–15 days are required to have a variable pitch and desirable compaction levels to prepare a Test match wicket.

Secondary consolidation is believed to come into play even in the range of primary consolidation, although its magnitude is small, because of the existence of a plastic lag right from the beginning of rolling. However, it is almost impossible to separate this component from the primary compression.

Block density is created within the pitch profile during a preseason rolling schedule with a heavy roller at proper moisture levels with 8–10 mm of grass height. In general, a total of 2–2.5 h of heavy roller rolling is completed over a three-week period, with an intermittent rolling holiday of a day or two taken to achieve simultaneous maximum primary compaction through initial rolling and secondary compaction with the thixotropic effect or rest hardening. The purpose is to regain loose density due to the over-saturation of the pitch profile during the monsoon season and the preseason renovation practices (such as top dressing, deep-choring, de-thaching, etc.) that decompact the pitch profile.

Primary and secondary compression, thixotropic rest hardening, and elastic rebound are the general scientific concepts that justify the rolling schedules and techniques being adopted on the cricket pitch profile.

f. **Why the medium and heavy rollers are deployed once again after a day or two of rest, in spite of the fact that the pitch profile has already achieved the maximum compaction levels at OMC through the initial rolling schedules?**

Air voids are developed due to the loss of moisture caused by the radiation of thermal energy on the pitch surface and transpiration through the grass roots (nearly up to 100–150 mm) deep within the pitch profile. These deep air voids within the pitch profile are to be further compacted by the rollers with deep vertical compactive potential (heavy rollers with larger drums).

Second, fine-grained soils (clay smectite) exhibit another type of volume rebound. Water held between platy (flaky) smectitic clay particles by certain forces (rollers) gets squeezed out under the compressive strength. When these stresses are removed, these forces cause the water to be sucked in again, resulting in the phenomenon of swelling. To compress this refilled water in the clay micelles, further intermittent rolling is desirable to maintain and sustain the desired levels of pitch surface density.

Vertical stress distribution or propagation in soil due to external loading occurs when the load is applied to the soil surface and increases the vertical and lateral stress within the soil mass. The increased stress is greatest directly under the loaded area and dissipates within the soil mass as a function of distance away from the loaded area, which is commonly called "spatial attenuation of applied stress."

The stress–strain behaviour in the cricket pitch soil profile is extremely complex and is determined by a variety of factors including drainage condition (generally, the pitch profile with clay-dominated soils has low hydraulic conductivity and permeability in compacted conditions), void ratio (depending on the pitch compactive potential), load level (weight and size of the rollers and rate (either the roller compactive potential)), and stress path (rolling schedules) (Fig. 8.15).

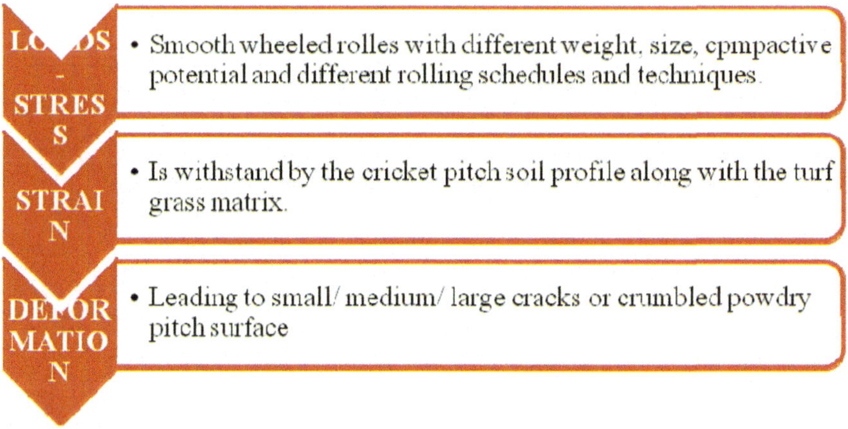

Fig. 8.15 Propagation and dissipation of stress within the pitch profile Conceptualised by Dr. S. B. Singh, Curator, BCCI

In other words, this means that most of the strain due to increased stress from the applied load will be experienced at relatively shallow depths below the load, generally known as the "depth of significant influence" (DOSI). According to the linear elastic theory, the size of the pressure bulb is proportionate to the size of the loaded area. This is a key concept that is used to evaluate the DOSI of an applied surface load. The depth Ds is a finite depth below which there are no significant strains in the soil mass due to the load imposed at the surface.

As an approximation to the exact solution given by the Boussinesq charts, the total loads at the surface of the soil mass may be distributed over an area of the same shape as the loaded area on the surface, but with dimensions that increase with depth at a rate of 1 horizontal: 2 vertical (1H: 1 V) units, generally known as the approximate (2:1) stress distribution concept.

An isobar is a line that connects all points below the ground surface with equal stress increments. In other words, an isobar is a stress-increment tour. Each isobar represents a fraction of the stress applied at the surface and delineates the zone of influence such that the area contained within two isobars experiences stresses greater than the lower isobar and less than the upper isobar. Since these isobars form closed figures that resemble the form of a bulb, they are also termed "bulbs of pressure" or "pressure bulbs," which give the feeling of the spread of stresses through a soil mass. The stress and strain developed below the pitch surface by the rollers are just the mirror image of each other and decrease with the square of the distance (Fig. 8.16, Table 8.6).

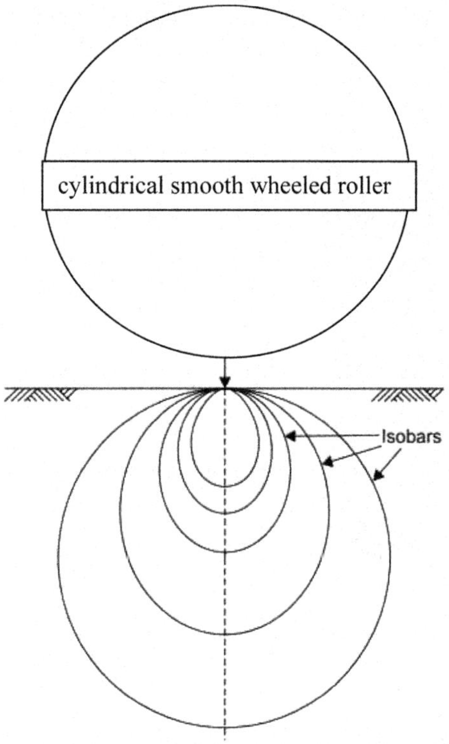

Fig. 8.16 Pressure bulbs (depth of significant load) for point load generated by a cylindrical smooth-wheeled roller below the pitch surface

It is generally considered that the rollers have maximum vertical impact up to 4 inches below the pitch profile. The point load exerted on the pitch surface by a roller weighing 1000 kg and measuring 80 cm in width is = 1000/80 = 12.5 kg/cm^2.

Table 8.6 Pressure exerted at 1 cm is only 1/100 of this pressure exerted at 100 mm.

stress at 1 cm (10) mm	= 5.97 kg/ cm^2,
Stress 5 cm (50) mm	= 0.238 kg/cm^2,
Stress 10 cm (100) mm	= 0.0597 kg/cm^2

Source McIntyre Keith and Don, Cricket Wickets, Science Vs Fiction, 2001

The total load at the surface of the soil mass may be distributed over an area of the same shape (pressure isobars) as the loaded area on the surface, but with a rate of 1 horizontal: 2 vertical (approximate stress distribution). The impact of rolling (the depth to which the roller compacts) is governed by a number of factors, including the available water in the pitch profile, the size of the peds, texture, structure, the weight and diameter (mass/width factor) of the roller, the number of passes, grass root growth, and depth.

Equations for stress and strains induced in a homogeneous, isotropic, weightless, linearly (free of initial stress and deformation) elastic half space, with a plane horizontal surface, by a point load perpendicular to the surface, were first provided by.

Boussinesq (1885) as

$$Pz = \frac{3PZ^3}{2\pi R^5} = \frac{3p}{2\pi R^2}\text{Cos}3\beta = \frac{Ip}{Z^2}$$

where I = is termed as influence factor,

Pz = change in vertical stress,

P/q = load per unit length,

Z = depth.

The thicker the layer of earth subjected to field compaction, the lower the energy output per unit weight of soil and, hence, the less compaction under each pass of the roller. A suitable thickness of soil in each layer is necessary to achieve uniform thickness. Layer thickness depends upon soil type, PAWC, roller type, weight, and the contact pressure of its drums (either compactive potential). Normally, 200–300 mm of layer thickness is optimum in the field for achieving homogeneous compaction. The load per cm of width and diameter determines the performance of smooth-wheeled rollers.

g. **Why smooth-wheeled rollers instead of vibratory rollers are used for pitch profile compression?**

Most of the cricket pitch soils are cohesive in nature, having a high quantity of clays that cannot be compacted by the vibratory force. Generally, 8 inches of drainage layers (4 inches of loamy sand and 4 inches of river-washed sand) lying just beneath the playing surface in 3-layered wickets are compacted with the help of vibratory rollers.

Static load produces very low or little compaction of loose sand. Medium or loose sand does not compact easily when moist because of the shear strength developed by capillary forces. When dry or submerged, all sands can be compacted by vibration. Today's cricket outfields are constructed with medium- to fine-grained sands and organic granules. This is generally done to have a softer outfield in which grass can grow well with good drainage levels.

Pressure is required to consolidate clays. The loose clay before compaction has a loose, random structure. Each clay particle has attached to its surface a thin layer of solidly adsorbed water and a thick layer of viscous diffused water. The thickness of the DDL depends on the water content of the soil mass. In compacting the clays, the position of the particles must be changed by forcing the contact points along adjacent surfaces to position themselves more nearly parallel and with reduced voids. When the clay to be compacted is saturated, compaction can only be accomplished by forcing compressible water out of voids. Because of the extremely low permeability of clays, considerable time (also known as hydrodynamic lag) is required for the water to flow out and allow the volume to decrease.

When the clay has a high water content that is less than saturation (as on the first day of pitch rolling), a thick layer of DDL water surrounds the particles, separating the contact points with low viscosity water. Under these conditions, only a small amount of pressure is required to force the grains into new positions. But a high degree of compaction cannot be produced with this high water content because the thick layer of DDL water prevents the particles from being forced close together before failure occurs. As the water content decreases, the thinner layer of attached water allows a high degree of compaction, but more work is required to move the particles in highly viscous media into new compact positions. For a given amount of compactive effort, there is a corresponding water content that will produce the densest state. The water content that will allow the densest compaction under a given compactive effort is known as OMC. When the water content is very low, a given compactive effort will produce a low degree of compaction because of the high resistance of the highly viscous water. This is the reason why dry pitches should not be rolled.

A high water content of low viscous attached water enables compactive effort to move particles into a denser state and acts as lubricant. The effort increases as the water content increases, until the water content is high enough that the large amount of water with the entrapped air prevents further compaction. As the compactive effort is increased, the water content at which the densest state will be produced is decreased. Under no circumstances can all the air be driven out of the pitch soil complex during the compaction so that the mass is completely saturated. This stage, theoretically referred to as "zero air void ratio (ZAVD)," is not achievable under underground conditions.

h. **Is there a different rolling technique or different rolling protocols for cricket pitch soils having different textures, morphological, structural, physical, chemical, and mineralogical compositions?**

Pitch soils with a high proportion of non-swelling and shrinking capacity need manual or mechanical means to decompact and aerate the over-compressed pitch surface after the match, as these types of pitches lack the natural ability to rejuvenate themselves due to their intrinsic non-swelling and shrinking behaviour. Pitch surface and turf grass recovery on these types of pitches are very poor (after the match, with an over-compressed pitch surface) and need mechanical or manual measures to recover to the desired pitch surface levels (Fig. 8.17).

Fig. 8.17 Mechanical and manual means (through spike rollers) of aerating and decompacting the overcompacted pitch surface (having non-swelling and shrinking fine clay soils) after the match or during the preseason renovation

The high compressibility of fine clay with a high proportion of sand, moderate silt and clay, and non-swelling and non-shrinking inactive (kaolinite dominant) pitch soils (having low water holding and retention capacity, moderate to high hydraulic capacity, etc.) is generally achieved by increasing the pressure on the pitch surface through mechanical means (rolling with rollers having different compactive potential).

The desirable compressibility of fine smectic-dominant swelling and shrinking active clay (having high water holding and retention capacity and low hydraulic capacity at alkaline to saline pH levels) is generally achieved not only by mechanical means (rolling with rollers of varying compactive potential), but also by slow desiccative drying of the pitch surface under the various climatic and weather variables available within and just above the pitch surface.

8.6 Summary

In this modern era of scientific research and development, a lot has been done in the fields of soil mechanics, civil engineering, geotechnical engineering, and soil science to understand the complex nature involved in soil compression. But the practical implications of the knowledge and skill developed by the above-mentioned institutions on the cricket pitches to attain the desired compression levels are still in their nascent stage. Though Cranfield University in the UK has conducted extensive research on "optimisation of cricket pitch rolling" in collaboration with the ECB (England and Wales Cricket Board), their research parameters are limited to climatic or weather variables and on pitch soil with a low percentage of less swelling and shrinking clay (illite). The entire research and guidelines are based only on mechanical aspects of compaction.

With different rolling schedules and techniques for cricket turf wickets, different pitch surface densities can be achieved due to compression and shrinkage drying, which are the main goals when preparing wickets for a Test match, an ODI, or a T-20. The desirable turf wicket behaviour can be attained and retained through different rolling schedules and techniques and by controlling and regulating the swelling, shrinking, and drying cycles of smectite-dominant clay pitch profiles along with the turf grass at variable moisture levels.

Cricket pitches are made out in the open air, where the temperature, relative humidity, solar radiation, evaporation, and other weather and climate factors change throughout the day. These factors have a direct effect on the density of the turf pitch surface by controlling how much water stays put, moves around, and moves out. Most of the time, the weight, diameter, and surface area of the rollers that touch the pitch surface determine how well they can roll. Every turf pitch soil, based on its particle size distribution, morphology, structure, physical (water holding and retention capacity, hydraulic capacity, etc.), chemical (soil pH levels with predominant cations), and predominant mineral composition, has different compressive potential. Turf pitch rolling protocols or different rolling schedules and techniques should be based on considering the format of the match, climatic and weather variables, and roller and turf pitch surface compaction potential as well.

Keeping in view the total compression levels achieved on the turf pitch surface through the compaction and consolidation processes, an integrated rolling management (IRM) programme needs to be developed. Rolling patterns, techniques, and schedules should be based on the particle size distribution, clay type, percentage, mineral content, turf grass leaf type, root area index, organic matter contents, pitch available water content, and weather variations in the form of evapotranspiration, temperature, relative humidity, dew point, fog, mist, altitude, etc. Though two soils with the same particle size distribution but different mineralogical types, such as swelling and shrinking, smectites or non- or limited swelling types like kaolinites or illites cannot have the same rolling schedules and techniques, At different optimal moisture contents, both pitch soils will achieve the highest dry densities. Also, the impact of primary and secondary compression, or the thixotropic effects, will be entirely different on pitch soils dominated by smectite clay or kaolinite clay.

Desired compression levels (DBD at an OMC) attained during the full swell-shrink cycle of the pitch profile at the time of pitch preparation or the decompression leading to the decrease in the DBD due to the development and further attenuation of cracks or crumbling (due to the desiccative loss of moisture from the pitch profile) need to be measured and quantified. The desirable variations in pitch surface during the game (especially in Test matches) are caused by decompression (a constant decrease or increase in DBD) with moisture variations (due to loss, migration, or retention of soil moisture) within the pitch profile. Optimum rolling (to attain the desired bulk density at optimum moisture content) and maximum drying and shrinkage (drying, rest hardening, desiccation due to evapotranspiration through the turf grasses, etc.) should be the integrated rolling management (IRM)[3] mantra for achieving the desired pitch behaviour of a cricket pitch profile.

Very little research has been done on how to make an integrated turf pitch surface dense through compaction or drying out and shrinking. Keeping in view the total compression levels achieved on the turf pitch surface through the compaction, shrinkage, and drying processes, an integrated rolling management (IRM) programme needs to be developed.

References

Abebe AT, Tanaka T, Yamazaki M (1989) Soil compaction by multiple passes of a rigid wheel relevant for optimization of traffic. J Terramech 26(2):139–148

Abouleid AF (1982) Measurement of swelling and collapsible soil properties "Foundation engineering." In: Pilot G (ed) Presses de le cole nationale des Ponts et chausses. Paris, France.

Adams WA, Baker SW, Carré MJ, Young RJ, James DM (2004) Pitch properties and performance. England and Wales Cricket Board, Lord's Ground, London

Adams WA, Gibbs RJ (1994) Natural turf for sport and amenity: science and practice. CAB International, UK

Adams WA, Tanavud C, Springsguth CT (1985) Factors influencing the stability of sports turf root zones. In: Proceedings of the fifth international turf grass conference, pp 391- 398.

Aitchinson GD, Holmes JW (1953) Aspects of swelling in the soil profile. Aust J Appl Sci 4:244–253

Albrecht BA (1996) Effect of desiccation on compacted clays. Msc Thesis, Department of Civil Engineering, University of Wisconsin, Madison, WI, USA

Altmeyer WT (1955) Discussion of Engineering properties of expansive clays. In: Proc. ASCE, vol 81, no SM2

Atique A, Sanchez M (2011) Analysis of cracking behavior of drying soil. In: Second international conference on environment science technology, IACSIT Press

Baker SW, Binns DJ, Cook A, Mooney SJ (2001) The performance of the cricket pitches in relation to soil type and moisture content. Int Turf Grass Soc Res J 9:469–478

[3] The idea of "Cricket Pitch Soil Rolling Management" is explained in Chapter 9, Pages 245–268, where integrated rolling management (IRM) is listed and discussed in detail through a case study 9.5.

Baker SW, Hammond LKF, Owen AG, Adams WA (2003) Soil physical properties of first class cricket pitches in England and Wales. 1. Classification system for characteristics. J Turf Grass Sport Surf Sci 79:2–11
Baker SWA, Binns DJ, Carre MJ, Haake SJ (1998) The effect of soil type and profile construction on the performance of cricket pitches. 2. Playing quality during the first season of use. J Sport Turf Res Inst 74:93–107
Batey T, McKenzie DC (2006) Soil compaction: identification in field. Soil Use Manag 22(2):123–131
Baver LD (1940) Soil physics. Wiley, New York
BCCI Curators Certification Course Manual (2013) BCCI, 2nd edn
Been K, Sills GC (1981) Self weight consolidation of soft soils: an experimental and theoretical study. Geotechnique 31(4):519–535
Biot MA, Clingan FM (1941) General theory of three-dimensional consolidation. J. Applied Physics 12
Bowles IE (1992) Engineering properties of soils and their measurement. McGraw Hill, New York
Brady NC, Weil RR (2008) Nature and properties of soil, 14th edn. Prentice Hall
Chertkov VY (2000) Using surface cracks, spacing to predict crack network geometry in swelling soils. Soil Sci Soc Am J 64:1918–1921
Conte E (2004) Consolidation analysis for saturated soils. Can Geotech J 41(4):599–612
Dexter AR (1986) Root penetration and compaction pans in relation to soil structure and strength. Soil Tillage Res 8:332
Dexter AR (1987) Compression of soil around roots. Plant Soil 97(3):401–406
Dexter AR (1988) Advances in characterization of soil structure. Soil Tillage Res 11(3–4):199–238
Dong Y, Lu N (2017) Measurement of suction stress characteristics curve under drying and wetting conditions. Geotech Test J 40(1):107–121
Dong Y, Lu N, Fox PJ (2020) Drying- induced consolidation in soil. J Geotechn Geoenviron Eng 146(9):n040200092
ECB Staff (2007) Recommended guidelines for the construction, preparation and maintenance of cricket pitches and outfields at all levels of the game. England and Wales Cricket Board, Lord's Cricket Ground, London, UK
Ekwue EI, Lall DZ, Stone RJ (2006) Engineering properties of major soils used in cricket pitches in Trinidad. West Indies J Eng 28(2), Technical Paper
Evans RDC (1991) Cricket grounds: the evolution, maintenance and construction of natural turf cricket tables and outfields. The Sports Turf Research Institute, Bingley, UK
Fleureau JS, Kheirbek- Saoud RS, Taibi S (1993) Behaviour of clayey soils on drying- wetting paths. Can Geotech J 30(2):287–296
Fredlund DG, Hasan JU (1979) One dimensional consolidation theory: unsaturated expansive soils. Can Geotech J 16(3):521–531
Gens A, Alonso EE (1992) A framework for the behaviour of unsaturated expansive clays. Can Geotech J 29(6):1013–1032
Gibson RE, England GL, Hussey MJL (1967) The theory of one- dimensional consolidation of saturated clays. I: Finite non- linear consolidation of thick homogeneous layers. Geotechnique 17(3):261–273
Gibson RE, Lo KY (1961) A theory of consolidation for soils exhibiting secondary compression. Publication No, Norwegian Geotechnical Institute, p 41
Gobat JM, Aragan M (2004) The living soil fundamental of soil science and soil biology. Wiley Mathey
Grim RE (1953) Clay mineralogy. Mc Graw-Hill, New York
Grim RE (1959) Physico-chemical properties of soils. ASCEJ Soil Mech 85(SM2):1–17
Haris JR (1978) Maintenance of soil structure under playing turf. In: Emerson WW, Bond RD, Dexter AR (eds) Modification of soil structure. Wiley
Henkel DJ (1959) The relationship between the strength, pore- water pressure and volume- change characteristics of saturated clays. Geotechnique 9:119–135

Holtz RD, Kovacs WD (1981) An introduction to geotechnical engineering. Prentice- Hall, Upper Saddle River, NJ
Horn R, Lebert M (1994) Soil compactactibility and compressibility. In: Soane BD, Van Ouwerkerk C (eds) Soil compaction in crop production. Elsevier Science B.V.
IOG/ECB (2006) Management and maintenance of the cricket pitch. The Institute of Groundsmanship, Milton Keynes, UK
Jaky J (1994) The coefficient of earth pressure at rest. J Soc Hungarian Arch Eng:335–358
Kay BD, Angers DA (2000) Soil structure. In: Summer M et al (eds) Handbook of soil science. CRC Press, Boca Raton, p A229
Koolen AJ, Kuipers H (1983) Agricultural soil mechanics. Springer- Verlag, Berlin
Lambe TW, Whitman VR (1969) Soil mechanics. Willey (SI Version)
Lolret A, Alonso EE (1980) Consolidation of unsaturated soils including swelling and collapse behaviour. Geotechnique 30(4):449–477
Lu N, Likos WJ (2004) Unsaturated soil mechanics. Wiley, New York
McIntyre K, Don (2001) Cricket wickets- science vs fiction, horticulture engineering consultancy, 5- Brimage Palace Kambah, ACT, Australia
McIntyre K, Jacobsen B (2000) Practical drainage for Golf: sportsturf and horticulture. Wiley, New Jersey
McKeen RG (1992) A model for predicting expansive soil behaviour. In: Proceeding in seventh Int. conference on expansive soils, 1–6, Reston, VA: ASCE
Means RE, Parcher JV (1965) Physical properties of soils. Prentice Hall, New Delhi
Mitchell JK (1993) Fundamentals of soil behaviour. Wiley, New York
Morris PH, Williams DJ, Graham J (1992) Cracking in drying soils. Can Geotechn J 29:263–277
Mun W, McCartney JS (2015) Compression mechanism of unsaturated clay under high stresses. Can Geotech J 52(12):2099–2112
Murthy VSN (2002) Geotechnical engineering: principles and practices of soil mechanics and foundation engineering. Marcel Dekker Inc., Madison Avenue, New York
Nagraj TS, Srinivasamurty SR, Vastalya A (1994) Analysis and prediction of soil behavior. Taylor and Francis
NSRI (2001) A guide to better soil structure. Cranfield University, Silsoe, UK, National Soil Resource Institute
Onafeko O, Reece AR (1967) Soil stresses and deformation beneath rigid wheels. J Terramech 4(1):59–80
Parsons AW (1992) Compaction of soils and granular materials: a review of research performed at the Transport Research Laboratory, Department of Transport. HMSO, London
Pearson J (2010) Basic guide to turf cricket pitch preparation, VCA
Perloff WH, Baron W (1976) Soil mechanics, principles and applications. Wiley, New York
Powai IIT (2008) Mumbai, physical, chemical, free swell index and Atterberg limit analysis reports
Ranjan G, Rao ASR (2002) Basic applied mechanics, 3rd edn. New Age International Private Limited
Raper RL, Jhonson CE, Bailey AC, Burt EC, Block WA (1995) Prediction of soil stresses beneath a rigid wheel. J Terramech 61:57–62
Reddy ES, Sastri, KR (2002) Measurement of engineering properties of soils. New Age Publication
Salini U, Girish MS (2011) Suitability of soil for cricket pitch preparation. College of Engineering Trivandrum, Technical paper at NATCON- 2011, Centre for Engineering Research and Development, Kerala, 28–29
Sharkley K (2002) Geotechnical engineering behaviour of Denver expansive clay shale. MS Thesis. Department of Civil Engineering, Colorado School of Mines
Shimpton P, James L (2009) Optimization of cricket pitch rolling. PhD Thesis, Cranfield University, School of Applied Science and Natural Resource Department
Shipton PMR (2004) The effect of moisture and grass species on turf grass root zone, physical conditions in winter games pitches. MSc thesis
Singh A (2009) Modern geotechnical engineering, 3rd edn. CBS Publication

Sivapullaiah PV, Sitaram, TG, Rao KSS (2004) Simple correlation between soil plasticity and expansion potential. Adv Eng Geol Geotechn 441
Skempton AW (1944) Notes on compressibility of clays. Q J Geol Soc 100(1–4):119–135
Spoor G, Tijink FGN, Weisskopf P (2003) Sub- soil compaction: risk, avoidance, identification and alleviation. Soil Tillage Res 73:175–182
Taiton NM Klug JR Edmondson D, Campbell RK, Deventer PWV, and Beer, MJD (1993) Principles and practice of pitch preparation. United Cricket Board of South Africa, May
TCCB and NCA (1989) Recommendations on the preparation of cricket pitches, Test and County Cricket Board and National Cricket Association, LORDS, London
Terzaghi K (1943) Theoretical soil mechanics. Wiley, New York
Terzaghi K, Peck RB (1967) Soil mechanics in engineering practice, 2nd edn. Wiley, New York
Venkatramaiah C (2006) Geotehnical engineering. New Age International, ISBN: 978–93–86649–70–6
Wood, C (2006) Recommended guidelines for the maintenance of cricket Pitches and outfields at all levels of game. England and Wales Cricket Board in Association with the Institute of Groundsmanship

Chapter 9
Cricket Pitch Behavioural Analysis Index (PBAI) and Pitch Behaviour Forecasting (PBF) Methods: Traditional and Conventional Methods *vs.* Scientific Methods Based on Recent Research and Developments

9.1 Introduction

Curators judge the pitch's behaviour through their senses based on scientific logic and experience. Traditionally, judging the pitch behaviour is done through core sampling (by observing the colour and moisture status of the cricket pitch profile), observing the initiation and propagation of cracks, cracking time, and rate; cracking or crumbling patterns (wear and tear) of the pitch profile; the nature, size, depth, width, and geometry of cracks; the status of the rolled-in grass and its colour; the colour and shine of the finished pitch surface; and ball rebound techniques. These are the most popular, indirect, and yet simple methods to judge pitch status and behaviour in India and elsewhere. These popular and indirect methods are simple but come only through the vast experience of pitching.

Judging and forecasting pitch behaviour based on the Pitch Quality Index, the Database Management System of a pitch profile, the Pitch Behaviour Analysis Index (index properties based on textural, particle size distribution, morphological, structural, physical, chemical, and mineral properties), cracking and crumbling patterns (wear and tear), and by the use of scientific gadgets, such as the Clegg Hammer, moisture metres, are the recent evolving trends. These recently available mechanical devices, which are extensions of our senses, enabled the pitch curators to analyse and interpret the pitch behaviour with speed and accuracy.

9.2 Different Methods of Analysing, Interpreting, and Predicting Pitch Behaviour

"Traditional and conventional methods vs. scientific methods based on recent research and developments"—Cricket pitches express their behaviour through cracks or crumbling (wear and tear) of the playing surface during the course of the game, especially the multi-day matches (Ranji and Test matches). The predominant

S. B. Singh et al., *Cricket Pitches*, Springer Transactions in Civil and Environmental Engineering, https://doi.org/10.1007/978-981-99-2913-9_9

morphological, structural, physical, and chemical (wear and tear) characteristics of the playing surface during the course of the game, especially the multi-day matches (Ranji and Test matches).

The predominant morphological, structural, physical, chemical, and mineralogical cations, along with the prevalent weather variables such as temperature, relative humidity, evapotranspiration, solar radiation, wind speed, the format of the match, etc., have a dominant role in controlling the pitch's behaviour. The smectitic (swelling and shrinking) type of clay soil gives a wide variation in pitch behaviour (through the initiation and propagation of cracks of different shape, size, and geometry during different phases of the day) as compared to the non-swelling kaolinitic type of clay soil. The presence of different exchangeable cations within the smectitic clays also has a dominant influence on controlling the pitch behaviour (Figs. 9.1 and 9.2).

Fig. 9.1 Cracking patterns of some of the pitch soils (WACA, GABBA, VISAG, VISAG-Red, Bangalore, and Pakistan Clay) used for international venues (*Data source* National Cricket Academy, BCCI, Bengaluru)

Fig. 9.2 Cracking patterns of some of the pitch soils (Centurian, Bendigo, Bermuda, Ongar, World-1) used for international venues (*Data source* National Cricket Academy, BCCI, Bengaluru)

9.2.1 Different Extreme Scenarios of Pitch Behaviour

Different extreme scenarios of pitch behaviour can be expected based on the composition of the soil and the extreme type of clay mineral present in it (as stated below).

Scenario 1: A pitch soil with high monovalent exchangeable cations like sodium in dominant smectite clays (as determined by having high Atterberg limits in the form of a liquid limit and plasticity index, a low shrinkage limit, a high cation exchange capacity (CEC), a high coefficient of linear expansion (COLE), a high free swell index/ratio (FSI/FSR), a low bulk density, high and high OMC and high compressibility) with moderate to low silt and sand contents and having high evapotranspiration rates express their behaviour by formation of cracks on the playing surface.

The time of crack initiation, timing, rate, angle, intensity of cracking (crack propagation rate), nature, shape, size, length, breadth, and depth of cracks developed during the course of the game determine the pitch behaviour. The majority of continental pitches from Australia and South Africa fall into this category (they have sodium, calcium, or magnesium-dominant fine smectitic clay with proportionate silt, sand, and humus) (Fig. 9.3).

Fig. 9.3 Large polygonal pitch surface cracks developed on the pitch surface (fourth and fifth day of Test match—India Vs South Africa, Test matches—19 to 23 October 2019, JSCA International Cricket stadium—curated by: Dr. S.B. Singh, Curator, JSCA/BCCI)

Scenario 2: Pitch soils with a high proportion of low or non-swelling and less shrinking clays (such as illite or kaolinite with low compressibility, consistency, liquid limits, plasticity index, CEC, COLE, FSI) and a high proportion of silt and sand with high evapotranspiration rates typically have limited or no pitch surface cracking or crumbling. Due to the low plasticity of these types of soils, the water holding capacity within the pitch profile will be less, and it will also have less volume changes with changes in the moisture of the pitch profile. As a result, the pitch surface of such soils may not crack but crumble due to pitch soil structure loss caused by low water holding capacity and a high proportion of silt or sand (silty clay). Most of the subcontinental pitches (pitches in India, Bangladesh, Pakistan, Sri Lanka, etc.) come under this category.

Fig. 9.4 Crumbling and powdering of kaolinite-dominant mineral fine silty clay pitch surface after the fifth day of Test match

Fig. 9.5 Different methods of testing pitch behaviour (*Data source* National Cricket Academy, BCCI, Bengaluru)

Fig. 9.6 Tony Hemming, ICC curator demonstrating the analysis of the pitch behaviour analysis through ball rebound and core sampling (showing different compression and moisture levels) method (*Data source* National Cricket Academy, BCCI, Bengaluru)

Fig. 9.7 Tony Hemming, ICC curator demonstrating the analysis of cricket pitch surface density level through Clegg Hammer test (*Data source* National Cricket Academy, BCCI)

Clegg Impact Hammer (CIH)—To measure pitch surface hardness and potential ball bounce using a 2.25 kg Clegg Hammer. Named after Dr. Baden Clegg, it was originally developed for engineering purposes. 1 CIV is equal to 10 gravities for readings in CIV units. Surface hardness has been defined as "the ability of a surface to absorb the impact energy created by any object striking that surface" (Rogers 1988). The instrument measures the rate of deceleration of a hammer dropped from a standard height, and the mass of the hammer varies in different models. Clegg Hammer readings have been found to relate well to the rebound height of a ball dropped from a standard height above the surface (Baker et al. 2001a, b), and the 2.25 kg model has been shown to be very useful on cricket pitches (McAuliffe and Hannan 2001). The instrument is designed to be used on flat surfaces, and vertical orientation of the instrument is recommended during measurements. The final prepared match pitch surface is relatively flat and well-suited for measurements with the Clegg Hammer. It is generally recommended that the slope on cricket pitches should not exceed 1% in any direction (McIntyre and McIntyre 2001; Tainton and Klug 2002). However, it is not expected that the pitch should be perfectly level (horizontal), and a small, designed, directional slope is beneficial to facilitate pitch surface water, which can otherwise lead to the development of wet spots. The hammer was dropped from a height of 45 cm (as shown in Fig. 9.7), and peak deceleration values were recorded by the instrument as Clegg readings in units of gravities (g) (as shown in Tables 9.1 and 9.2) following four successive hits of the hammer on the same spot. For all measurements, the instrument was placed squarely on the surface, so that the instrument's orientation to the vertical was determined by the surface slope at each location. The angle between the pitch surface of the falling hammer and that of the cricket pitch surface immediately below was assumed to be zero. Measurements with this instrument were terminated whenever there were obvious surface depressions created by the falling hammer.

Possible errors in measurements with the Clegg Hammer include frictional forces between the hammer and the barrel of the instrument, a build-up of air pressure below the falling hammer, and an inconsistent angle of contact between the hammer and the surface (Carre et al. 2006). There are no defined or approved guidelines (issued by either the BCCI or ICC playing conditions guidelines) regarding the Clegg Hammer readings of the finished match pitch surface. Generally, it is based on the desirable pitch surface target densities attained during various phases of the match and the available CH/EC readings at that point in time.

Mean surface hardness (Clegg) readings have been used to indicate pitches that are slow (below 200 g), medium-fast (200–350 g), or fast (350 g) based on the measurements with the 2.25 kg Clegg instrument (Lopez and Chinnery 2009).

9.2.2 *Conclusions and Interpretation of Clegg Hammer Readings*

The smaller difference between the CH extreme values and moderate EC values shows the moderate uniformity in pitch behaviour related to pitch surface hardness and density levels. The larger the difference between the CH and EC values, the greater the uniformity in pitch behaviour related to pitch density and compaction levels. It is reasonable to conclude that the greater the difference between the CH maximum and minimum, the lower the EC values, and vice versa.

Higher EC values will have uniform density and compaction levels throughout the pitch surface, while lower EC values will have non-uniform density and compaction levels on the pitch surface, resulting in variable bounce and turns.

This equilibrium or uniformity in pitch surface density and compaction levels is generally achieved by uniform wetting (at least 4 inches) and uniform drying of the pitch profile by the judicious use of rollers with their full compactive potential along with the uniform grass and roots developed on the pitch surface. Non-uniform wetting of the pitch surface with non-uniform grass and root density, bare patches on the pitch surface, high thatch (organic matter content), non-judicious and excessive use of heavy rollers, and wrong rolling techniques and schedules lead to a wide difference in CH maximum and CH minimum values with lower EC values.

Table 9.1 Clegg Hammer and equilibrium coefficient readings of sample pitch number1

Pitch #	Table showing the CH and EC values for pitch No-1		
Date:	26.08.2016		
Tested by:	Mr Tony Hemming and Level-1 Team		
NORTH END			
East			**West**
	1	**2**	
1	**192**	**306**	**1**
2	**215**	**243**	**2**
3	**360**	**262**	**3**
4	**330**	**264**	**4**
5	**335**	**253**	**5**
6	**305**	**294**	**6**
7	**277**	**237**	**7**
8	**316**	**244**	**8**
9	**303**	**240**	**9**
	1	**2**	
	AVG. Gravitates	276.44	
CH (Max)	360	CH (Min)	192
EC:	CH (Min) / CH (Max)	0.53	

Note Difference between CH maximum to CH minimum = 360 − 192 = 168. Larger the difference in Between CH maximum and minimum values lesser the EC values

Table 9.2 Clegg Hammer and equilibrium coefficient (EC) readings of sample pitch number 2

Location:	NCA for demonstration purpose only		
Pitch #	2		
Date:	26.08.2016		
Tested by:	Mr Tony Hemming and Level-1 Team		
NORTH END			
East			**West**
	1	**2**	
1	**252**	**253**	**1**
2	**341**	**225**	**2**
3	**290**	**277**	**3**
4	**303**	**316**	**4**
5	**335**	**311**	**5**
6	**390**	**263**	**6**
7	**356**	**294**	**7**
8	**262**	**269**	**8**
9	**276**	**241**	**9**
	1	**2**	
	AVG. Gravitates	291.89	
CH (Max)	390	CH (Min)	225
EC:	CH (Min) / CH (Max)	0.58	

Note Difference between CH maximum to CH minimum = 390 – 225 = 165. Larger the difference in between CH maximum and CH minimum values Lesser the EC values

9.3 Pitch Behaviour Analysis by Interaction with Experts

An Interview with Shri Mahendra Singh Dhoni (MAHI) about His Interpretation and Analysis of Cricket Pitching Behaviour.

In order to understand how the captains comprehend and analyse the cricket pitch behaviour (days before and during the matches) and whether the pitch behaviour can be standardised and quantified (benchmarking or forming a Pitch Quality Index) by examining and analysing its mineralogical, chemical, physical, or morphological compositions, weather variables, and different pitch preparation methods and techniques (Fig. 9.8).

Fig. 9.8 Pitch interview session along with the author

Following are a few interesting questions posed by the first author (Dr. S. B. Singh) during the interview with Shri M. S. Dhoni about his experiences analysing pitching behaviour and the responses given by him.

Question 1**:** As a captain, how do you comprehend and analyse the pitch behaviour by visual examination just a day before the match, before the toss, or during the match? What different parameters do you consider in arriving at those conclusions about pitch behaviour that can assist and alter match decisions?

Answer: It is the integrated effect of all the physical observations of the pitch playing surface (its colour, surface shine, crack shape, size, evenness, and its binding with the surface), climatic and weather conditions (temperature, humidity, and moisture in the atmosphere and pitch profile), altitude, wind speed, and direction, the format of the game, and many other minute variations (such as soil pH, predominant minerals and exchangeable ions, and humus) that are considered to judge and analyse the pitch behaviour.

The pitch should not be allowed to behave solely according to the conditions on and within the profile, but should make intelligent use of the available weather variables (such as temperature, relative humidity, evapotranspiration, solar radiation, wind speed, and so on) and conditions such as pitch soil morphological, physical, chemical, mineralogical, and engineering properties, available moisture, grass vegetative cover and roots, natural weather and climatic conditions, and so on. Either way, a technique based on available conditions and resources, skill, experience, research, and developments based on scientific logic should be such that the pitch profile should not be let free to act by itself, but one should make it behave as desired at any point in time. This can be achieved by integrating and synergizing the cumulative effects of all available pitch profile resources, skill, and experience of the curator to achieve the desired variations in pitch behaviour and offer equal opportunities to fast bowlers, batters, and spinners during different sessions of the match. All of the available resources for the complex pitch profile should be tamed and put to use to achieve the match's goals.

This judgement comes from intuition and also from experience gained by not only playing as many matches as possible on continental and subcontinental pitches but also by observing the behaviour of pitches during the course of the matches. The best team accepts the above-mentioned challenges and variabilities as a part of the game and tries to acclimatise, accommodate, and perform without complaining. The pitch available for play, as well as the surrounding climate and weather conditions, cannot be changed, but they can be wisely and intelligently judged about their expected pitch behaviour and plan match strategies accordingly. For this, a wide experience of playing on all possible continental and subcontinental pitches with good intuition is a must.

Question 2: How do you analyse and judge the pitch's moisture levels, and how is that useful in predicting the pitch's behaviour?

The moisture level of a pitch is generally (traditionally) judged by observing the colour of the pitch and the way it feels when the sides of the pitch are pierced (by hardly up to 6–12 mm) with the spikes of the shoe worn by the cricketer or sportsperson, and observing the pattern formed by it and how the spikes of the shoe come out of the soil. If the shoe spikes pierce down and come out easily, forming a clear hole pattern in the field soil, which indicates that the pitch profile is dry. On the other hand, if the spikes come out with a sticky substance and with a slight strain, then it indicates the pitch is moist. If the spike comes out easily but with a spongy feeling, i.e. some part of the pitch soil comes out along with the spikes, then it indicates that the top portion of the pitch is dry but there is moisture beneath. The pitch surface's dark colour with less shine indicates that there is enough moisture in it and that the pitch surface is compacted and wet at its best. The ash-coloured and dull appearance of the pitch and grass indicates that there is less moisture present and that the pitch surface is compacted dry rather than optimised.

The pitch soil moisture, along with the soil and grass, provides an idea of the various compaction levels during the game. By controlling the moisture levels during the pitch preparation process and during the game, the behaviour of the pitches can be regulated in a desirable pattern.

This desirable pitch behaviour can be achieved by controlling and regulating the pitch soil moisture just above and below the surface, similar to the techniques used in wineries which create an artificial water stress in the berry plant to achieve a particular, alkaline flavour. In some wine industries, to achieve a particular taste and alkalinity in wine, the grape plants are forced to wilt by creating an artificial drought. These plants start sucking water from the grape berries in order to maintain the turgidity and other metabolic activities of the vines, thereby making the fruit berries dehydrated and alkaline in taste (which is desired). This period of induced water stress encourages the vine to concentrate its limited resources into lower yields of smaller berries, creating a favourable skin-to-juice ratio that is often desirable in quality wine production.

A similar analogy can be used within the cricket pitch profile to regulate pitch soil moisture during preparation and during the game. The available capillary water in liquid form and the unavailable hygroscopic water present in vapour form within the micropores and tightly adsorbed on the clay micelle forming the double diffused layer can be regulated and managed in different permutations and combinations (or proportions) to achieve the desired pitch behaviour. By regulating the pitch soil moisture levels, the swelling and shrinking capacities of the pitch profile can be regulated, which facilitates the regulation and management of the pitch.

Question 3: Does the pitch behaviour of cricket pitches prepared with different coloured soils, such as red, brown, yellow (coarse textured), or black (fine textured), differ? If yes, then how?

Answer: Red or brown (coarse textured) pitch soils provide favourable conditions for a wider range of play by giving equal opportunity to pacers and seamers through good bounce; spinners through good carry and grip; and batsmen through good opportunity for stroke play. It cracks, crumbles, and powders early and provides equal opportunities for the spinners to operate. Nonetheless, it tests the batsman's endurance and ability to perform in wide pitch variable behaviour. There is a wide intensity of variations in red soil (coarse textured) pitch behaviour as compared to black soil (fine textured) pitches, which are generally known for their pace, bounce, and stroke playing opportunities. Red or brown soil wickets provide a space for the spinners to breathe and show their art and importance, which are generally dying on fast and bouncy wickets (Fig. 9.9).

Fig. 9.9 Different pitch surface colour and their wear tear during different phases of days matches (five days Test matches or four days Ranji matches)

Question 4: Do you believe that the physical, chemical, or mineralogical report of the match pitch, as well as the weather report certified and published by reputable institutions, are reliable for analysing and comprehending the nature and behaviour of pitches? Also, can the pitch behaviour during the matches be indexed based on such scientific analysis and findings?

Answer: Based on scientific research and developments, no such concept has been developed or any initiatives taken in India to study and analyse its effect on pitch behaviour. These theoretical- and laboratory-tested, scientifically approved methods and techniques should first be practically tried and tested on ground conditions before their approval and final application.

Question "5": How do you comprehend and read the behaviour of the pitch through its cracking patterns?

Answer: The size, depth, and evenness of the pitch cracks indicate the moisture level and the compaction done with the rollers. Subcontinental pitches typically have small, superficial cracks or no cracks. Such pitches breakdown and crumble early and get powdered, favouring the spinners. Such a small but large number of cracks are often judged by applying thumb pressure to the ridges and periphery of the cracks. Small but multiple cracks get displaced on application of pressure, while large plate-like cracks remain firm on application of pressure. Pitches with small (but numerous) cracks or no cracks are generally flat and slow. These kinds of pitches have variable bounce and pace, which aid the batsman in playing shots and strokes. Such pitches, if the edges or the periphery are stiff and do not move or dislocate from their original position by applying thumb pressure, will have a good bounce and carry the ball to the batsman (Fig. 9.10).

Fig. 9.10 A large-sized deep and wide pitch surface cracks developed on the pitch surface at the J.S.C.A. International Stadium, Ranchi and on the cricket pitch surface at Perth, Australia

Large, deep, and wide pitch surface cracks are generally observed in Australian (Perth wickets) and some South African pitches, in which even the bat can penetrate such wide cracks (Fig. 9.10). These wide cracks induce variability and extra bounce and pace in this type of wicket. Even if the number of cracks is few but large in size, bisecting each other (as shown in Plate 9.10b), they are useful to analyse the pitch behaviour. Balls pitched on cracks, behind cracks, or at crack intersections behave differently. The behaviour of the cracks between the intersecting lines should also be monitored. Two or more adjacent cracks should be at a uniform horizontal level so as to achieve uniform bounce and pace.

There should not be any ripple formations on the pitch surface, as these undulations or ridges may produce variable and unpredictable bounce, which is dangerous for the batsman to play.

Question 6: All formats of cricket, whether a Test match, an ODI, or a Twenty20 match, are played on the same type of surface. Should there be different types of pitches for different formats of games?

Answer: No, the Test match pitch is the mother of all available formats of the game of cricket, as the conditions available on either the second or third day of the Test match are what are desirable for the ODI and T-20. Pitches for national or international venues should be multidimensional; either a good Test wicket should be good for ODI and T-20, and vice versa. Generally, second- or third-day pitch conditions for Test match wickets are considered good for ODI and T-20 cricket.

Question: 7: How do you assess the importance and impact of rollers before and during the match?

Answer: The weight and width of the rollers matter, as well as the way they interact with the type of soil surface, grass cover, and moisture available on the pitch surface. In India, the weight of the roller is much more important: either light or heavy rollers. It is desirable to conduct research on rollers and their impact on pitch behaviour under localised and acclimatised conditions. The choice of a light or heavy roller is determined by the current moisture conditions, grass conditions, pitch soil type, ball strength combination, and innings to bat on. If given to bat on the third, fourth, or fifth day of the Test match, I will go for heavy rollers to accentuate the cracks, wear, and tear, or powder the pitch so as to assist the bowlers, especially the spinners. Later in the game, further wear and tear on the pitch will provide a rough surface, friction, and grip for the ball. In the case of pitches with heavy cracks, the use of heavy rollers will evenly settle down the cracks and make batting slightly comfortable.

A uniformly rolled playing surface without any undulations or ridges is a must. Though looking like a grassy and good wicket, one of the international matches played in India was called off due to an ill-prepared, uneven playing surface having undulations and ripples. This match was called off due to an uneven, unpredictable bounce that was extremely dangerous for the batsman to play.

The rolling speed of the roller is also an important factor that should be considered while rolling. I have observed a very low roller speed in Australia while preparing their pitches. The slow rolling speed compacts the pitches much more than the fast one.

A large or small contact angle between the rollers and the pitch surface is also an important factor that should be looked at while rolling to get the best shallow, vertical, or horizontal compaction.

Question 8: How do you visualise the impact of grass on the pitch's behaviour?

Answer: A uniformly distributed grass cover in the turf wicket provides an anchoring effect that is critical in providing consistent pitch pace and bounce. For fast and bouncy pitches, green grass with fresh shoots and no lateral spread of thick stolons is preferred. Brown-coloured, evenly rolled grass is generally considered good for batting.

Question 9: The use of heavy plastic pitch covers overnight produces a greenhouse-type environment on the covered pitch surface that regulates the non-available pitch moisture in vapour form. This phenomenon is generally considered to be thermo-osmosis and directly affects the pitch's behaviour during the course of the Test match. How do you visualise this?

Answer: If the above phenomenon is scientifically proven, traditional heavy plastic pitch covers should be replaced with the Hover Craft (an elevated, dome-shaped, lightweight pitch cover used in some venues) to reduce the heating caused by large, thick, and heavy pitch covers.

Question 10: Drop-in pitches (DIP) were used for the majority of the 2015 World Cup matches. Only Australia and New Zealand have such technology and know-how to produce such pitches. DIP is a transplanted pitch prepared in unapproved laboratory conditions that are unregulated, untested (both materially and process-wise), and more likely to be tempered. In-situ pitches, where a natural equilibrium exists between the pitch soil, grass, and ambience, are missing in DIP. There exists a natural equilibrium between the soil profile and the - and macroclimatic conditions surrounding it, which directly affects the nature and behaviour of the pitches. Don't you think that these types of pitches will create a technological gap among the cricket-playing nations that do not have them or cannot financially or technologically afford them?

Answer: In 2015, when world cup matches were played, no such thought entered the minds of players or the cricket governing bodies of different nations. The process and techniques for making the DIP should be made public and should be certified and verified among the playing nations. Basically, these types of pitches are generally developed by Australia and New Zealand, where the cricket game is competing with other popular parallel sports, such as rugby and football, being played on the same ground. To minimise the wear and tear on the pitches from these games, the concept of DIP was developed. Other reasons for choosing DIP include a lack of stadium space and the inability to fully utilise the stadiums built throughout the year. Since the nature and behaviour of the DIP resemble the Indian subcontinental pitches, it suits our core competencies. They are not able to produce the same wild cracks, pace, and bounce as their counterparts (Perth wickets) under DIP conditions (Fig. 9.11).

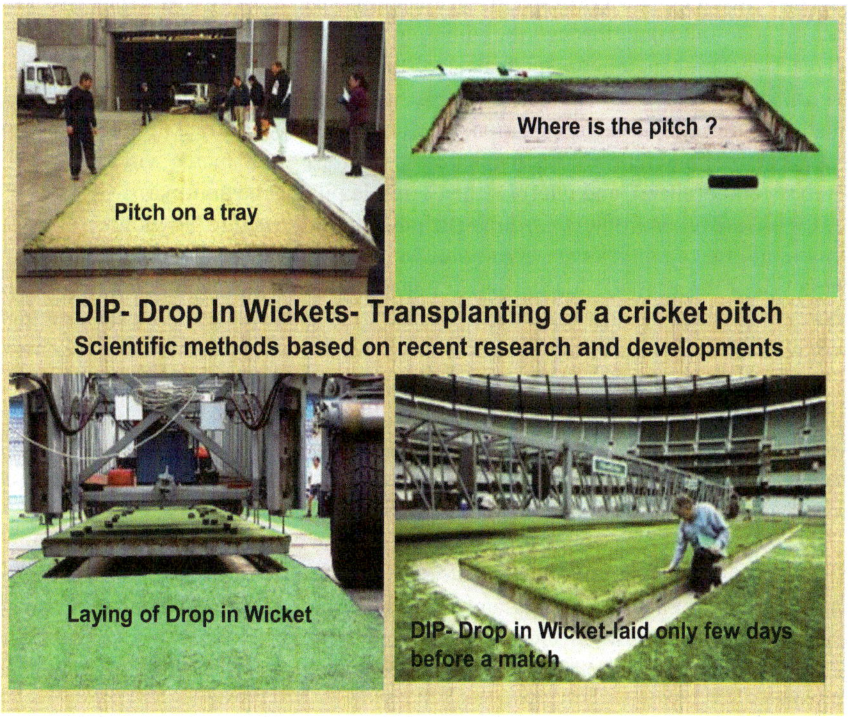

Fig. 9.11 Laying of drop-in pitches in Australia

All the available knowledge based on skill, experience, and scientific innovations on pitch making should be channelled so as to improve and enhance the level of our pitch standards, by analysing, judging, and comprehending their character and behaviour, keeping in mind our natural legacy and core competencies.

9.4 Cricket Pitch Behaviour Analysis Through the Measurement of Cracking or Crumbling Patterns

Cricket pitch behaviour analysis through the measurement of cracking or crumbling patterns (wear and tear) of the pitch surface during multi-day (four-day Ranji matches, five-day Test matches), ODI, or T-20 matches.

Every cricket pitch (based on existing climatic and weather variables such as temperature, relative humidity, solar radiation, evapotranspiration, and wind speed available within and just above the pitch profile), their construction methods, pitch soil structure, morphology, physical, chemical, and predominant minerals, express their behaviour through the cracking, non-cracking, or crumbling patterns of the pitch surface. Based on the above-mentioned predominant factors, every pitch surface follows a definite path through the initiation and propagation of cracks of different shape, size, depth, geometry, intensity, or crumbling of the pitch surface, etc., during the different sessions of multi-day matches (Ranji or Test matches). Mapping and analysing the initiation, propagation, rate, and intensity of cracking or crumbling patterns (wear and tear) during the different sessions of the match will enable us to analyse, comprehend, and interpret the desirable pitch behaviour narrowly.

Cracks are fissures other than those attributed to soil structure (also known as extra-structural cracks). Cracks are commonly vertical, subplanar, or polygonal and are the result of desiccation, dewatering, or consolidation of earthy material (Figs. 9.12 to 9.23). Cracks are most common in clayey soils and are more noticeable in high shrink-swell soils. Cracks can be crust-related cracks or trans-horizon cracks. Crust-related cracks are shallow, vertical cracks related to the crust that may be reversible or irreversible in nature. Reversible crust-related cracks are very shallow (0.5–0.7 cm), very transient (generally persisting for less than a few weeks), and formed by drying from the surface down. Irreversible crust-related cracks are shallow (0.5–2.0 cm) and seasonal (not present all year or every year).

Trans-horizon cracks are deep vertical cracks that commonly extend across more than one horizon and may extend to the surface; they are derived from wetting and drying or the original dewatering and consolidation of parent material. The primary cause of transient (seasonal; closes when rewetted) reversible trans-horizon cracks is soil wetting and drying. Irreversible trans-horizon cracks are permanent (they exist all year).

Desiccation cracking or crumbling has been quite a challenging puzzle for the cricket fraternity. The presence of cracking in clay-dominant cricket pitch soils can significantly alter its mechanical, hydrological, physio-chemical, and thermal properties, which are commonly desirable on the cricket pitch surface during the different sessions of the match, providing variations in the pitch's behaviour.

9.4.1 Benefits of Cracking and Crumbling (Wear and Tear) of the Pitch Surface

Desiccation cracking can have a negative impact on cohesive soils by reducing the strength, stiffness, and stability of the overall soil mass while increasing compressibility and soil hydraulic properties (Chertkov 2000). Cracking of soil mass also causes changes in its engineering properties (for example, shear strength, bearing capacity, overall stability, and so on) and affects various mechanisms such as the transport of gases and waste containment in the landfill, the loss of moisture and nutrients from the root zone, and the formation of zones for water infiltration into the top covers of waste storage (Kelpe and Olson 1985; Miller et al. 1998; Benson et al. 1999; Puppal 1999). The cracks of different shape, size, depth, rate, geometry, etc., initiated and propagated during different sessions of multi-day matches provide the desirable variations in pitch behaviour.

Cracks create zones of weakness in a soil mass, reducing overall strength and stability while also increasing the compressibility of the swelling and shrinking-dominant pitch surface (due to moisture and air migration generating air voids, resulting in high compressibility).

Cracks also provide pathways for moisture movements (by the formation of air entry and exit points within the pitch profile, leading to better aeration of the turf grass) and help in the diffusion of entrapped poisonous gases in the root zones (carbon dioxide, carbon monoxide, etc.) within the over-compacted pitch surface under anaerobic conditions.

9.4.2 Mechanism of Shrinkage and Crack Formation

It is a generally occurring phenomenon that shrinkage often leads to cracking in clay soils as a result of consistent moisture loss. Cracking usually occurs when soils are restrained while undergoing shrinkage, which generates high negative pore pressure, otherwise known as soil suction, within the drying mass (Mitchell 1993; Holtz and Kovas 1981).

Expansive soil exposed to a slight change in moisture content can lead to a volumetric change in the crack surface area. These soils expand during precipitation by absorbing water into their structure and also shrink during desiccation; during a period of dryness, moisture is also lost via evapotranspiration (Rao et al. 2004). The volumetric changes are determined by factors like the type of clay mineral present, the compaction of soil, and the depositional environment, which is responsible for the arrangement of particles, the overburden pressure, and the expected weathering (Mishra et al. 2008).

In general, it is believed that shrinkage is due to water loss by evaporation, and if drying continues from the surface down, the dried-out upper layer shrinks first and tensile stress develops, leading to a negative pore water pressure and resulting in cracking meaning that shrinkage leads to cracking (Kodikara et al. 2002).

The primary requirement for the formation of desiccation cracks is restrained shrinkage leading to tensile stress development exceeding the tensile strength. If the drying soil body can be considered free from any restraints, the full potential shrinkage free from cracks would develop, which would be the function of the drying environment and soil properties. The restraints against free shrinkage can be internal, external, or a combination of the two. Internal constraints are usually due to soil structure, such as differing soil fabric units and moisture gradients. Displacement boundary conditions and interface friction can be regarded as external constraints.

Crack initiation is mainly controlled by two factors, first is the distribution of tensile strain due to the restraint of shrinkage strain within the soil mass. Studies have shown that cracks will initiate at the centre of the soil drying mass, where the tensile stresses are concentrated most (Nahlawi and Kodikara 2006). The second factor is related to the initiation of cracks as a result of flaws or imperfections on the soil surface; the initiation of cracks is affected by the tensile stress at these locations. Flaws can be due to microcracks, air bubbles, large particles, and surface texture, as stated by Nahlawi and Kodikara (2006).

The initiation and propagation of cracks, as observed by Tang et al. (2010), are due to the effect of capillary suction and rearrangement of constituent particles. As this suction advances, units of soil at the surface are exposed to tensile stresses, which continue as evaporation proceeds. It gets to a point where the tensile stress exceeds the tensile strength of the soil, at which point cracking will occur (Figs. 9.12, 9.13, 9.14, and 9.16).

Fig. 9.12 Cracking patterns (initiation and propagation) of fine red clay soil (being used at the JSCA cricket stadium complex, Ranchi, Jharkhand) under artificial light, 30-W lamp, the distance of 15 cm and 48 h of shrinking time

Fig. 9.13 Cracking patterns (initiation and propagation) of fine swelling and shrinking-dominant clay soil (being used at the JSCA cricket stadium complex, Ranchi, Jharkhand) under artificial light (30 W lamp, distance of 15 cm and 48 h of shrinking time)

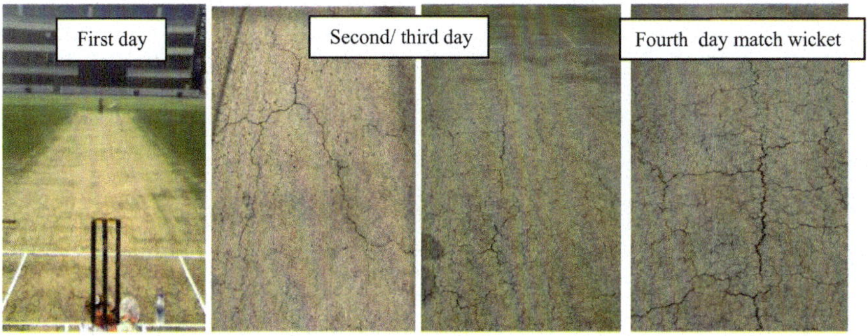

Fig. 9.14 Cracking patterns (initiation and propagation during the second, third, and fourth day of the match under available weather and climatic variations (temperature, relative humidity, evapotranspiration, solar radiation, dew, etc.) of fine swelling and shrinking smectite-dominant clay soil during the Ranji match—23–26 November 2015 at the JSCA cricket stadium complex, Ranchi, Jharkhand

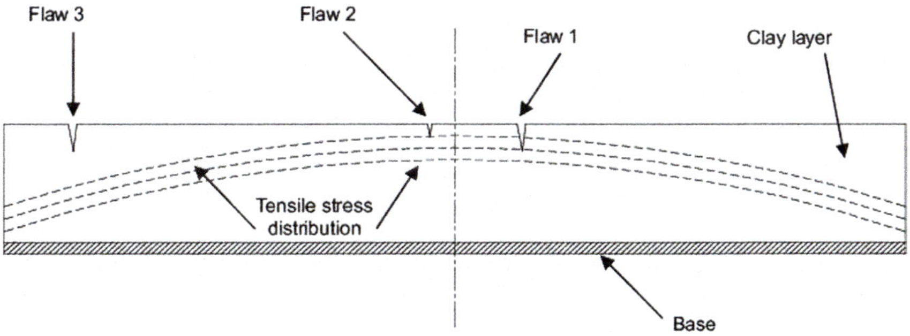

Fig. 9.15 Effect of flaws on crack initiation using hoop stress in circular test (Costa et al. 2012). According to fracture mechanics, the tensile stress that is needed to activate a flaw is inversely proportional to the square root of the flaw size. Hence, it is possible for cracks to occur at locations away from the points where the tensile stress is maximum, if flaws at those locations are sufficiently large to get activated by the prevailing tensile stress

Fig. 9.16 Uniform, semi-uniform (**A** and **B**—fine smectite Bolangir soil at the JSCA International Stadium, Ranchi, Jharkhand and non-uniform turf grass cover with patches (**C**—Fine red soil pitch at Polo Ground, Shilong, Meghalaya) having different WHC and loss of pitch surface moisture due to evapotranspiration leading to variations in pitch surface compression and density levels

Uneven watering of pitch surfaces, large-sized granules, or pitch soil with non-uniform PSD (a low proportion of swelling and shrinking clay with a high proportion of undesirable silt and sand), the high proportion of silt and sand, uneven pitch surfaces with different WHC, and shallow or too deep pitch soil depths with different water movement, holding, and retention capacities.

George (1968) concluded that as a result of shrinkage, tension stresses would be set up in the soil. Shrinkage stress reaches its maximum value in the early stages of drying when it is highly localised at the exposed surface and decreases rapidly with depth. This localised stress can be relieved only by surface cracking or the plastic flow of materials. Thus, the mechanism of desiccation cracking may be explained as the failure of the material in tension when the tensile stresses developed in the material due to shrinkage exceed the tensile strength of the soil. According to G. D. Towner, the tensile strength and the tensile stress increase with decreasing water content (or increasing suction), at least when the clay is saturated.

Shrinkage through drying is an important natural process contributing to the disintegration of clay aggregates. Sand-sized particles distort the shrinkage of the surrounding clay matrix, causing the latter to shrink anisotropically until cracking ensues.

Hvorslev (as referred to by Lambe 1958) found that the pattern of cracking in soil upon drying tended to be perpendicular to the direction of applied force. It has been concluded that cracking is parallel to particle orientation.

The shrink-and-swell nature of clay is self-ameliorating in two ways. As the soil dries, cracks form in the soil. The soil fractures along lines of weakness. These cracks increase the surface area of the soil, inducing more drying and further cracking. Secondary cracks can form normal to the primary cracks, and tertiary cracks can form normal to the secondary cracks (Dexter 1991).

The size and spacing of the cracks are determined by the rate of drying; fast drying produces smaller cracks with reduced spacing, and slower drying produces more widely spaced, border cracks (as shown in Fig. 9.18). These cracks form potential channels for root growth, water infiltration, and gas diffusion through the soil (Dexter 1991).

The abundance of smectite present in soil results in a high coefficient of linear expansion (COLE), coefficient of linear contractibility (COLC), and volumetric shrinkage potential (VSP). In the upper horizons, low overburden pressure and cracks prevent the development of high lateral stresses. However, in subsoils, the difference between the lateral and vertical stresses is very large, and sphenoids and or slickensides are developed (Yaahen and Kalmar 1978; Knight 1980).

Fang (1994) reported that when water is lost from the surface mass, tensile forces are established in the drying surface layer. Due to the water loss, soil loses its ability to relieve tensile forces. These stresses are finally relieved by the formation of shrinkage cracks. As particles move closer, the surface layer is broken up into pieces of more or less distinct geometric shapes. This geometric shape of the cracks depends on the clay mineral composition, the heating process (either loss of moisture through evapotranspiration), and the pore fluids. MI (1995) showed that crack geometry was a function of the duration of desiccation and corresponding soil moisture suction. Yong and Warkentin (1975) stated that the cracking cycle may attract water and become a preferential zone of cracking. Cracking will occur at the location with the lowest cohesion, which corresponds to the wettest location in the soil. These broken bands may attract water and become a preferential zone of cracking during the first drying cycle (Fig. 9.17).

Fig. 9.17 Ped structure as retained within the pitch profile during pitch construction

Turf grass in cricket pitches increases the combined tensile and shear strength of the cricket pitch profile, controlling and regulating pitch surface cracking or crumbling. The cracking of the pitch surface is a very important feature in controlling pitch behaviour during a cricket match, especially a Test match. The shape, size, depth, number, initiation, propagation, and wear and tear of cracks on the pitch surface during the various sessions of the Test matches define the desirable pitch behaviour.

9.4.3 Classification of Cracks in the Cricket Pitch Soil

Gray (1989) classified the cracks in soils into two groups according to their mechanism of occurrence as follows:

1. Mechanically induced cracks (cracks induced through the drying and desiccation of turf pitches), and
2. Physicochemical cracks (cracks induced due to the interaction of clay minerals, organic matter, water, and air present in soil).

These types of cracks may be subdivided into three groups:

1. Syneresis cracks: Syneresis cracks are induced by changes in interparticle forces resulting from the replacement of interstitial water with a low-dielectric organic solvent or a highly saline aqueous solution (Brown and Anderson 1983).
2. Cracks induced by freeze–thaw cycles: Cracks formed as a result of freeze–thaw cycles result in crack formation and have the net result of increasing the hydraulic conductivity of the soil.
3. Cracks induced by desiccation: desiccation cracks are induced by the evapotranspiration of water and the consequent shrinkage. Turf cricket pitch soils having calcium, magnesium, or sodium-dominant smectite clays with high swelling and shrinking capacities generally induce cracks due to the phenomena of syneresis and desiccative cracks.

Fig. 9.18 Large polygonal desiccative cracks on the turf surface due to evapotranspiration. Homogeneous crack leads to heterogeneous crack after initial wetting and drying cycles generating more desirable wear and tear on the pitch surface

It is generally observed that the initial homogeneous system was broken into further heterogeneous systems after the first drying cycle; therefore, the second drying cycle treated each of the islands as a separate homogeneous system, which now cracks and further subdivides depending upon the flaws at the edges of the various boundaries that have now been created (as shown in Fig. 9.18). This continues in cycles until it reaches the smallest possible level where it can no longer be subdivided; i.e. where further drying will only result in evaporation and not cracking. At that point, the system is stable because it has become so weak that no more tensile strength can be developed, which will lead to further cracking.

Soil macrocracks due to pore water suction are more readily produced in fine-grained soils than in coarse-grained soils. This is because fine-grained soils have a smaller particle size, a very high specific surface area, and hence smaller intergranular voids. The smaller voids support large-pore water suction. These wide cracks are generally developed in pitch soils having a dominant calcium, magnesium, or sodium smectite clay with a low proportion of silt and sand. The conditions for crack propagation are more favourable at the ground surface, where pore water suctions are the largest and self-weight stresses are zero. The depth to which the crack extends is ultimately constrained by the increasing stresses due to the self-weight of soil, and the planar length of any crack is limited by its intersection with other cracks.

According to Fag and Daniels (2006), cracks can be classified into four major categories: shrinkage, tensile, thermal, and fracture. Shrinkage cracks are the most common, and they are formed due to moisture loss at the soil surface. During this process, cracking occurs when the tensile stress generated is greater than the tensile strength of the soil.

Thermal cracks take place when the soil mass experiences a rapid change in the thermal stress state of the soil. This can only happen in areas that experience high temperatures during the day and low temperatures at night, which cause a freeze–thaw effect on the soil mass.

Fracture cracks are a result of preexisting cracks within the soil that formed due to deviations in precipitation, evaporation rate, and the water table, which directly had an impact on the pore pressure.

9.4.4 Measurement of Cracks

Crack dimensions are usually measured using approximate methods. The initiation and propagation of shrinkage cracks cannot be categorised as pure orthogonal or pure non-orthogonal (e.g. hexagonal) patterns. Generally, non-orthogonal patterns seem to occur when the "effective" thickness of the drying soil is small, which leads to a high concentration of strain energy in comparison with the fracture energy required for crack propagation (Corte and Higashi 1960; Costa et al. 2012). Fractures tend to occur predominantly simultaneously under these conditions, with the best crack configuration to dissipate energy. The final state of the crack pattern can be a mixture of orthogonal, non-orthogonal, simultaneous, and sequential cracks. However, in clay cracking, sequential, orthogonal cracks dominate the final pattern (Costa et al. 2012).

The irregular shape and complex geometry of cracks prevent accurate measurement of length, width, and depth. The width and depth of a crack are not uniform along its length. Maximum length, width, and depth are commonly recorded using measurements taken with rulers and gauge wires.

Kleppe and Oslon (1985) developed a scale ranging from 0 to 4 to describe the severity of cracking. A crack severity of 0 indicates no cracking, whereas a crack severity of 4 describes cracks with greater than 20 mm and significant depths.

Al Wahab and El-Kedrah, Mi (1995) and Miller et al. (1998) developed a cracking index to quantify the amount and severity of cracking.

Cracking Index (CI) = $\left(\sum \text{B.L}\right)/\text{A}$,

where

N = number of cracks deeper than 2 mm on the sample surface area,

B = width of the cracks,

L = length of the cracks,

A = total surface area of the sample.

Fig. 9.19 Cracks of different shapes and sizes (single interconnected and several small cracks) developed on the pitch surface (fourth and fifth day of Test match—India Vs South Africa, Test matches—19–23 October 2019, JSCA International Cricket stadium)—curated by: Dr. S.B. Singh, Curator, JSCA/BCCI

Fig. 9.20 Depth of cracks developed on the pitch surface (fourth and fifth day of Test match—India Vs South Africa, Test matches—19–23 October 2019, JSCA International Cricket stadium)—curated by: Dr. S.B. Singh, Curator, JSCA/ BCCI

Fig. 9.21 Widening, propagation and wear and tear of the pitch surface on the fifth day of Test match India Vs South Africa, Test matches—19–23 October 2019, JSCA International Cricket stadium—curated by: Dr. S.B. Singh, Curator, JSCA/ BCCI

Based on the above definition of CI, the width and length of 2 mm-deep or more cracks were measured at the end of each shrink cycle. This method does take the length and width of cracks into account for its determination. This potentially leads to overlooking the effects of the various irregular shapes and sizes of cracks in the calculation of the CI (Fig. 9.19).

Another method to analyse the cracking patterns of the soil surface is through the determination of the crack intensity factor (CIF), which is the percentage of cracked area to the total surface area of the sample. CIF is the most useful and effective parameter for the quantification of desiccation cracks ("Application of Laser Microscopy for Studying Crack Characteristics of Fine-Grained Soils," Singh, D.N., and Uday, K.V. 2013). The cracks included in the CIF analysis are usually larger than 1 mm. Nowadays, digital image processing is being also considered for calculation of CIF which was introduced as a descriptor of the extent of surface cracking. A computer-aided image analysis programme is used to determine the CIF values. The cracking patterns are determined using photographs of a drying soil. In a photograph of a drying soil, cracks appear darker than the remaining uncracked soil surface in photograph of a drying soil. The contrast between the colours of the cracked and uncracked soils is used to calculate the CIF values. Scanned photographs of soil surfaces are analysed using MATLAB to determine the CIF values.

More research is required to determine the crack width, length, shape, rate, and timing of crack formation and propagation on cricket pitches during different phases (sessions) of day matches (four-day Ranji or five-day Test matches). While it is easy to develop pictures of the soil crack network, examination of the crack network is a non-trivial matter and highly software-driven. Algorithms exist that can count the length and width (as shown in Figs. 9.20, 9.21 and 9.22) of the cracks, but logically defining a single crack is difficult; as the cracking process continues, eventually there is generally just one highly connected (with itself) crack and a few small individual cracks (as shown in Fig. 9.20). A more informative view would have been gained had the total crack length over time been determined and examined in relation to the extent of shrinkage, as is suggested for future work.

9.4.5 Wetting and Drying Cycles of Cricket Pitches

Tang et al. (2011) worked on a laboratory experiment to investigate the effects of wetting and drying on the initiation of cracks on a clay layer from a slurry state. For this laboratory experiment, four duplicate slurry samples were prepared and subjected to five wetting and drying cycles. The rate of evaporation, the initiation of cracks, and the changes in the structure of the four samples were monitored, and the crack patterns that formed were analysed using image processing.

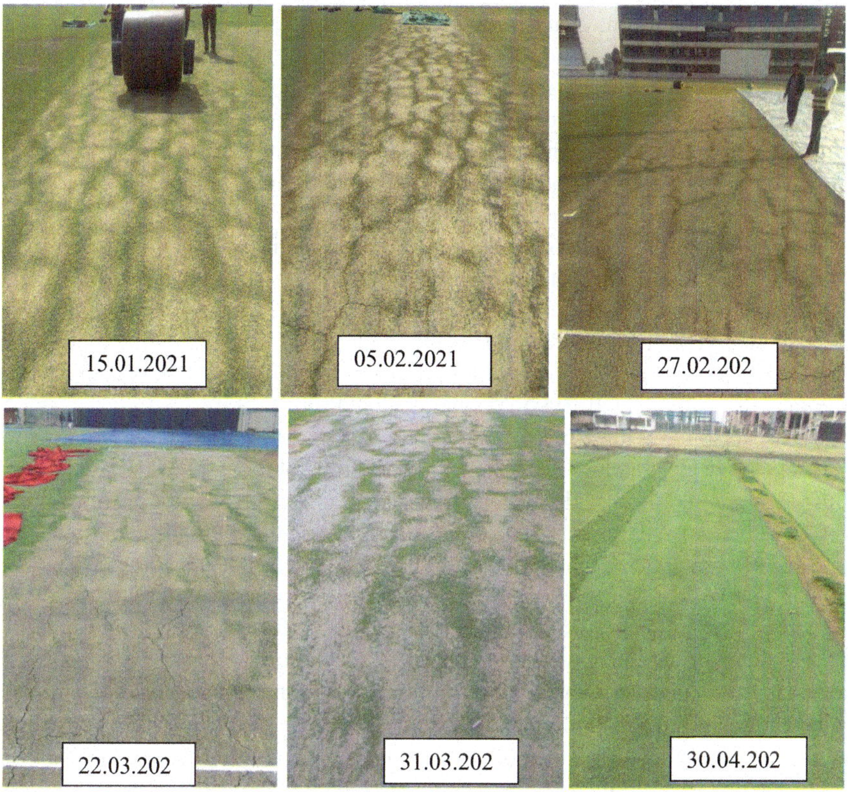

Fig. 9.22 Cracking patterns and geometry of Bolangir pitch soil (at the JSCA International stadium complex) during different wetting and drying cycles (different match pitch preparation on 15.01.2021, 05.02.2021and 27.02.2021) and regeneration of new grass turf on the boundaries of developed cracks (on 22.03.2021)

It was observed and recorded that the measured cracking moisture content, the ratio of crack surface, and the thickness of all four samples increased after the first three cycles and then approached equilibrium. The observed cracks formed after the second wetting and drying cycle were more uneven compared to the first cycle.

The relative increase in the surface crack caused the pore volume to shrink during desiccation. In addition to the results, it was observed that the wetting and drying cycles led to the rearrangement of the sample's structure, as the original structure was altered to a clear and aggregated structure with noticeable interaggregate pores.

The crack dimensions changed slightly with each subsequent cycle until the third cycle. This agrees with studies conducted by Tang et al. (2011), who indicated that an equilibrium state can be achieved after certain cycles, which in this case is the third cycle. After the third cycle, it seems that there is a rearrangement of the sample structure into more stable configurations. The subsequent fourth cycles experienced little or no significant variation in the crack formation, and the major cracks reopened again with similar dimensions and geometry. This is assumed to occur as a result of the rearrangement of the sample structure diminishing and approaching a more stable structure (Kodikara et al. 2002; Tang et al. 2011). Equilibrium states tend to occur during multiple wetting and drying cycles, while this largely depends on the percentage of clay minerals in the soil. High plasticity soils, on the other hand, required more cycles to achieve equilibrium than sandy soils (Al Wahab and El-Kedrah 1995; Omidi et al. 1996; Yesiller et al. 2000).

A cricket pitch with a life expectancy of 10–15 years goes through numerous wetting and drying cycles. Every match (pitch preparation and match day) has one full wetting and drying cycle, which varies depending on the predominant quantitative and qualitative nature of pitch soil, the existing weather/climatic conditions, and the desired format of the match (T/20, ODI, or day matches). Depending on weather/climatic variations and pitch profile conditions (soil structure, morphology, physical, chemical, mineralogical conditions, etc.) shorter format of the matches (T-20, ODI) generally have narrow wetting and drying cycles as compared to the multi-day matches (Ranji and Test matches), where wide variations in pitch surface behaviour are desirable. Once the soil cracks, the cohesive force within the soil is destroyed and cannot be rebound even if the crack closes; this leads to an increase in pores, which in turn increase the permeability of the soil (Anderson et al. 1982).

This phenomenon is generally observed in a cricket pitch surface (Fig. 9.23) having predominant swelling and shrinking clay in which the wide cracks (of different shape, size, depth, dimensions, etc.) during the pitch preparation and the matches are not evenly closed even after watering or saturating the pitch surface (swelling). Fresh grass regenerates on the boundaries or edges of the closed cracks first due to better aeration and high moisture contents during later stages of the match (generally the 3rd, 4th, or 5th day of the match), generating the desirable variations in pitch behaviour. The ball pitching in between the crack plates generates high pace and bounce (due to the high pitch surface dry density), and a wide deviation is observed when the ball is pitched in between the cracks. Crack shape, size, depth, structure, boundary strength, etc., generated during the different phases of the match, provide the desirable variations in pitch soils dominated by fine swelling and shrinking clay.

Fig. 9.23 Same cracking patterns appearing after repeated wetting and drying cycles on days match pitches (Ranji match) for smectite-dominant fine clay of Bolangir soil. cracking patterns observed at the days matches during Ranji match at the JSCA International Stadium, Ranchi, Jharkhand and 2—Final cracking pattern achieved during the Ranji match (Baroda Vs Bengal, Dated 20.02.2022, from 20 February 2022) at Barabati, Cuttack, Odisha)

During the absorption of water, the mechanism of closure was observed as a result of the breaking off and falling in of weak particles of soil as they lost their cohesion at the edge of the cracks. The soil particles shed into the gap, leading to its gradual closure as the depth of the cracks reduces. This is generally observed during the later stages of a match in the pitch soils having a high proportion of clay with silt or sand proportions (silty clay), in which the edges of the cracks wither away or crumble due to medium to low WHC and a high proportion of undesirable silt assisting the spinners.

9.5 Case Study

The goal of this case study is to understand the basic scientific and logical principles underlying the art of match pitch preparation, measuring and analysing pitch surface behaviour through measurement of cracking or crumbling (wear and tear) of the pitch surface during one complete wetting and drying cycle (for days matches—Test or Ranji matches) and how to develop, analyse, and interpret the Database Management System (DBMS), Pitch Quality Index (PQI) and pitch behaviour analysis index (PBAI) of a cricket pitch profile.

Pitch Behaviour Analysis Index (PBAI)

Based on Mineralogical, Physical, and Chemical, Engineering properties, PQI/PQS and DBMS of a pitch profile—For India Vs South Africa, Test matches—19–23 October 2019, JSCA International Cricket stadium Curated by: Dr. S.B. Singh, Curator, JSCA/BCCI (Table 9.3).

Pitch profile:	Three-layered wicket of 12 years of standing
Turf grass	Bermuda, Selection 1 variety, uniformly dense with nearly 7–8 mm height

Match pitches objective—2/5 wicket, result oriented, providing equal opportunity to all classes of play.

Table 9.3 Soil properties of JSCA Int. Stadium (main pitch square), Ranchi, Jharkhand

1	Mineral type	Smectite	72.8
		Illite	12.00
		Kaolinite	15.2
2	Exchangeable	Calcium	22.92
	Cations	Magnesium	4.8
	Cmol (p^+) Kg^{-1}	Sodium	1.09
		Soil p H	6.09
		FSI—Free Swell Index	20
3	Atterberg	LL—Liquid limits	71
	Limits	PL—Plastic limits	29
		PI—Plasticity Index	42
4	Particle size	Soil Texture	Silty clay (CH)
	Distribution (%)	Total clay	59
		Silt	34
		Total sand	7

Expected Pitch Behaviour: Calcium-dominant smectitic silty fine clays with moderate sand but a high level of silica, a medium level of cracks, and the turf grass will generate medium pace and bounce, but the surface cracks will crumble during the later stages of the game, favouring the spinners (due to very high levels of little or non-swelling illitic or kaolinitic clays and high silica contents).

Weather forecast during pitch preparations and match days: Due to the retreating monsoons, frequent thunderstorms and rain are expected during the pitch preparation period. The weather during the match is expected to be normal.

Preparation schedules: Prematch wicket preparation and match wicket preparation prematch wicket preparation is adopted to maintain the base compaction of the pitch profile with 7–8 mm of grass and rolling with the heaviest rollers intermittently. The purpose is to attain the density of the pitch profile by retaining much of the grass density and drying the bottom of the pitch profile through evapotranspiration.

Match pitches with nearly 5–8 days of pitch preparation time (depending upon the weather forecast) with optimum rolling patterns and minimum use of heavy rolling are planned so as to have a balance between the grass cover, moisture levels, and the density of the pitch profile.

1	Pitch commenced on	23rd October
2	Events accruing during preparation	Practice was held on the square
3	Total number of days preparation	17 days
4	Total hours of rolling	4 h and 45 min
5	Initial grass coverage	100% light coverage
6	Rollers Weights to be used during pitch preparations	250, 500, 750, 2000 kg rollers

Prematch wicket preparation, Day 1, 02 Oct 2019.
Climatic Conditions—Cloudy.

Temperature:	Max: 30.0 C. Min: 25.0 C. Ave: 22.0 C
Humidity:	Ave: 65% Evapotranspiration: 3.7 mm
Irrigation:	Nil Rainfall: Nil

Pitch Daily Maintenance.

Dew Removal:	No
Mowing Height:	7 mm double cut
Irrigation:	Not required as nearly 4 inches of pitch profile was uniformly saturated by last one week of rainfall. Covered the whole pitch square with light breathable covers overnight

Pitch Rolling. 10 AM to 11.30 AM)—Straight and cross roll (2 passes each) with 500 kg roller to smoothen and firm up the surface followed by sprinkling of 4 cans of water and covered with hessian (Fig. 9.24).

Fig. 9.24 Different prematch wicket preparations, Day 1, 02 Oct 2019

Day 2, 03 Oct 2019

Climatic Conditions—Cloudy, showers in the afternoon (2.30 PM)

Temperature:	Max: 30.0 C. Min: 25.0 C. Ave: 22.0 C
Humidity:	Ave: 60%
Evapotranspiration:	3.7 mm Irrigation: Nil Rainfall: Nil

Pitch Daily Maintenance

Pitch Rolling. (10 AM to 10.45 AM)—Rolling with heavy roller (5 passes, 30 min) to attain density at the bottom of the pitch profile and attain maximum drying of the pitch surface through evapotranspiration followed by sprinkling of 4 cans of water and covered with hessian (Fig. 9.25).

Fig. 9.25 Different prematch wicket preparations, Day 2, 03 Oct 2019

Day 3, 04 Oct 2019

Weather forecast—Cloudy and T storm in spots

Temperature:	Max: 29.0 C. Min: 21.0 C
Humidity:	Ave: 64%
Evapotranspiration:	3.7 mm, Irrigation: Nil Rainfall probability: 44- 55%

Pitch Daily Maintenance

Pitch Rolling. (10 AM to 10.30 AM)—Rolling with heavy roller (3 passes, 20 min) to attain density at the bottom of the pitch profile and attain maximum drying of the pitch surface through evapotranspiration followed by sprinkling of nearly 8 cans of water and covered with hessian (Fig. 9.26).

Fig. 9.26 Different prematch wicket preparations, Day 3, 04 Oct 2019

Day 4, 05 Oct 2019, Saturday

Weather forecast—An afternoon T storm or two

Temperature:	Max: 29.0 C. Min: 21.0 C.
Humidity:	Ave: 65%
Evapotranspiration:	3.7 mm, Rainfall probability: 63- 68%

Pitch Daily Maintenance

Pitch Rolling. (10 AM to 10.30 AM)—Rolling holiday, pitch mowed at 7 mm (with GR1000) so as to attain maximum drying of the pitch surface through evapotranspiration followed by sprinkling of nearly 4 cans of water and covered with hessian (Fig. 9.27).

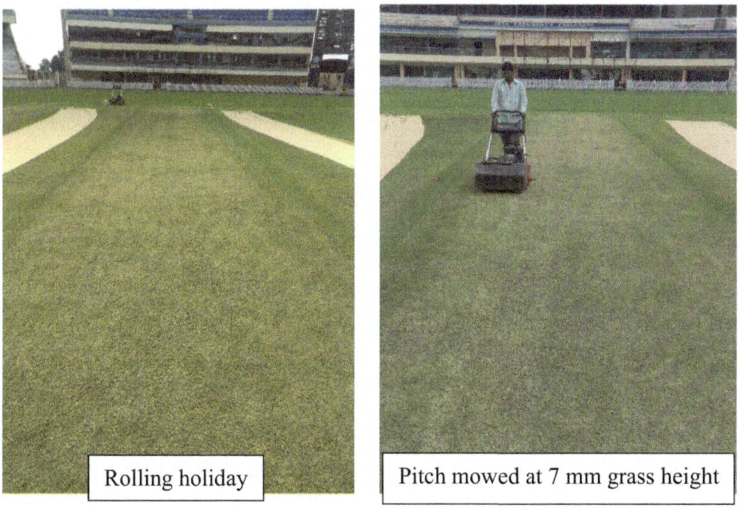

Fig. 9.27 Different prematch wicket preparations, Day 4, 05 Oct 2019

Day 5, 06 Oct 2019, Sunday

Weather forecast—An afternoon T storm at noon (rained heavily for 15 min—3.30 PM)

Temperature:	Max: 28.0 C. Min: 21.0 C
Humidity:	Ave: 66%
Evapotranspiration:	3.7 mm, Rainfall probability: 73- 41%

Pitch Daily Maintenance—Rolling with heavy roller (3 passes, 20 min) to attain density at the bottom of the pitch profile and attain maximum drying of the pitch surface through evapotranspiration followed by light irrigation the pitch with hose pipe. Considering the T storm, the whole pitch square was covered with light breathable covers (Fig. 9.28).

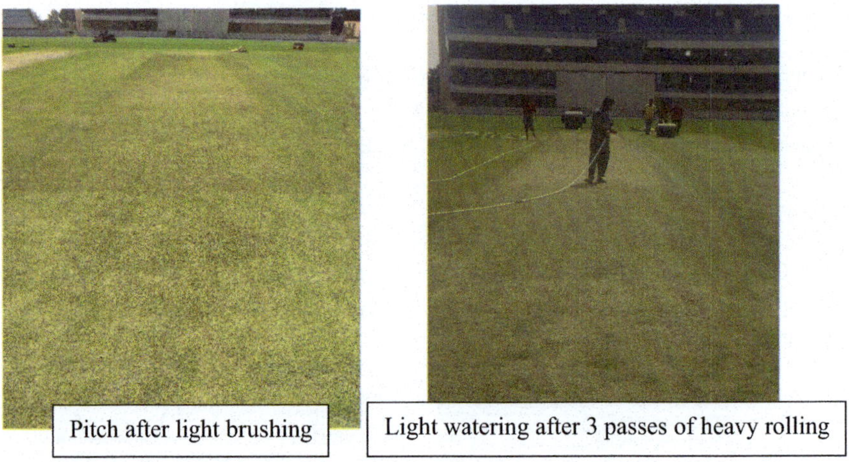

Fig. 9.28 Different prematch wicket preparations, Day 5, 06 Oct 2019

Day 6, 07 Oct 2019, Sunday

Weather forecast—Cloudy with a stray afternoon T storm

Temperature:	Max: 29.0 C. Min: 21.0 C
Humidity:	Ave: 62%
Evapotranspiration	0.7 mm, Rainfall probability: 41- 40%

Pitch Daily Maintenance—Rolling holiday. Frequent light irrigation the pitch with hose pipe morning and evening so to saturate the pitch profile up to 2–3 inches. Considering the T storm, the whole pitch square was covered with light breathable covers in evening (Fig. 9.29).

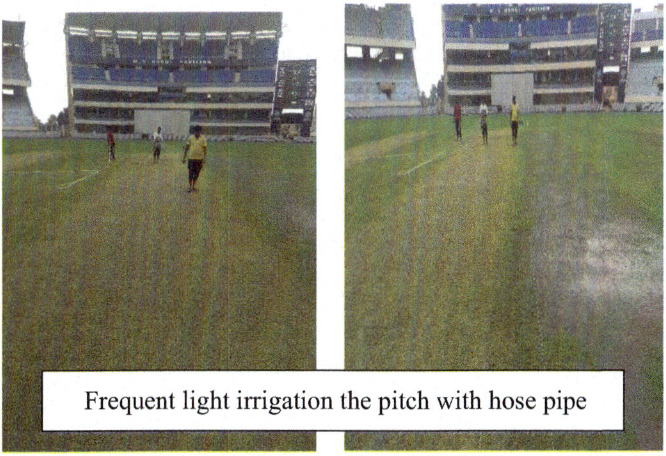

Fig. 9.29 Different prematch wicket preparations, Day 6, 07 Oct 2019

Day 7, 08 Oct 2019, Sunday
Weather forecast—Cloudy with a stray afternoon T storm

Temperature:	Max: 30.0 C. Min: 21.0 C
Humidity:	Ave: 70%
Evapotranspiration:	3.7 mm, Rainfall probability: 47- 58%

Pitch Daily Maintenance—Rolling holiday. Pitch square covered with breathable covers (Fig. 9.30).

Fig. 9.30 Different prematch wicket preparations, Day 7, 08 Oct 2019

Day 8, 09 Oct 2019, Sunday

Weather forecast—Mostly cloudy with light shower in afternoon

Temperature:	Max: 25.0 C. Min: 20.0 C
Humidity:	Ave: 58%
Evapotranspiration:	3.7 mm, Rainfall probability: 53–40%

Pitch Daily Maintenance—Pitch was brushed twice and mowed at 6 mm grass height followed by 2 passes of straight–cross–straight rolling with 250 kg and 500 kg rollers. Considering the T storm, the whole pitch square was covered with light breathable covers from noon onwards (Fig. 9.31).

Fig. 9.31 Different prematch wicket preparations, Day 8, 09 Oct 2019

Day 9, 10 Oct 2019, Thursday

Weather forecast—Dense fog early in the morning, intermittent clouds with bright sunlight

Temperature:	Max: 25.0 C. Min: 21.0 C
Humidity:	Ave: 70%
Evapotranspiration:	3.7 mm, Rainfall probability: 51- 41%

Pitch Daily Maintenance—Straight rolling with 750 kg roller three passes. Considering bright sunlight and high ET 4 cans of water were sprinkled followed by covering with the hessian. Considering the T storm, the whole pitch square was covered with light breathable covers overnight (Fig. 9.32).

Fig. 9.32 Different prematch wicket preparations, Day 9, 10 Oct 2019

Day 10, 11 Oct 2019, Friday

Weather forecast—Intermittent clouds with bright sunlight

Temperature:	Max: 26.0 C. Min: 19.0 C
Humidity:	Ave: 74%
Evapotranspiration:	3.7 mm, Rainfall probability: 67- 19%

Pitch Daily Maintenance—Straight rolling with 500 kg roller for two passes followed by 4 passes of heavy rolling with 2000 kg in defined area. Observing bright sunlight and heat the pitch was sprinkled with 3 cans of water and covered with hessian. The whole pitch square was covered with light breathable covers overnight (Fig. 9.33).

Fig. 9.33 Different prematch wicket preparations, Day 10, 11 Oct 2019

Day 11, 12 Oct 2019, Saturday
Weather forecast—Bright sunlight Intermittent clouds with

Temperature:	Max: 29.0 C. Min: 18.0 C
Humidity:	Ave: 73%
Evapotranspiration:	4.4 mm, Rainfall probability: 60–55%

Pitch Daily Maintenance—Pitch was gently brushed followed by mowing at 6 mm. Straight rolling with 500 kg roller for two passes followed by 3 passes of heavy rolling with 750 kg in defined area. Observing bright sunlight after rolling session the pitch was sprinkled with and heat the pitch was sprinkled with 6 cans of water and covered with hessian. Observing heavy clouds and T storm (rained heavily for 15 min at 2.30 PM) the whole pitch square was covered with light breathable covers in afternoon and overnight (Fig. 9.34).

Fig. 9.34 Different prematch wicket preparations, Day 11, 12 Oct 2019

Day 12, 13 Oct 2019, Sunday
Weather forecast—Bright sunlight

Temperature:	Max: 27.0 C. Min: 18.0 C
Humidity:	Ave: 67%
Evapotranspiration:	5.5 mm, Rainfall probability: 56–55%

Pitch Daily Maintenance—Pitch was gently brushed followed by mowing at 6 mm. Straight rolling with 500 kg roller for 3 passes followed by sprinkling with 6 cans of water and covered with hessian. The pitch square was covered with light breathable covers overnight.

Day 13, 14 Oct 2019, Monday
Weather forecast—Bright sunlight before noon followed by cloudy weather

Temperature:	Max: 28.0 C. Min: 18.0 C
Humidity:	Ave: 67%
Evapotranspiration:	4 mm, Rainfall probability: 55- 14%

Pitch Daily Maintenance—Pitch was gently brushed followed by mowing at 6 mm. Straight rolling with 500 kg roller for two passes followed by brushing at whole pitch and mowing at 4 mm grass height. 3 passes of heavy rolling with 2000 kg in defined area were done. Observing bright sunlight after rolling session the pitch was sprinkled with 6 cans of water and covered with hessian. The whole pitch was covered with the hessian at noon so as to minimise the excessive desiccation of pitch surface. To regulate and maintain the moisture levels and replenish the moisture losses due to ET the whole pitch square was once again sprinkled with 6 cans of water in evening. Observing cloudy weather afternoon the whole pitch square was covered with light breathable covers in evening and overnight (Fig. 9.35).

Fig. 9.35 Different prematch wicket preparations, Day 13, 14 Oct 2019

Day 14, 15 Oct 2019, Tuesday
Weather forecast—Partly sunny

Temperature:	Max: 28.0 C. Min: 17.0 C
Humidity:	Ave: 69%
Evapotranspiration:	4 mm, Rainfall probability: 11–8%

Pitch Daily Maintenance—No rolling and watering in first session. Rolled with 2000 kg roller for 3 passes followed by brushing and sprinkling of 2 cans of water on pitch surface. Whole pitch square was covered with light breathable covers in evening and overnight (Fig. 9.36).

Fig. 9.36 Prematch wicket preparations, Day 14, 15 Oct 2019

Day 15, 16 Oct 2019, Wednesday
Weather forecast—Mostly sunny

Temperature:	Max: 31.0 C. Min: 20.0 C
Humidity:	Ave: 73%
Evapotranspiration:	4 mm, Rainfall probability: 11–10%

Pitch Daily Maintenance—No rolling and watering at noon and at 4 PM. Repeated light watering was done so as to regulate the excessive cracks on the pitch surface due to desiccation and to destress the low-cut grass from rolling stress. Pitch was brushed twice and mowed uniformly at 4 mm grass height. Whole pitch square was covered with light breathable covers in evening and overnight (Fig. 9.37).

Fig. 9.37 Different prematch wicket preparations, Day15, 16 Oct 2019

Day 16, 17 Oct 2019, Thursday
Weather forecast—A couple of pm T storm/mostly sunny

Temperature:	Max: 29.0 C. Min: 20.0 C
Humidity:	Ave: 78%
Evapotranspiration:	4 mm, Rainfall probability: 63–55%

Pitch Daily Maintenance—No rolling and watering evening with 6 cans of water. Pitch was brushed twice and mowed uniformly at 3 mm grass height. Whole pitch square was covered with light breathable covers in evening and overnight (Fig. 9.38).

Fig. 9.38 Different prematch wicket preparations, Day 16,17 Oct 2019

Day 17, 18 Oct 2019, Friday
Weather forecast—An am T storm, partly sunny

Temperature:	Max: 27.0 C. Min: 20.0 C
Humidity:	Ave: 81%
Evapotranspiration:	4 mm, Rainfall probability: 80%

Pitch Daily Maintenance—No rolling and watering evening with 6 cans of water. Pitch was brushed twice and mowed uniformly at 3 mm grass height. Whole pitch square was covered with light breathable covers in evening and overnight (Fig. 9.39).

Fig. 9.39 Different prematch wicket preparations, Day 17, 18 Oct 2019

Day 18, 19 Oct 2019, Saturday, First match day
Weather forecast—An afternoon T storm

Temperature:	Max: 27.0 C. Min: 20.0 C
Humidity:	Ave: 80%
Evapotranspiration:	4 mm, Rainfall probability: 82%

Pitch Daily Maintenance—No rolling before the match, cloudy up to noon, rained heavily at 3 PM. Match day first hour dominated by pacers followed by good to bat upon. Considering the T storm whole ground was covered with covers (Fig. 9.40).

Fig. 9.40 Different prematch wicket preparations, Day 18, 19 Oct 2019

Day 19, 20 Oct 2019, Sunday, Second match day
Weather forecast—An afternoon T storm

Temperature:	Max: 27.0 C. Min: 20.0 C
Humidity:	Ave: 83%
Evapotranspiration:	4 mm, Rainfall probability: 80%

Pitch Daily Maintenance—Cloudy in morning. One and half hours for outfield uncovering and mopping of the whole outfield. Whole pitch square was covered with light breathable covers overnight (Fig. 9.41).

Fig. 9.41 Different prematch wicket preparations, Day 19, 20 Oct 2019

Day 20, 21 Oct 2019, Monday, Third match day
Weather forecast—An afternoon T storm

Temperature:	Max: 25.0 C. Min: 19.0 C
Humidity:	Ave: 87%
Evapotranspiration:	4 mm, Rainfall probability: 80%

Pitch Daily Maintenance—Cloudy in morning. Showers by 6 AM. One and half hours for outfield uncovering and mopping of the whole outfield. Whole pitch square was covered with light breathable covers overnight (Fig. 9.42).

Fig. 9.42 Different prematch wicket preparations, Day 20, 21 Oct 2019

Day 21, 22 Oct 2019, Tuesday, Fourth match day
Weather forecast—Isolated T storm, Cloudy

Temperature:	Max: 27.0 C. Min: 18.0 C
Humidity:	Ave: 77%
Evapotranspiration:	4 mm, Rainfall probability: 30%

Pitch Daily Maintenance—Cloudy in morning. One and half hours for outfield uncovering and mopping of the whole outfield (Figs. 9.43 and Table 9.4).

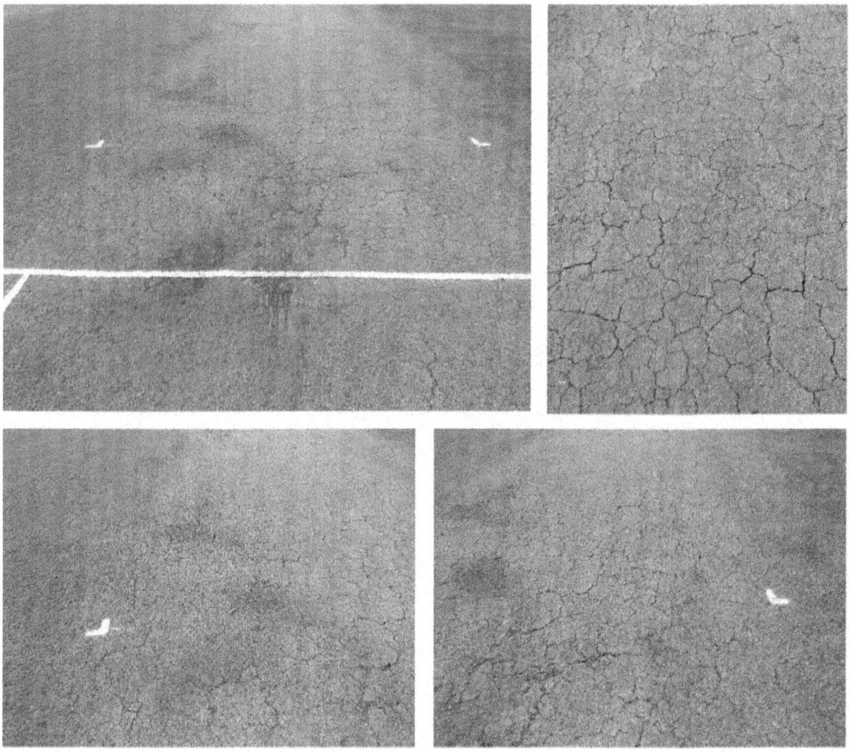

Fig. 9.43 These figures are for days 21, 22 October 2019, Fourth match day

Fig. 9.44 After match wear and tear on the Test match whole pitch surface (Fifth day 23.10.2019) From South to North End

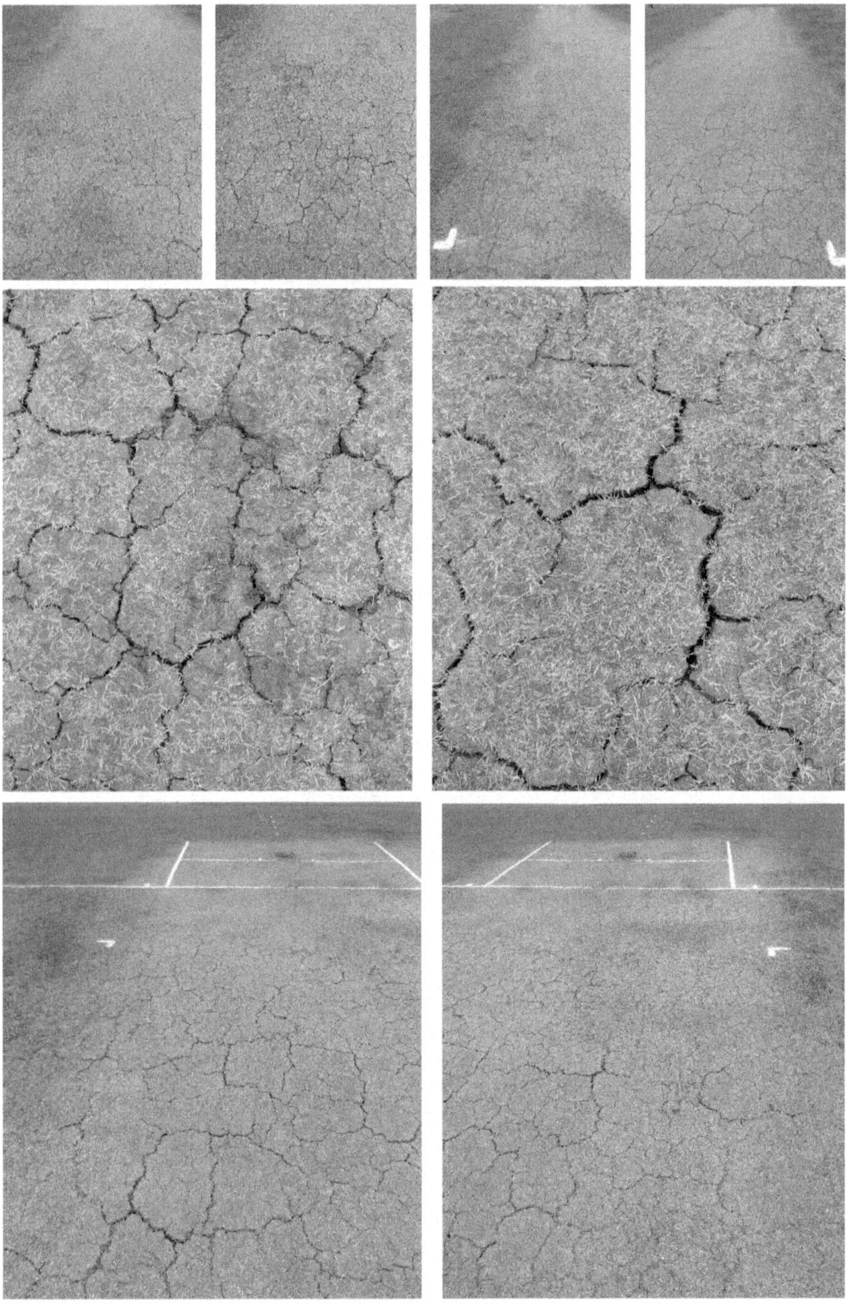

Fig. 9.45 Pitch wear and tear on fifth day of match (23.10.2019) From North end to South end

Fig. 9.46 Initiation, propagation, and development (one complete wetting and drying cycle- from day one of match pitch preparation- 02.10.2019 to final completion of the Test match 23.10.2019—17 days) during India Vs New Zealand Test match held at the JSCA International Stadium, Ranchi, Jharkhand- 19- 23 October 2019

Table 9.4 Total rolling and drying schedules during 17 days of pitch preparation time (PPT) during India Vs New Zealand Test match held at the JSCA International Stadium, Ranchi, Jharkhand—19–23 October 2019

			Rolling objective	Minimum rolling and maximum drying
				Prematch rolling schedules
Day	Date	Roller	Rolling	Comments
		weight (Kg)	duration (Min)	
1	02-Oct	500	30	Straight and cross roll-
				2 passes each
2	03-Oct	2000	30	Heavy roller with large drums to
				attain the density at the bottom surface
3	04-Oct	2000	30	- do-
4	05-Oct			Rolling holiday, grass mowed at 7 mm and gently brushed so as to increase the evapotranspiration rates and dry the pitch surface
5	06-Oct	2000	30	Heavy rolling followed by brushing the grass
6	07-Oct	Rolling holiday		Light frequent sprinkling of water so as to saturate the pitch profile up-to 2 to 3 inches
7	08-Oct	Rolling holiday		
	Total prematch rolling time		Light roller	30 min
			Heavy roller	1 h and 30 min
		Total prematch rolling	time	2 h
				Match pitch rolling schedules
8	09-Oct	250	20	Cross rolling

(continued)

Table 9.4 (continued)

			Rolling objective	Minimum rolling and maximum drying
		500	10	straight rolling
9	10-Oct	750	15	straight rolling with medium large drum roller
10	11-Oct	500	10	straight rolling
		2000	25	straight rolling in defined area
11	12-Oct	500	10	straight rolling
		750	15	straight rolling with medium large drum roller
12	13-Oct	500	15	straight rolling
13	14-Oct	500	10	straight rolling
		2000	20	straight rolling in defined area
14	15-Oct	2000	15	straight rolling in defined area
15	16-Oct			Rolling holiday
16	17-Oct			Rolling holiday
17	18-Oct	First match day		No rolling before the toss
	Total match	pitch rolling time	Light roller	1 h 15 min
			Medium roller	30 min
			Heavy roller	1 h
			Total	2 h and 45 min
	Total rolling time	during pitch preparation	time of 17 days	4 h and 45 min

9.6 Needs Further Clarification or Research

1. "Vegetation effect" or "turf effect": the rate at which moisture is lost within the soil gives rise to soil suction resulting in crack initiation. The moisture is lost either through evaporation or evapotranspiration. Since vegetation (turf grass on cricket pitches) reduces the moisture within the soil due to the process of evapotranspiration, it is necessary to study and understand the rate at which this occurs and its influence on the initiation and development of cracks using different types of grass.
2. Compaction effect: Several studies show that compaction of soil invariably improves the strength parameters, which in turn limit the potential for cracking. Compaction increases the soil density, reduces porosity, and dissipates the excess pore pressure (Batey 1990). As it has been established that cracking occurs when the tensile stress generated by negative pore pressure (suction) exceeds the soil's tensile strength, compaction will enhance the soil's tensile strength, thereby reducing its susceptibility to cracking.
3. The development of cracks in the laboratory under artificial conditions (a few mm in a rounded or rectangular petri dish) is entirely different from the conditions on an outdoor pitch. Conditions such as a turf pitch surface, the desiccation approach (under different weather and climatic variables such as temperature, relative humidity, evapotranspiration, wind, and solar radiation), and the wetting approach, all of which are difficult to stimulate appropriately in the laboratory.
4. Cracks occur when the tensile stress developed during desiccation exceeds the tensile strength of the soil. The main cause of cracking is not moisture content, but a second derivative known as suction. Many authors seem to use water content directly to explain desiccation cracks, probably because suction is difficult to measure in the soil. Meanwhile, its quantification can shed more light on the mechanism of cracking; therefore, it needs to be investigated further.
5. More drying and wetting cycles should be performed to confirm or invalidate some previously known concepts, as well as to introduce other variables such as temperature, drying rate, clay mineralogy, wetting approach, drying time, compaction effect, vegetation, and other worst-case conditions in preparation for the future of extreme cyclic events as predicted.
6. Future investigation should look at the image processing techniques, which would permit crack measurements at different moisture contents with greater accuracy. Though different scientific methods and techniques are available to map, scan, measure, or interpret the initiation, propagation, rate, and intensity of cracking of fine swelling and shrinking-dominant pitch soils, methods, and techniques to map, scan, measure, or interpret the initiation, propagation, rate, and intensity of crumbling of fine non-swelling kaolinite-dominant fine clay having a high proportion of silt and sand are yet to be certified and quantified. The involvement of highly nonlinear material behaviour that varies with the environmental factors (such as temperature, RH, ET, wind, and solar radiation) and cyclic wetting and drying makes the initiation, propagation, and development of cracks more

complicated. Meanwhile, the mechanism of cracking or crumbling of the pitch surface under wetting and drying cycles, including crack initiation, advancement, closure, and other resulting structures, is still vague and needs to be researched.

7. Cracking has been studied by several researchers over time, but not much has been established with respect to the connection between temperature, wind speed, and the exact rate at which cracking develops (Tang et al. 2010). According to research, there is a link between the rate of cracking and the effect of wind speed, which is directly related to the rate of moisture loss. The higher the wind speed, the greater the moisture loss, which quickly accelerated the rate of evaporation (Ellah 2009).

9.7 Conclusions

Integration of both the indirect (conventional and traditional style based on skill and experience) and direct methods (by the use of scientific data and gadgets) of pitch behaviour analysis and forecasting during the pitch preparation would be the best and prudent way to achieve the desired objectives for different formats of the game of cricket.

Many research efforts have been made in the areas of soil mechanics and soil science in this modern era of scientific research and development to understand the complex nature involved in soil behavioural analysis. However, the practical application of the understanding of soil behaviour on cricket turf pitches so as to attain the desired behaviour is still in its nascent stage.

Cracks on the pitch surface are either mechanically (via rolling) or physio-chemically (via pitch profile desiccation) induced by changes in interparticle forces. This needs to be scientifically evaluated.

Of late, the initiation of cracks, time, rate, intensity, or cracking or crumbling patterns (nature, shape, size, length, breadth, and depth of cracks) developed in the cricket pitch profile can be scientifically measured and quantified with the help of digital imaging techniques and computer software such as MATLAB Version 7. Cracking or Crumbling Intensity Factor (CIF) is the most useful and effective parameter for evaluation and quantification of cracks developed on the cricket pitch soils dominant in smectite clay content or non-swelling or shrinking clays (such as illite or kaolinitic soils), which can be analysed and correlated with the variations in the pitch behaviour during the course of the game, especially the Test matches.

9.8 How to Analyse the Pitch Behaviour of Subcontinental and Continental Pitches

Australian pitches have traditionally been known for producing consistent fast bowlers due to the amount of bounce that can be generated on this surface. Because of the high percentage of Smectite (>80%), Ca^{+2}, Mg^{+2}, H^{+}, and Na^{+}-dominant exchangeable fine clay (82%) with a lower percentage of silt (6%), and fine sand (only 2%), the West Australia Cricket Association (WACA) ground in Perth is regarded as possibly the quickest pitch in the world. Brisbane pitch soil is also known to aid fast bowlers with bounce and to have similar morphological, physical, chemical, and mineralogical pitch soil composition to Perth pitch soil.

Adelaide Oval and the Sydney Cricket Ground (SCG) have been known to assist spinners more as these pitches have more dust covers. Adelaide Oval's Ca^{+2} smectites, intermediate kaolinite, illite, and quartz, moderate clay (47%) and intermediate silt (st covers. Adelaide Oval's Ca^{+2} smectites, intermediate kaolinite, illite, and quartz, moderate clay (47%) and intermediate silt (11%), and a high proportion of fine sand are the main reasons for providing spinner-friendly wickets. The SCG wicket had hardly any cracks because of the large amount of non-swelling clay (clay: 48%) pitches have more dust covers. Adelaide Oval's Ca^{+2} smectites, intermediate kaolinite, illite, and quartz, moderate clay (47%) and intermediate silt (11%), and a high proportion of fine sand are the main reasons for providing spinner-friendly wickets. The SCG wicket had hardly any cracks because of the large amount of non-swelling clay (clay: 48%, silt: 15%), and fine sand: 12%, with fewer amounts of smectites and predominant kaolinite (20–30%) and quartz (20–30%).

The Melbourne Cricket Ground (MCG) generally assists seam bowlers initially, but it has a tennis-ball bounce. The Wamberal wickets are hard due to the high percentage of quartz as well as the 20–30% of smectites, which also add to the hardness. The surface of these wickets also tends to powder when dry because of the high level of quartz and silt (21%). These types of wickets are famous for being hard and becoming powdery and dusty, which allows the ball to grip for the spinners (Figs. 9.47, 9.48, 9.49, 9.50 and Table 9.5).

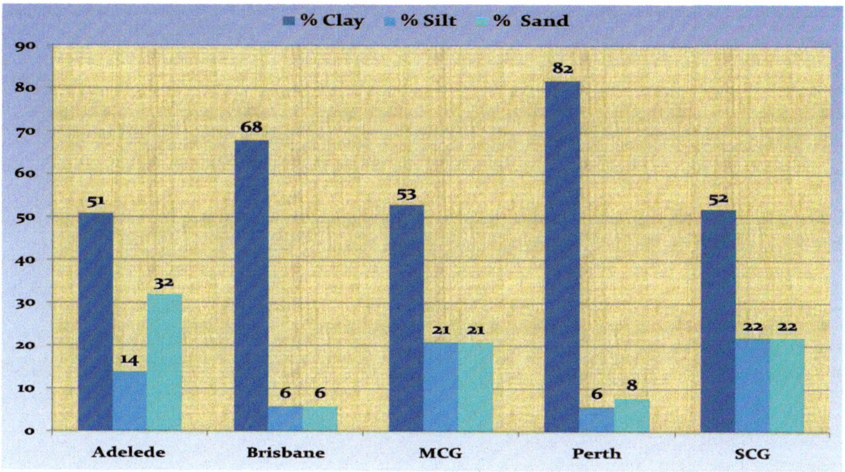

Fig. 9.47 PSD of some of the major Test playing cricket pitches in Australia (*Source* McIntyre Keith and Don, Cricket Wickets—Science Vs Fiction, 2001, p 68)

Table 9.5 Physical, chemical, and mineralogical composition of some of the major Test grounds around Australia in 1980s

Soil Name	Adelaide (Oval)	Brisbane	MCG	Merri Creek	Perth (WACA)	Harvey	SCG	Wamberal
Mineralogy								
Smectite (%)	40–50	> 80	65–80	65–80	> 80	65–80	50–65	20–30
Illite (%)	20–30	5–10	10–20	10–20	Nil	Nil	5–10	10–20
Kaolinite (%)	10–20	Nil	5–10	10–20	10–20	10–20	20–30	20–30
Quartz (%)	5–10	Nil	10–20	10–20	Nil	Nil	20–30	30–40
Clay (%)	47	68	50	50	82	78	48	55
Silt (%)	11	16	16	15	6	11	15	21
Fine sand	28	15	24	23	2	7	12	15
pH	8.1	8.2	7.4	6.3	5.8	4.5	5.9	5.5
EC (dsm^{-1})	0.19	0.21	0.4	0.54	1.13	1	0.56	1.02
CEC	46.2	84	54.8	58.2	67.1	64	30.1	29.4
Ca^{+2}	36.6	48.3	24	21	11.8	7.6	9	1.59
Mg^{+2}	7.5	24.5	15.4	15.7	24.6	22.7	1.8	2.2
Na^{+}	1.2	0.54	2.6	1.1	6	6.8	1.2	1.1
H^{+}	Nil	9.5	11.6	19.5	24.2	26.3	17.4	25.2
ESP (%)	2.6	0.6	4.8	1.9	8.9	10.6	4	3.7
Ca^{+2}/Mg^{+2}	4.9	2	1.6	1.3	0.5	0.3	5	3.72
Cs	11	1	6	6	2	2	22	4
DI	0	0	9	3	10	12	4	10

Note Soil mineralogy (smectite, illite, kaolinite, Quartz) and clay, silt, sand is in percentage (%), Electrical conductivity (EC is in (dsm^{-1}), Cation exchange capacity (CEC) and exchangeable cations ($Ca^{+2}/Mg^{+2}/Na^{+}/H^{+}$) is in cmol ($p^{+}$) Kg^{-1}, Exchangeable Sodium percentage (ESP is in %)., Dispersion Index (DI) (*Source* McIntyre Keith and Don, Cricket Wickets—Science Vs Fiction, 2001, p 68)

Sources McIntyre, Keith, and Don, "Cricket Wickets: Science vs. Fiction," 2001; and Wikipedia, the free encyclopedia: https://en.wikipedia.org/wiki/Cricket_pitch

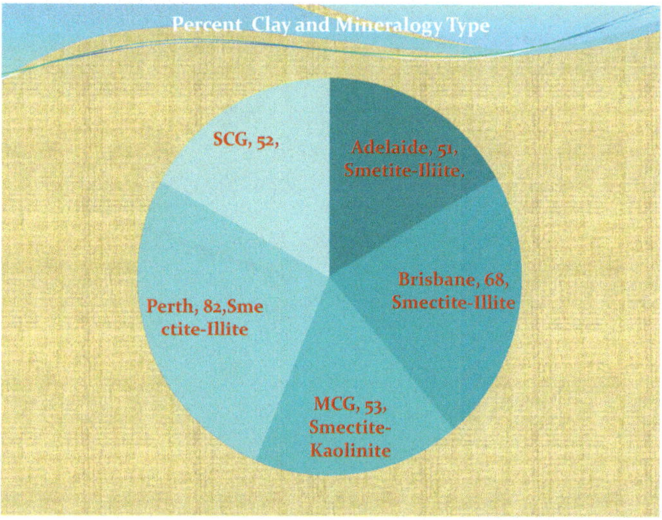

Fig. 9.48 Clay percentage and mineral types of some of the important pitches of Australia (*Source* McIntyre Keith and Don, Cricket Wickets—Science Vs Fiction, 2001, p 68)

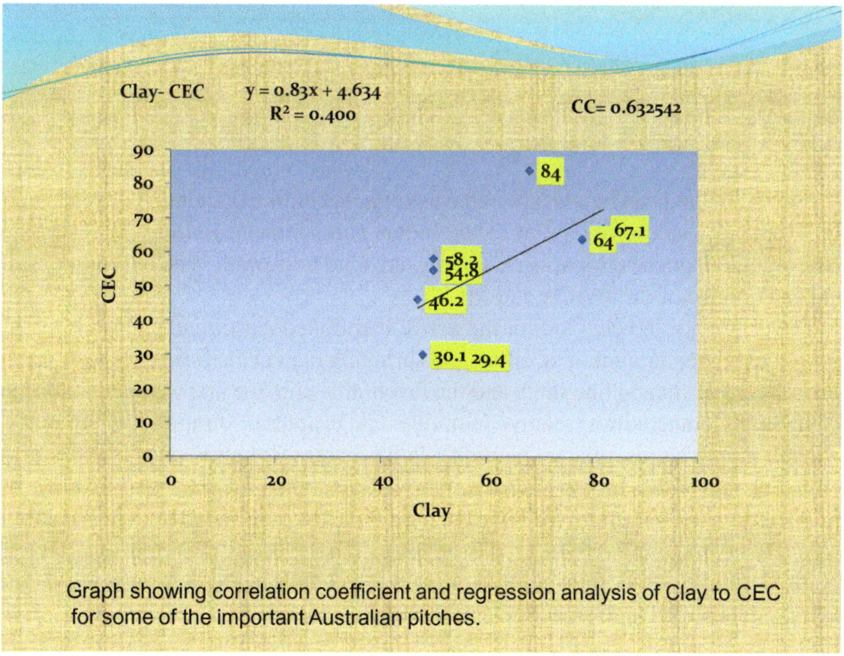

Fig. 9.49 Correlation coefficient and regression analysis of Clay to CEC for some of the important Australian pitches (*Source* McIntyre Keith and Don, Cricket Wickets—Science Vs Fiction, 2001, p 68)

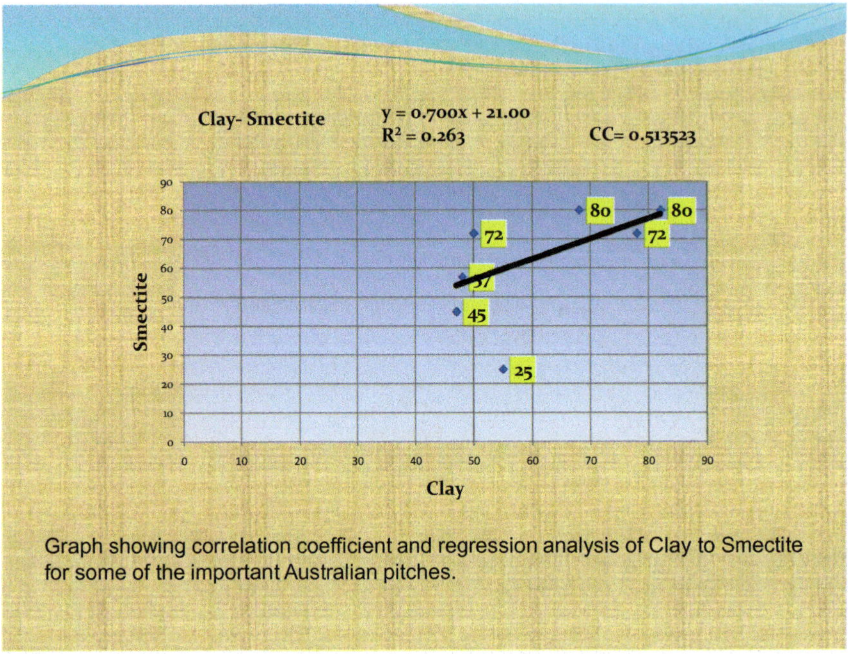

Fig. 9.50 Correlation coefficient and regression analysis of Clay to Smectite for some of the important Australian pitches (*Source* McIntyre Keith and Don, Cricket Wickets—Science Vs Fiction, 2001, p 68)

Swing bowling can be a weapon in Australia, but unlike England, it depends upon the overhead conditions, similar to the Indian subcontinent. Batting in Australia is easier on most pitches. Most backfoot players tend to do well. The only issue is the unusual bounce of the WACA and MCG.

The difference in behaviour of the above-mentioned Australian pitches is mainly due to the higher amount of swelling and shrinking clay in the form of smectites, the low per cent of silt and fine sand, and the favourable climatic and weather conditions having high temperatures, relative humidity, and evapotranspiration rates.

High temperatures, relative humidity, and evapotranspiration rates that desiccate a pitch profile with a high percentage of swelling and shrinking smectitic clays are the prerequisite conditions for laying the type of fast and bouncy pitches available in Australia and South Africa. Due to adverse climatic conditions (low temperature, relative humidity, evapotranspiration rates, etc.) available in England or New Zealand, laying pitches with high smectitic clay is not feasible.

9.8.1 Pitches in South Africa

Pitches in South Africa resemble those in Australia with added swing (lateral movement) and comparatively lesser bounce (due to high silt content). Spinners gain less or negligible assistance on such a pitch surface (Fig. 9.51 and Table 9.6).

Fig. 9.51 Average Particle Size Distribution (PSD) of South Africa cricket pitches soils

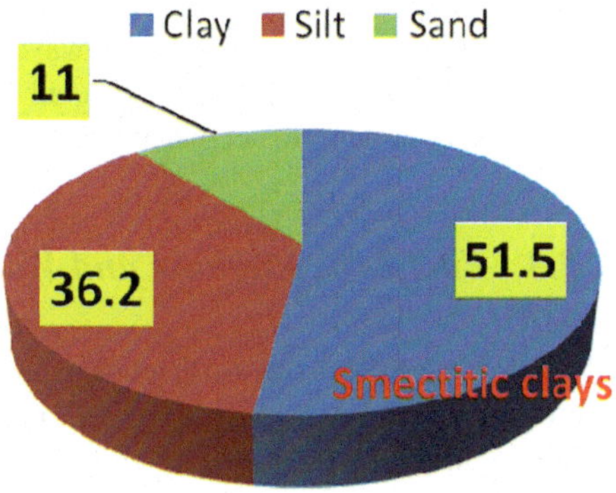

Table 9.6 PSD, chemical, and mineralogical composition of one of the often-used cricket pitch soils in South Africa

PSD		in %		Exchangeable cation (mg/L)
Clay		51.5		Ca^{+2}—7132
Silt		36.2		Mg^{+2}—2087
Sand	Fine	9.1		K^{+}—109
	Medium	2.2		N^{+}—2
	Coarse	0.7		Total—53.2
pH (water)		5.9		
pH (KCL)		4.9		
Clay minerals (%)				
Smectites		81		
Kaolinite		14		
Mica		5		

Source Principles and practices of pitch preparation, United Board of South Africa, May 1983

9.8.2 Pitches in Zimbabwe

Pitches in Zimbabwe closely resemble those in South Africa, the only difference being in the bounce. South African pitches have a fast bounce, whereas Zimbabwean pitches have a spongy, tennis-ball bounce, making hitting on the up a risky proposition. Most pitches in Zimbabwe have a slower bounce, so batting is more favourable on those pitches. Conditions at the Queens Sports Club tend to aid batsmen, with spin coming into the game in a big way in the latter stages. With the temperature touching 28 degrees, the strip is expected to dry out quickly and flatten into a batting beauty. The seamers' best chance will be with the new ball.

9.8.3 Pitches in England and Wales

The main characteristics of English wickets are green turf, swing promoting low temperatures and evapotranspiration rates, high relative humidity, a low or non-swelling and shrinking type of clay (basically illites or kaolinites), high silt, and moderate sand.

Soil physical, chemical, and mineral composition in cricket pitches in the UK (Tables 9.7, 9.8, Figs. 9.52, 9.53, 9.54 and 9.55).

Table 9.7 Particle size distribution (PSD) of cricket pitch soils used in UK

	Clay	BC	OL
Clay	47.2	34	30.1
Silt	24	29.9	43.2
Fine sand	6.5	17	12.1
medium sand	18.6	17.3	12.9
Coarse sand	3.7	1.7	1.7
Total sand	28.8	36	26.7

Note BC—Boughton County soil, OL—Ongar loam soil, C—Clay soil

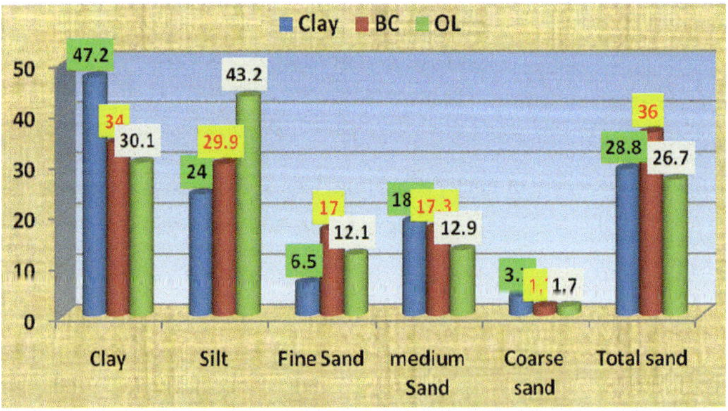

Fig. 9.52 Chart showing the particle size distribution of major soils used in cricket pitches of UK

Table 9.8 Whole soil quantitative mineralogical analysis of cricket pitch soils used in UK

	Clay	BC	OL
Quartz	42.4	52.1	58.7
Kaolinite	7.3	6.5	3.3
Mica/Illites	6.9	4.8	2.6
Illite/Smectite	28.6	19.3	19.7

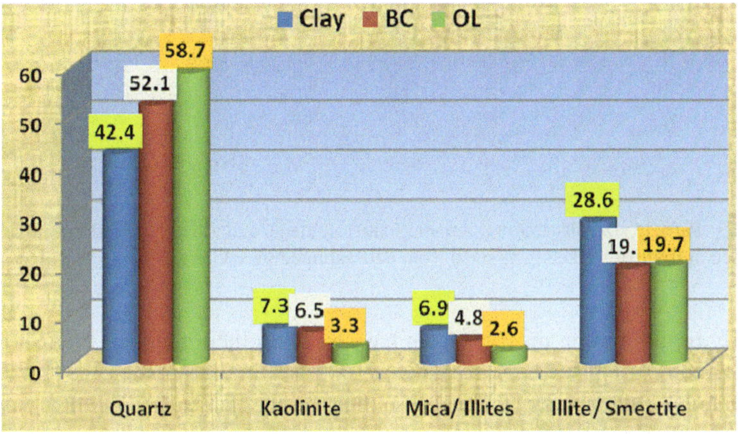

Fig. 9.53 Whole soil quantitative mineralogical analysis of major soils used in cricket pitches of UK

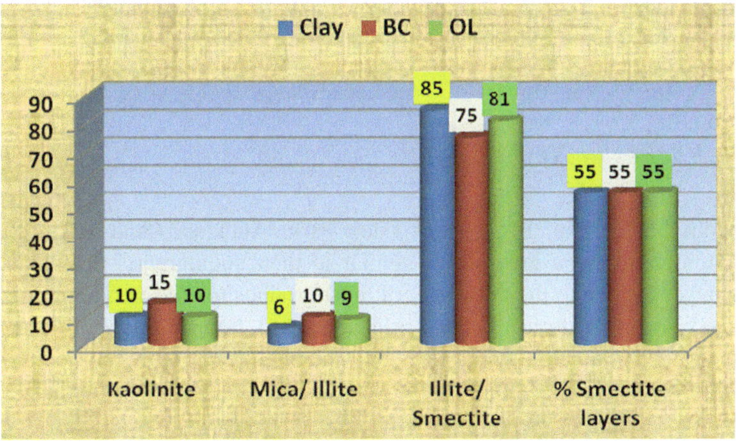

Fig. 9.54 Relative proportion of clay minerals in clay size fraction of major soils used in cricket pitches of UK (*Source* Simon Parsons, The aeration of clay soil in cricket, School of Applied Sciences, Cranfield University, PHD thesis, 2008–12, p 91)

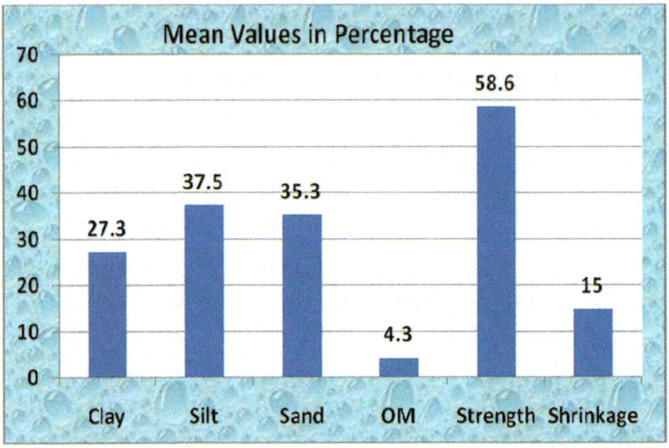

Fig. 9.55 Particle size distribution, organic matter contents and ASSB strength mean values for soils used in English wickets (extracted from data supplied by E.C.B 2004)

Early in the season, most batsmen have to be on their guard as English pitches prove to be fickle, like the country's weather. Later in the summer, the pitches tend to get harder but crack as well and lose their green. This makes the task easier for batsmen, and only genuine fast bowlers like those bowling in the range of 130–150 km/h and spinners can prevent the batsmen from scoring runs. Spinners prove less effective in the first half of the season and tend to play their part only in the second half of the day or innings. The humid conditions and little dust make the grounds an ideal place to practice reverse swing with a 50-over ball. The Oval is the most dangerous of all fields because the ball tends to reverse swings.[1]

9.8.4 *Pitches in New Zealand*

The pitches here, like the ones at Eden Park, Auckland, and Basin Reserve, Wellington, can have a green tinge similar to their counterparts in England. The ball swings a lot due to the proximity of most grounds to the sea. These pitches are often the best-rounded ones in the world, with hard, bouncy pitches offering seam movement for the quick bowlers early on. The pitches resemble the South African ones, with added wind. Batting can be very trying, and batsmen often take time to adjust to the conditions. As they are often in open stadiums, bowlers can frequently be assisted by winds that increase the speed and make it difficult for the batsmen to face the ball at one end of the pitch. At the other end, the wind acts as a wall and

[1] Peter Shipton and Lain James, Guidelines for Rolling in Cricket, 2009; and Simon Parsons, The Aeration of Clay Soils in Cricket, Cranfield University, 2008–12.

decreases the pace on the ball, and as a result, adjusting to this is often the most trying part for visiting teams.

9.8.5 Pitches in India and the Subcontinent

Most of the Indian subcontinental pitches are silty clay in texture, having less or no swelling illitic or kaolinitic type clay minerals, aided by high temperatures, high relative humidity, and high evapotranspiration rates. Silts are generally very fine, non-cohesive soil particles, which reduce the cohesiveness of the pitch soil. This reduction in the cohesiveness of the soil causes the pitches with high silt contents to powder and crumble. Clay along with the humus (clay-humus interaction) present in the soil acts as a binding agent and holds all the aggregates together.

The pitch surface strength (the desirable pitch surface density) at a point in time and durability (consistency) thus depends a lot on the amount and type of clay present. Historically, Indian pitches have been known for their spin bowling rather than seam or swing. The main reason behind this behaviour is due to the pitch soils having a low percentage of swelling and shrinking clays and a very high proportion of silt and sand. Hot and humid climates with high evapotranspiration rates aggravate pitch surface conditions (by reducing pitch surface moisture and structure early in the game) and favour spinners later in the game.

They outwitted opposing batsmen not only through line, length, and trajectory variations but also by physically and psychologically exploiting rough spots resulting from wear and tear on the playing surface and cracks from increasing surface dryness as the game progressed (Figs. 9.56 and 9.57).

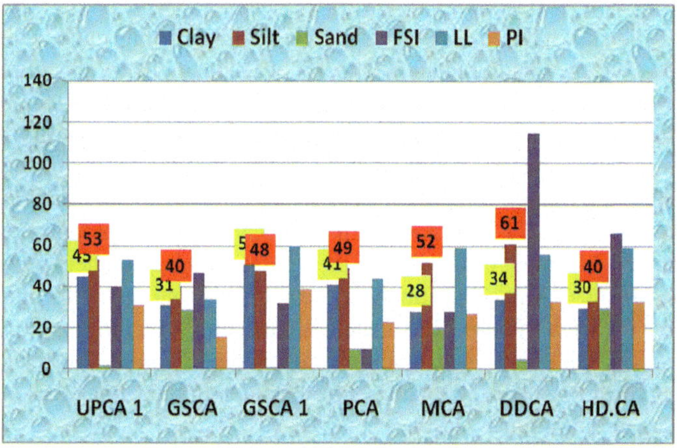

Fig. 9.56 Some of the major Test playing cricket pitch venues in India having moderate non-swelling and shrinking clay with high proportion of silts and sand (*Source* IIT Powai, Bombay 2008)

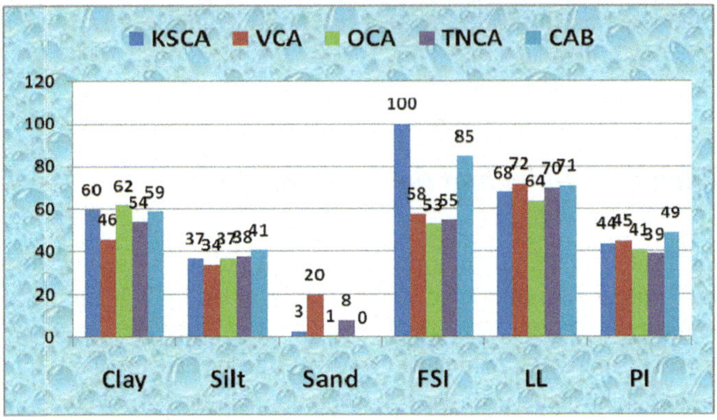

Fig. 9.57 Some of the major Test playing cricket pitch venues in India having high swelling and shrinking clay with moderate proportion of silts and sand (*Source* IIT Bombay 2008)

Indian pitches and attitudes towards pitch making have changed considerably in the past few years, though. The induction of several newer "green top" venues (such as the ones at Mohali and Dharamshala), the emergence of "genuine" Indian fast bowlers, plus the development of domestic league cricket (the Ranji Trophy), have resulted in a greater variety of pitches. Some contemporary pitches provide good support for pace, bounce, and swing, giving visiting teams from countries such as South Africa and Australia the feeling of being at home.

9.8.6 *Pitches in Pakistan*

Pitches in Pakistan have historically supported spin bowling rather than seam or swing. However, the conditions in most parts of Pakistan, like Rawalpindi, Lahore, and Peshawar, have also supported the reverse-swing capabilities of local bowlers in the past, like Imran Khan, Sarfraz Nawaz, and Wasim Akram. The dry and windy conditions usually lend good support to the faster bowlers as well. Such pitches, having virtually no grass, afforded little assistance for pace, bounces, or lateral air movement, but created a very good turn. In decades past, legendary off spinners such as Saqlain Mushtaq, credited with the doosra, and leg spinners such as Abdul Qadir and Mushtaq Ahmed routinely toyed with visiting teams to plot dramatic victories for Pakistan in home Test matches, particularly on turning pitches in hot, humid conditions at Arbab Niaz Stadium and Gaddafi Stadium. Pitches in Pakistan are flat and considered batting paradises for batsmen in the winter season, and they suit spinners in the summer season.[2]

[2] Wikipedia, the free encyclopedia, https://en.wikipedia.org/wiki/Cricket_pitch.

9.8.7 Pitches in Bangladesh

The Bangladeshi wickets receive a lot of rain in a short period of time, which reflects their soggy nature. The conditions vary on grounds like Sher-e-Bangla Cricket Stadium and the Chittagong Divisional Stadium.[3]

9.8.8 Pitches in Sri Lanka

Generally dusty and shorn of grass, the rain here also makes it a "sticky wicket." Wickets are flat and don't offer much bounce, unlike the pitch at Asgiriya Stadium; Kandy in Sri Lanka offers generous bounce and favours fast bowling. Spin is the key in these conditions (Fig. 9.58).

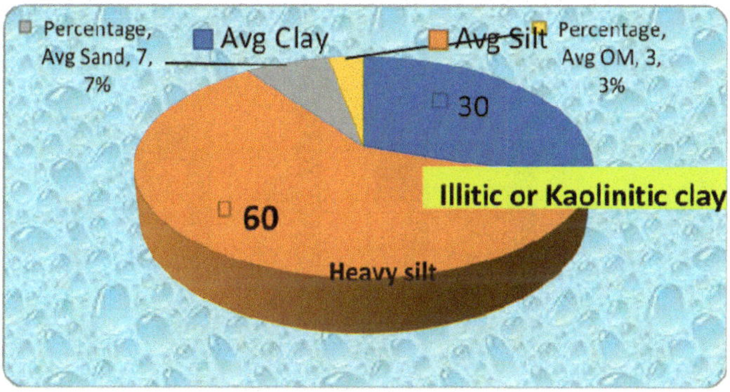

Fig. 9.58 Average PSD of Sri Lankan cricket pitches soils (*Source* U. P. Nawagamuwa et al., "Improvement of local soils in order to make fast and bouncy cricket pitches," in Wikipedia, the free encyclopedia", https://en.wikipedia.org/wiki/Cricket_pitch)

Spinners have fine records on the pitches in Sri Lanka. The heat requires an extreme level of fitness, and the sweaty clothing doesn't quite shine. Reverse swing, off-spin, and leg-spin are all effective for a good player.

9.8.9 Pitches in the United Arab Emirates (UAE)

The UAE features dry and batting-friendly pitches. A new ball helps the bowler and bowler's eye reverse swing after 34 overs of mandatory refurnishing of the ball. UAE conditions differ significantly from those of Pakistan due to the Gulf's sandy soils. Grounds are not that hard. Dubai Cricket Stadium offers some grass, and fast bowlers like Shoaib Akhtar loop the ball on to the ground. Sheikh Zayed Stadium is batting-friendly, and the cracks come very late into play.

[3] Wikipedia, the free encyclopedia, https://en.wikipedia.org/wiki/Cricket_pitch.

9.8.10 Pitches in the West Indies

The West Indies tend to produce balanced pitches. Neither the bounce is disconcerting nor the movement extravagant. It also does not assist spin like the subcontinental pitches, and hence, for quality batsmen, these pitches could be a batting paradise. However, bowlers who are willing to bend their backs will find some assistance from these pitches. Pitches here have earned a reputation of helping the fast bowlers, mainly because of the era gone by when the West Indies used to have some of the fastest bowlers in cricket, and hence, the pitches appeared to be faster than they are. Tall bowlers like Joel Garner, Michael Holding, Andy Roberts, Colin Croft, Curtly Ambrose, and Courtney Walsh produced bounce and speed even on the most docile pitches that wreaked havoc on any side, and they used to run through the line-up (Figs. 9.59, 9.60 and 9.61).

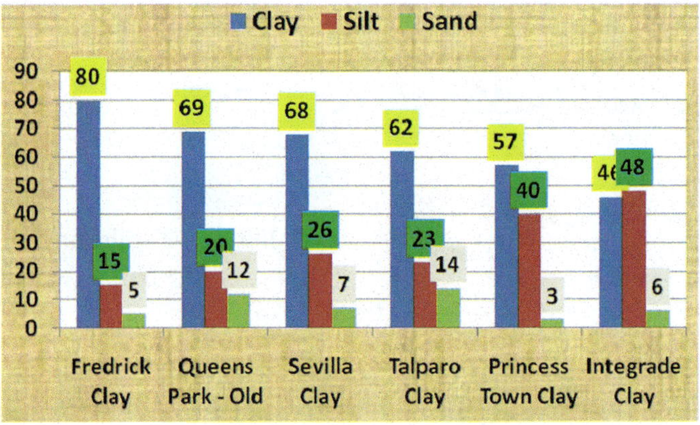

Fig. 9.59 Particle size distribution for some of the commonly used soils in Trinidad, West Indies

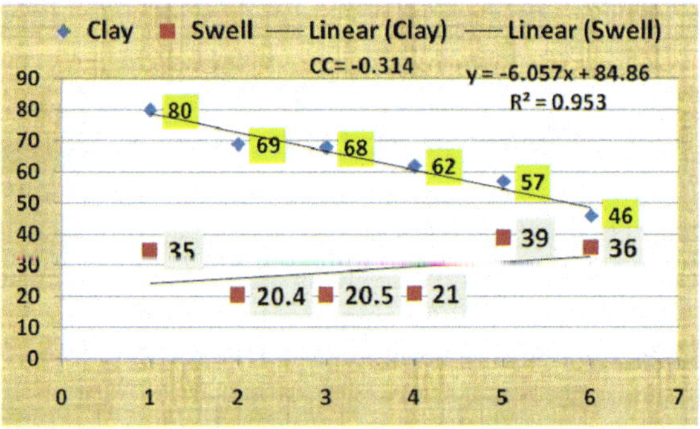

Fig. 9.60 Correlation coefficient and regression in between the clay and swell percentage for some of the commonly used soils in Trinidad, West Indies

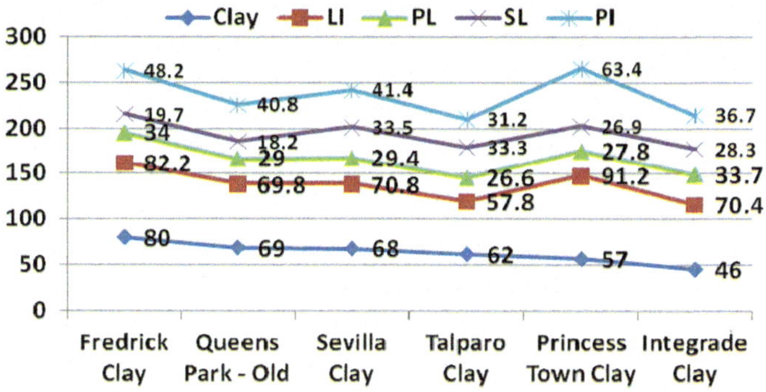

Fig. 9.61 Relationship between clay and the Atterberg limits for some of the commonly used soils in Trinidad, West Indies

However, some of the best batsmen have arisen from the Caribbean too, like Vivian Richards, Clive Lloyd, Rohan Kanhai, and Brian Lara. Gary Sobers is generally considered the greatest all-rounder of all time. Spinners also have something in the pitches, as they offer a little dust and cracks from the fourth day on.[4]

9.9 Cricket Pitch Behavioural Analysis Through the Database Management System (DBMS) of Pitch Profile

Database Management Systems (DBMSs) for cricket pitches are desired for cricket pitch preparation, analysis, interpretation, comparisons, and setting benchmarks and standards to achieve the desired objectives and pitch behaviour. DBMS for preparing a cricket pitch can be divided into four stages:

1. DBMS for preseason pitch preparation and renovation details such as dates of scarifying, dethatching, top dressing, grass planting, fertiliser, fungicide, and pesticide doses and applications, pitch profile (new or old pitch), soil (morphological, physical, chemical, mineralogical, and engineering properties), grass and water test reports, equipment and labour availability, and maintenance reports. The complete details required can be extracted and analysed through the pitch questionnaire (refer to Annexure 1).
2. DBMS prior to pitch preparation, such as the nature and type of pitch soil and grass, match type, format, and time available for pitch preparations, as well as at least a 15-day weather forecast for rainfall, temperature, relative humidity, dew points, evapotranspiration rates, and so on. The Pitch Quality Index (PQI) (refer to Annexure 2) can extract and analyse all required details.

[4] Ekwue EI, Lall DZ, Stone RJ (2006) Engineering properties of major soils used in cricket pitches in Trinidad. West Indies J Eng 28(2) and Wikipedia, the free encyclopedia, https://en.wikipedia.org/wiki/Cricket_pitch.

3. DBMS related to the pitch preparation methods and schedules as pitch soil wetting, rolling patterns and schedules, roller types and their compactive potentials, grass shoot and root density and height, compactive or density of pitch surface measured through Clegg Hammer or EC values or indirect methods through spikes, core sampling, size, shape, and depth of pitch surface cracks, shine and colour of pitch surface, colour of rolled-in grasses, etc. The full set of required details can be extracted and analysed through the pitch questionnaire and the following three case studies.
4. DBMS related to the post-match results as the pitch surface status in terms of wear and tear, cracks, and crumbling of the pitch surface, as well as the analysis and interpretation of match data with the desired and actual outcomes.

9.10 Classification of Cricket Pitches Based on the Database Management System of a Pitch Profile, the Pitch Behavior Analysis Index (PBAI)

Classification of cricket pitches based on the Database Management System of a pitch profile, the Pitch Behaviour Analysis Index (index properties based on pitch soil texture, particle size distribution either proportion of clay, silt, sand, or humus, morphological, structural, physical, chemical, and mineralogical), cracking and crumbling patterns (wear and tear) in the form of their initiation, rate, and intensity of propagation on the pitch surface, depth, width, shape, size, geometry, prevalent day and night weather variations (in the form of evapotranspiration, temperature, relative humidity, solar radiation, wind, dew).

1. Swelling and Shrinking Clay.
a. At a moderate to strong alkaline pH range (pH > 8.5, EC 4, high ESP (> 5), SAR, BS, CEC, WHC, dispersion index, and a low proportion of silt, fine sand, and air voids, very fine swelling and shrinking Na^{+}/Mg^{+2} and moderate Ca^{+2} and Mg^{+2}-dominant fine clay smectites with a very high dry density of pitch surface. Very high dry surface density due to deflocculation (dispersion of clay particles in Na^{+}-dominant saline conditions) and hydration of clay particles soil structure is lost, and the soil is dispersed, resulting in sticky soil when wet and very hard soil when dry. High WHC and low air voids result in a dense pitch surface. Regeneration of grass and the pitch surface is poor due to low porosity (air voids), dispersion of clay particles, and the presence of slats at toxic levels proportions, e.g. the Perth and Harvey cricket pitch soil (Table 9.5, Fig. 9.47) very wide surface cracks with a very high water holding capacity (due to the hydration of clay) and dry density of the pitch surface. By interacting with the prevalent weather and climatic variations (in the form of evapotranspiration, temperature, relative humidity, solar radiation, wind dew, etc.), these types of pitches express their pitch surface behaviour through large (polygonal or hexagonal) cracks and very high pitch surface dry density with very low air voids within the pitch

profile provides a narrow spectrum of pitch behaviour (due to high water holding capacity and high dry density with a well-structured pitch surface), favouring pacers early in the match, followed by batsmen, while spinners have little scope on this type of pitch. After matches, recovery of the pitch surface and the turf grass is moderate.

b. High swelling and shrinking Ca^{+2}-dominant fine clay smectites at alkaline pH range (pH neutral to slightly alkaline, providing good flocculated aerated pitch surface, EC > 4, moderate ESP, SAR, high BS, CEC, WHC, good flocculated structure (stable soil aggregate and structure due to clay- calcium- humus bridge) and regeneration of grass, the pitch surface during and after the match is good, with proportionate silt and fine sand proportions- Example- The Brisbane (Table 9.5, Fig. 9.63), MCG and Merri Creek (Table 9.6 and Fig. 9.51), Kingsmead, South Africa - (Table 9.5, Fig. 9.47) cricket pitch soil. wide surface cracks due to the pitch surface's high dry density by interacting with the prevalent weather and climatic variations (in the form of evapotranspiration, temperature, relative humidity, solar radiation, wind, dew, etc.), they generally express their behaviour by generating a uniformly high pace and bounce with good regeneration of the pitch surface and turf grass.

c. Moderate swelling and shrinking Ca^{+2}-dominant fine silty clay, illite, and smectites (mixed minerals) with a neutral to moderately alkaline pH range (EC > 4, moderate ESP, SAR, high BS, CEC, WHC, and flocculated structure) with non-proportionate silt and fine sand (as shown in Table 9.13 and Fig. 9.66. By interacting with the prevalent weather and climatic variations (in the form of evapotranspiration, temperature, relative humidity, solar radiation, wind, dew, etc.), the pitch surface expresses its behaviour through multiple medium-sized cracks during the initial stages of the match, but crumbles and assists spinners during the later stages of the match due to a very high proportion of silt having a moderate WHC of the pitch surface And provides a wide range of pitch surface behaviour in which pacers, batsmen, and spinners all have an equal opportunity to perform in a limited number of match days.

2. Non-swelling and less shrinking kaolinite, an Al^{+3} or H^{+} ion-dominant fine clay, has a low pH range, CEC, BS, WHC, and compressibility, with proportionate or non-proportionate silt and high to medium sand proportions. Generally, such pitch surfaces, by interacting with the prevalent weather and climatic variations (in the form of evapotranspiration, temperature, relative humidity, solar radiation, wind dew, etc.), express their behaviour as moderate pace and bounce with crumbling (powdery) pitch surfaces during the later stages of the match. These types of pitch surfaces generally provide a narrow moisture range (due to low water holding capacity) with a loss of pitch surface structure during the early phases of multi-day matches.

9.11 Categorisation of Cricket Pitches Based on DBMS, Pitch Behaviour Analysis Index (PBAI) / Pitch Behaviour Forecasting (PBF) Methods

1. Pitch Behaviour Analysis Index (PBAI) / Pitch Behaviour Forecasting (PBF) methods.

Table 9.9 PQI, PQS, and DBMS of a pitch profile based on mineralogical, physical, chemical, and engineering properties for Test matches (day matches). Name of the Ground/Country-WACA- The Perth/Australia in percent

1	Mineral Type (%)	Smectite	80
		Illite	Nil
		Kaolinite	10–20
		Quartz	Nil
2	Exchangeable	Calcium	11.8
	Cations	Magnesium	24.6
	cmol (p^+) Kg^{-1}	Sodium	6
		ESP—Exchangeable sodium percentage (%)	8.9
		Calcium/Magnesium ratio	0.5
		Soil pH	5.8
	cmol (p^+) Kg^{-1}	Cation exchange capacity (CEC)	67.1
3	Particle size (%)	Soil texture	Clay (CH)
	Distribution	Total clay	82
		Silt	6
		Total sand	2

Expected Pitch Behaviour: Magnesium and sodium-dominant (high ESP) smectitic clay; highest achievable pitch surface dry density and compression levels due to deflocculation of clay particles with the least air voids; extra-large, wide, over consolidated, and consistent cracks generating extra pace, bounce, and desirable variation in pitch behaviour (Fig. 9.62, Table 9.9).

Fig. 9.62 Predominant pitch behavioural patterns of the Perth (WACA), Australia—based on their predominant morphological, mineralogical, physical, chemical, and engineering properties

2. Pitch Behaviour Analysis Index (PBAI)/Pitch Behaviour Forecasting (P.B.F).

Table 9.10 PQI, PQS, and DBMS of a pitch profile based on mineralogical, physical, chemical, and engineering properties for Test matches (days of matches). Name of the Ground/Country-Brisbane/Australia in percent

1	Mineral Type (%)	Smectite	80
		Illite	5–10
		Kaolinite	Nil
		Quartz	Nil
2	Exchangeable	Calcium	48.3
	Cations	Magnesium	24.5
	cmol (p^{+}) Kg^{-1}	Sodium	0.54
		Exchangeable sodium percentage (ESP)	0.6
		Calcium/Magnesium ratio	2
		Soil pH	8.2
		CEC—Cation exchange capacity	84
4	Particle size	Soil texture	Clay (CH)
	Distribution (%)	Total clay	68
		Silt	16
		Total sand	15

Expected Pitch Behaviour: Calcium and magnesium-dominant smectite clay with optimum levels of silt and sand, high CEC, and high levels of pitch surface dry density and compression levels due to good flocculated pitch soil structure with lesser air voids and large, wide, over consolidated, consistent polygonal cracks generating pace, bounce, and desirable variation in pitch behaviour (Fig. 9.63, Table 9.10).

Fig. 9.63 Predominant pitch behavioural patterns of the Brisbane, Australia—based on their predominant morphological, mineralogical, physical, chemical, and engineering properties

3. Pitch Behaviour Analysis Index (PBAI)/Pitch Behaviour Forecasting (PBF).

Table 9.11 PQI, PQS, and DBMS of a pitch profile based on mineralogical, physical, chemical, and engineering properties for Test matches (days of matches). Name of the Ground/Country- Sydney Cricket Ground (SCG)/Australia in percent

1	Mineral type (%)	Smectite	50–65
		Illite	5–10
		Kaolinite	20–30
		Quartz	20–30
2	Exchangeable	Calcium	9
	cations	Magnesium	1.8
	cmol (p^+) Kg^{-1}	Sodium	1.2
		Exchangeable sodium percentage (ESP)	4
		Calcium/Magnesium ratio	5
		Soil pH	5.9
		Cation exchange capacity (CEC)	30.1
4	Particle size	Soil texture	Clay
	Distribution (%)	Total clay	48
		Silt	15
		Total sand	12

Expected Pitch Behaviour: Calcium and Magnesium-Dominant Smectite in Moderate Quantity with a Medium Proportion of Non- or Less-Swelling Illite Kaolinites, Moderate Levels of CEC, Medium Levels of Silt, Sand, and Quartz, and Less or Negligible Cracks That Further Crumble, Assisting the Spinners in the Later Part of the Game (Fig. 9.64, Table 9.11).

Fig. 9.64 Predominant pitch behavioural patterns of the Sydney Cricket Ground (SCG), Australia - based on their predominant morphological mineralogical, physical, chemical, and engineering properties

4. Pitch Behaviour Analysis Index (PBAI)/Pitch Behaviour Forecasting (PBF).

Table 9.12 PQI, PQS, and DBMS of a pitch profile based on mineralogical, physical, chemical, and engineering properties for Test matches (days of matches). Name of the Ground/ Country- South Africa bulli in percent

1	Mineral type	Smectite	81
	(%)	Illite	14
		Kaolinite	5
		Quartz	
2	Exchangeable	Calcium	7.132
	Cations	Magnesium	2.087
	cmol (p^{+}) Kg^{-1}	Sodium	2
		Soil pH	5.9
4	Particle size	Soil texture	Clay (Smectitic)—CH
	Distribution (%)	Total clay	51.5
		Silt	36.2
		Total sand	12

Expected Pitch Behaviour: Calcium-dominant smectitic clay with moderate sand but a high level of silt will generate pace by generating high pitch surface dry density and compression levels (due to flocculated soil structure) but lesser bounce and carry due to high levels of silt (Fig. 9.65, Table 9.12).

Fig. 9.65 Predominant pitch behavioural patterns of the South Africa bulli—based on their predominant morphological, mineralogical, physical, chemical, and engineering properties

5. Pitch Behaviour Analysis Index (PBAI)/Pitch Behaviour Forecasting (PBF).

Table 9.13 PQI, PQS, and DBMS of a pitch profile based on mineralogical, physical, chemical, and engineering properties for Test matches (days of matches). Name of the Ground/ Country- Eden Garden- Kolkata in percent

1	Mineral type	Smectite	40.8
	(%)	Illite	28.2
		Kaolinite	31
2	Exchangeable	Calcium	22.69
	Cations	Magnesium	8.13
	cmol (p^{+}) Kg^{-1}	Sodium	2.13
		Soil pH	6.49
		Free swell index (FSI)	85
3	Atterberg	LL—Liquid limits	71
	Limits (%)	PL—Plastic limits	22
		PI—Plasticity index	49
4	Particle size	Soil texture	Silty clay (CH)
	Distribution (%)	Total clay	59
		Silt	41
		Total sand	0

Expected Pitch Behaviour: Calcium-dominant smectitic silty fine clays with moderate sand but a high level of silt, a medium level of cracks, and the turf grass will generate medium pace and bounce, but the surface cracks will crumble during the later stages of the game, favouring the spinners (due to very low levels of little or non-swelling illitic or kaolinitic clay and silt contents) (Fig. 9.66, Table 9.13).

Fig. 9.66 Predominant pitch behavioural patterns of the Eden Garden- Kolkata—based on their predominant morphological, mineralogical, physical, chemical, and engineering properties

6. Pitch Behaviour Analysis Index (PBAI)/Pitch Behaviour Forecasting (PBF).

Table 9.14 PQI, PQS, and DBMS of a pitch profile based on mineralogical, physical, chemical, and engineering properties for Test matches (days of matches). Name of the Ground/ Country- Barabati Stadium, Cuttack, Odisha in percent

1	Mineral type	Smectite	89.4
	(%)	Illite	0
		Kaolinite	10.6
2	Exchangeable	Calcium	37.58
	cations	Magnesium	7.10
	cmol (p^+) Kg^{-1}	Sodium	2.99
		Potassium	0.38
		Soil pH	6.62
		FSI—Free swell index	53
3	Atterberg	LL—Liquid limits	64
	Limits (%)	PL—Plastic limits	23
		PI—Plasticity index	41
4	Particle size	Soil texture	Silty clay (CH)
	distribution	Total clay	62
	(%)	Silt	37
		Total sand	1

Expected Pitch Behaviour: Calcium-dominant smectitic silty fine clay with moderate sand but a high level of silt, medium to large levels of cracks, and turf grass will generate high levels of pitch surface density and compression levels due to flocculation of clay particles with lesser air voids and extra-large, wide, consolidated, and consistent cracks generating extra pace, bounce, and desirable variation in pitch behaviour (Fig. 9.67, Table 9.14).

Fig. 9.67 Predominant pitch behavioural patterns of the Barabati Stadium, Cuttack, Odisha—based on their predominant morphological, mineralogical, physical, chemical, and engineering properties

7. Pitch Behaviour Analysis Index (PBAI)/Pitch Behaviour Forecasting (PBF).

Table 9.15 PQI, PQS, and DBMS of a pitch profile based on mineralogical, physical, chemical, and engineering properties for Test matches (days of matches). Name of the Ground/ Country- JSCA, Ranchi, Jharkhand in percent

1	Mineral type	Smectite	44.7
	(in total clay)	Illite	30.8
	(%)	Kaolinite	21.2
		Quartz	Nominal
		Soil pH	6.09
		Free swell index (FSI)	20
3	Atterberg	LL—Liquid limits	71
	Limits (%)	PL—Plastic limits	29
		PI—Plasticity index	42
4	Particle size	Soil texture	Silty clay (MH)
	Distribution	Total clay	59
	(%)	Silt	34
		Total Sand	07

Expected Pitch Behaviour: Calcium-dominant smectitic silty fine clay with moderate sand but a high level of silt and a medium level of cracks along with the turf grass will generate medium pace and bounce, but the surface cracks along with their boundaries will crumble during the later stages of the game, favouring the spinners (due to the high levels of silt contents) (Fig. 9.68, Table 9.15).

Fig. 9.68 Predominant pitch behavioural patterns of the JSC A, Ranchi black soil, Jharkhand—based on their predominant morphological, mineralogical, physical, chemical, and engineering properties

8. Pitch Behaviour Analysis Index (PBAI)/Pitch Behaviour Forecasting (PBF).

Table 9.16 PQI, PQS, and DBMS of a pitch profile based on mineralogical, physical, chemical, and engineering properties for Test matches (days of matches). Name of the Ground/ Country- Wankhede, Mumbai in percent

1	Mineral type	Smectite	Nominal
	(%)	Illite	NA
		Kaolinite	Prominent
		Quartz	Prominent
		Soil pH	5.76
		Free swell index (FSI)	28
3	Atterberg	LL—Liquid limits	59
	Limits (%)	PL—Plastic limits	32
		PI—Plasticity index	27
4	Particle size	Soil texture	Silty clay loam (MH)
	Distribution (%)	Total clay	28
		Silt	52
		Total sand	20

Pitch Behaviour Expected: Kaolinitic-dominant non-cohesive silty clay loam with non-swelling and shrinking clay and no surface cracks. Pace and bounce early in the game due to high dry densities achieved (due to high sand and quartz contents), but crumbles (loses its granular structure due to very low water holding capacity and low amounts of binding material such as clay and humus) later in the game, favouring spinners (Fig. 9.69, Table 9.16).

Fig. 9.69 Predominant pitch behavioural patterns of the Wankhede, Mumbai - based on their predominant morphological, mineralogical, physical, chemical, and engineering properties

9.12 Conclusions: Integrated Pitch Management (IPM)

There are hundreds of ways of preparing a wicket profile or a match wicket, but there are thousands of controllable and uncontrollable macro- and microfactors through which the pitch surface expresses its behaviour.

It is the integrated and synergistic effect of the pitch soil complex either the pitch soil (quantitative and qualitative composition), soil organic matter (SOM - humus), pitch grass, soil moisture, voids, soil salts (and their interaction with each other as soil-humus, soil-moisture, soil-grassroots along with flora and fauna within the rhizosheaths) with the micro- and macroweather and climatic variations available within and just above the pitch soil complex (in the form of radiative energy of sun (evapotranspiration), temperature, relative humidity, dew, etc.); along with the mechanical compactive effort of pitch rollers (in a defined rolling techniques and different pitch preparation methods) defines the cumulative nature and behaviour of pitches.

During pitch preparation, the curator always attempts to manage controllable factors (via various pitch preparation methods and techniques) while minimising the effects of uncontrollable factors (climatic and weather variations) via various pitch-making schedules and techniques. By controlling and regulating various controllable and non-controllable factors available on and within the pitch profile based on skill and the scientific reasons behind the pitch preparation schedules and techniques, the desired behaviour of the pitch can be achieved. This thought can be enumerated and visualised as a concept called "integrated pitch management," which is a combination of controllable and non-controllable factors involved in pitch preparation. Controllable factors of pitch preparations are in the hands of curators, but the impact of non-controllable factors (weather and climatic variables) on pitch preparations can only be minimised. The important non-controllable factors that are not in the hands of the curators are the atmospheric and weather parameters in the form of solar radiation, temperature, relative humidity, radiation, wind speed, rainfall, ET rates, dew, fog, mist, the type of soil and grass on the pitch, etc. Curators cannot eliminate the direct and indirect impacts of non-controllable factors in pitch preparation but can minimise their impact on pitch behaviour through various pitch preparation methods and techniques that they have mastered through their skills and experience.

It is the integrated, synergistic, and cumulative effect of all the above-mentioned controllable and non-controllable factors available on and within the pitch profile that defines the pitch's behaviour. No single factor affecting the cricket pitch soil complex is responsible for the desirable nature and variable behaviour of the cricket pitches. Even the soil organic matter (SOM), though available in a minuscule amount (less than 5%), or the exchangeable cations (as calcium, magnesium, or sodium) at microlevels, affect the pitch behaviour massively (e.g. the Perth wicket is considered the world's quickest wicket due to its magnesium and sodium-dominated smectitic clay only). Pitch soil clay quantity and quality (basically the smectites) are much more important among all the above-mentioned factors as they affect the variations in desired pitch surface density by interacting with the available moisture and the clay-humus interactions (Fig. 9.70).

Fig. 9.70 Concept of Integrated Pitch Management (IPM) and the soil (pitch), plant (turf grass), atmosphere (weather and climatic variables available within and above the pitch surface), continuum (SPAC) concept of the pitch profile

It is the SPAC of the pitch profile as pitch soil, plant (pitch grass), and atmosphere (all climatic and weather variables available above and within the pitch profile as solar radiation, evapotranspiration, temperature, relative humidity, dew, fog, mist, etc.) on a continuum with predetermined pitch preparation objectives, schedules, and techniques of the pitch preparation process in a dynamic state that defines the nature and character of the pitch profile, which varies from hour to hour, day to day, weather to weather, country to country, or in between the continents too) and every cricket pitch, by interacting with the existing available weather and climatic variations within and just above the pitch profile, has its own unique and specific language to express its behaviour during different phases of the match. To understand and comprehend these unspoken languages of pitch behaviour, a long understanding and experience of pitch making (under the existing climate and weather) with some knowledge of natural phenomena based on science is required.

Even when all of the pitch-making materials, climatic conditions, and processes are the same, the pitch behaviour varies from place to place. No single variable is responsible for defining the pitch behaviour, as each of the above-mentioned controllable and non-controllable factors interacts and regulates the cumulative behaviour of the pitch profile complex.

Even with the best pitch soil (in terms of structure, texture, and mineralogy), deep-rooted, evenly grown pitch grass, and the best available compacting schedule and effort, the pitch profile can't work at its best if the soil is dry. The pitch can be compacted to its best at the available optimum moisture content (OMC). It is the quantitative and qualitative presence or absence of the pitch soil complex, either the pitch soil, SOM, pitch grass, soil moisture and air voids, soil salts, and their interaction with each other (as soil humus, soil moisture, soil-grass roots, and flora and fauna within the rhizosheaths) with the micro- and macroweather and climatic variations in the form of radiative energy from the sun, temperature, and relative humidity that defines the nature and behaviour of any pitch.

Soil is a nanocolloidal particle that took several years of weathering to develop. The different clay-dominant soils that we are using on the pitch surface may have been created thousands of years ago in the womb of mother earth. It may take several years or generations to understand and decipher the complex nature of the soils that we are using on the pitch surface. The natural weathering forces (temperature, wind, water/ice, radiation, and time) that took lakhs or millions of year to transform the basaltic/granitic mother rocks into flour type powered fine sand, silt, and colloidal clay are the same forces (weather and climatic variables) that control the nature and behaviour of our cricket pitch soils.

The cricket pitch soil transported on the playing pitch surface will behave in the same way as it did at the source (swelling, shrinking, cracking or crumbling patterns, bulk density, water holding capacity, etc.) as the weathering natural forces in the form of temperature, relative humidity, wind, evapotranspiration rates, radiation, moisture (at their origin and excavation site and the pitch surface) are the same natural forces that controls the pitch behaviour also.

The more the groundsman or curator knows about the basic morphological, physical, chemical, and mineralogical properties of the soils that they are using on their pitches and weather variables (temperature, relative humidity, evapotranspiration, etc.), the better they will be able to comprehend and analyse or predict the nature of their pitches. The groundsmen and curators should not only concentrate on the above-mentioned soil reports but also on the origin, source, and topography of the areas from which they have excavated and transported their pitch soils.

References/Bibliography

(2016) "Soil Resource Information of Different Agro- Eco Sub-Regions (ASER) of India for crop and soil carbon modelling," Technical Bulletin, NBSS and LUP Publication 1035, ICAR, Nagpur

Adams WA, Baker SW, Carre MJ, Young RJ, James DM (2004) Pitch properties and performance. England and Wales Cricket Boards, London

Adams WA, Baker SW, James DM, Young RJ (2005) Measuring and modeling the bounce and pace of county championship cricket pitches. Int Turf Grass Soc Res J 10:1021–1026

Adams WA, Gibbs RJ (1994) Natural turf for sports and amenity, 1st edn, p 404. CAB International, Wallingford, UK

Adams WA, Tanavud C, Springsguth CT (1985) Factors influencing the stability of sports turf root zones. In: Lemaire F (ed) Proceedings of the fifth international turf grass conference, p 391. Avignon, France, Institutional de la recherche agronomique

Aitchinson GD, Holmes JW (1953) Aspects of swelling in the soil profile. Aust J Appl Sci 4:244–253

Albrecht BA (1996) Effect of desiccation on compacted clays. Msc Thesis, Department of Civil Engineering, University of Wisconsim, Madison, WI

Aldous DE (2016) Handbook on international turf management. CRC Press

Al-Khafaji AW, Andersland OB (1992) Geotechnical engineering and soil testing. Saunders College Publishing, New York

Al-Zoubi MS (2008) Swell characteristics of natural and treated compacted clays. Electron J Geotech Eng 13(D)

Anand JP, Balakrishna K, Laureano RH (2004) Volumetric shrinkage strain measurements in expansive soils using digital imaging technology. Geotechn J 27(6):1–10

Atique A, Sanchez M (2011) Analysis of cracking behavior of drying soil. In: Second international conference on environment science technology, IACSIT Press, Singapore

Baker SW, Binns DJ, Cook A, Mooney SJ (2001a) The performance of cricket pitches in relation to soil type and moisture content. Int Turf Grass Soc Res J 9:469–478

Baker SW, Cook A, Adams WA (1998a) Soil characteristics of first class cricket pitches and their influence on playing performance. J Turfgrass Sci 74:63–79

Baker SW, Cook A, Binns DJ (1998b) The effect of soil type and profile construction on the performance of cricket pitches. I. soil properties and grass cover during the first season of use. J Turfgrass Sci 74:80–92

Baker SW, Cook A, Binns DJ, Carre MJ, Haake SJ (1998c) The effect of soil type and profile construction on the performance of cricket pitches II. playing quality during the first season of use. J Sport Turf Res Inst 74:93–107

Baker SW, Hammond LKF, Owen AG, Adams WA (2001b) Influence of Clegg hammer weight on the relationship between surface hardness and ball rebound for cricket. J Turf Grass Res Inst 77

Basma AA, Al-Homoud AS, Husein Malkawi AI, Al-Bashabsheh MA (1996) Swelling-shrinkage behavior of natural expansive clays. Appl Clay Sci 11(2–4):211–227

Bosscher PJ, Connell DE (1988) Measurement and analysis of jointing properties in fine-grained soils. J Geotechn Eng, ASCE 114(7):826–843

Boward JF, Vallejo LE (1996) Clay liner crack propagation. In: Geotechnical Special Publication: Engineering Contaminated Soil and Interaction of Soil, Geomembranes (Proceedings of the 1996 ASCE).

Bowles JE (1979) Physical and geotechnical properties of soils. Wiley, New York
Brady NC, Weil RR (2008) Nature and properties of soil, 14th edn. Prentice Hall
Brain AA, Craig HB (2001) Effect of desiccation on compacted natural clays. J Geotechn Geo Environ Eng 127(1):67–75
Burdet L (1986) Turf cricket pitches, South Australian Cricket Association, Adelaide, p 28
Bureau of Indian Standards (1970) Indian Standard Classification and Identification of Soils for General Engineering Purposes, BIS, New Delhi, (Reaffirmed 1987), IS 1498
Chertkov (2002) Characteristic crack dimension of saturated drying soils. J Sci Res Dev IV:1–9
ECB Staff (2007) Recommended guidelines for the construction, preparation and maintenance of cricket pitches and outfields at all levels of the game. England and Wales Cricket Board, Lord's Cricket Ground, London, UK
Ekwue EI, Lall DZ, Stone RJ (2006) Engineering properties of major soils used in cricket pitches in Trinidad. West Indies J Eng 28(2), Technical Paper
El-Shoby MA, Rabha EA (1981) Some factors affecting swelling of clayey soils. J South East Soc Soil Eng 12:19–23
Fang H-y (1994) Cracking and fracture behaviour of soil. Geotechnical Special Publication Fracture Mechanics Proceedings of the ASCE National Convention, No 43:102–117
Fredlund DG, Rahardjo H (1993) Soil mechanics for unsaturated soils. Wiley, New York
Gobat JM, Aragan M (2004) The living soil fundamental of soil science and soil biology. Wiley
Grim RE (1953) Clay mineralogy. Mc Graw-Hill, New York
Gumbs FA, Griffith SM, Shillingford G (1981) The deign and physical features of the recently re-laid cricket pitch at the Queens Park Oval, Trinidad. J Sport Turf Res Inst 58:16–20
Harris JR (1961) The crumbling of cricket pitches. The Australian Scientist 01 April:73–74
Holtz RD, Kovacs WD (1981) An introduction to geotechnical engineering. Prentice- Hall, Upper Saddle River
Holtz WG, Gibbas HJ (1956) Engineering properties of expansive clays. In: ASCE Transaction, Paper no. 2814, vol. 121
IOG/ECB (2006) Management and maintenance of the cricket pitch. The Institute of Groundsmanship, Milton Keynes, UK
James DM, Carre MJ, Haake SJ (2005) Predicting the playing character of cricket pitches. Sport Turf Eng 8:193–207
Jayanthi PNV, Kuntikana G, Singh DN (2017) Stabilization of fine-grained soils against desiccation cracking using sustainable materials. Adv Civ Eng Mat 6(1):36–66. https://doi.org/10.1520/ACEM20160037
Johnson K, Hamilton S (1990) Background to excessive cracking of pitches at the Western Australian Cricket Association Ground, Turf Craft Australia, pp 28–32
Kotler P, Keller LK, Koshy A, Jha M (2008) Marketing management, 12th edn. Pearson
Lotse EG, Datta NP, Tomar KP, Motsara MR (1974) Mineralogical composition of some red and black soils of India. Nuclear Research Laboratory, IARI, PUSA, New Delhi, pp 216–226
Lush WM, Cummings DJ, McIntyre DS (1985) Turf cricket wickets. Search 16:142–145
McIntyre DK, Jackobsen B (1998) Drainage for sportsturf and horticulture, Horticultural Engineering Consultancy, Canberra, Australia
McIntyre DS (1985) Problems of the Melbourne test cricket pitch and their relevance to Australian turf pitches. J Turf Res Inst 61:90–91
McIntyre K (2001) Don, Cricket Wickets—Science vs fiction, horticulture engineering consultancy, 5- Brimage Palace Kambah, ACT Australia
Means RE, Parcher JV (1965) Physical properties of soils. Prentice Hall, New Delhi
Mitchell K (2010) Cricket wickets, Copyright ©Queensland Cricket, pp 1–37
Mitchell PH, Graham J, Williams DJ (1972) Cracking in drying soils. Can Geotechn J 29:263–277
Morris PH, Graham J, Williams DJ (1992) Cracking in drying soils. Can Geotechn J 29:263–277
Nagraj TS, Srinivasamurty SR, Vastalya A (1994) Analysis and prediction of soil behavior. Taylor and Francis

Naser Abu-Hejleh A, Dobroslav Z (1995) Desiccation theory for soft cohesive soils. J Geotechn Eng 121(6):493–502

Palaparthy V, Susha Lekshmi SU, John J, Sarik S, Baghini MS, Singh DN (2013) Soil moisture measurement system for DPHP sensors and in situ applications. In: 4th international symposium on electronic system design, ISED 2013, December 12–13, Singapore

Parsons S (2008) The aeration of clay soils in cricket, Engineering Doctorate (Academic year 2008–2012). Cranfield University, School of Applied Sciences

Patent No.: *277862*, Date of filing: 06/09/2012, Publication Date: 01/12/2016, A novel method for Determining cracking characteristics of fine-grained soils using laser microscopy (Indian Patent Application No. 2578/MUM/2012)

Patent No.: *358655*, Date of filing: September 4, 2013, Publication Date: February 16, 2021, An apparatus and method to determine the cracking characteristics of fine-grained soils under varied environmental conditions (Indian Patent Application No. 2873/MUM/2013)

Pearson J (2010) Basic guide to turf cricket pitch preparation, VCA

Peron H, Laloui L, Hueckel T, Hu L (2006) Experimental study of desiccation of soil. Proceedings of the Fourth International Conference on Unsaturated Soils, UNSAT 2006:1073–1084

Powai IIT (2008) Mumbai, physical, chemical, free swell index and Atterberg limit analysis reports

Prathyusha J, Uday VK, Singh DN (2013) Determination of cracking characteristics of soils for various geotechnical engineering problems, pp 219–224. In: Ulitsky VM, Lisyuk MB (eds) Proceedings of the Technical Meeting TC207—Workshop on Soil-Structure Interaction and Retaining Walls. 2–6 September 2013, Paris

Ranjan G, Rao ASR (2002) Basic applied mechanics, 3rd edn. New Age International

Rao BH, Venkataramana K, Singh DN (2011) Studies on determination of swelling properties of soils from suction measurements. Can Geotech J 48(3):375–387

Reddy ES, Sastri KR (2002) Measurement of engineering properties of soils. New Age Publication

Salini U, Girish MS (2011) Suitability of soil for cricket pitch preparation. College of Engineering Trivandrum, Technical paper at NATCON-2011, Centre for Engineering Research and Development, Kerala, pp 28–29

Sharma JK (2014) Fundamentals of business statistics, 2nd edn. Vikas Publishing

Shimpton P, James L (2009) Optimization of cricket pitch rolling. Ph.D. Thesis, Cranfield University, School of Applied Science and Natural Resource Department

Shinde SB, Kala VU, Kadali S, Tirumkudulu MS, Singh DN (2012) A novel methodology for measuring the tensile strength of expansive clays. Geomech Geoeng 7(1):15–25

Singh A (2009) Modern geotechnical engineering, 3rd edn. CBS Publication

Singh DN (2008) Physical, chemical, free swell index and Atterberg limit analysis reports of cricket pitch soils under the BCCI, IIT, Mumbai

Singh DN, Uday KV (2013) Application of laser microscopy for studying crack characteristics of fine- grained soils. Technical Note, Geotechn Test J 36(1):1–9

Sreedeep S, Singh DN (2006) Methodology for determination of osmotic suction of soils. Geotech Geolog Eng 24(5):1469–1479

Sreedeep S, Singh DN (2011) A critical review of the methodologies employed for soil suction measurement. Int J Geomech 11(2):99–104

Sridharan A, Prakash K (2000) Classification procedures for expansive clays. Inst Civ Engrs Geotech Eng Thomas Telford Press Ltd, pp 235–240

Stewart VI, Adams WA (1969) Soil factors affecting the control of pace on cricket pitches. In: Proceedings of the First International Turf grass Conference, pp 533–544

Stirk GB (1954) Some aspects of soil shrinkage and the effect of cracking upon water entry into the soil. Australian J Agri Res 5:279–290

Taiton NM, Klug JR, Edmondson D, Campbell RK, Deventer PWV, Beer MJD (1993) Principles and practice of pitch preparation, United Cricket Board of South Africa

TCCB and NCA (1989) Recommendations on the preparation of cricket pitches, Test and county Cricket Board and National Cricket Association, Lords, London

Tej PR, Singh DN (2013) Estimation of tensile strength of soils from penetration resistance. Int J Geomech 13(5):496–501

Thakur VKS, Sreedeep S, Singh DN (2006) Laboratory investigations on extremely high suction measurements for fine-grained soils. Geotechn Geolog Eng 24(3):565–578

Thornthwaite CW (1948) An approach toward a rational classification of climate. Geog Rev 38:55–94

Towner GD (1987) The mechanics of cracking of drying clay. J Agri Engg Res 36:115–124

Uday KV, Jayanthi NVP, Singh DN (2014) A generalized relationship for determination of tensile strength of fine-grained soils from shrinkage characteristics. Dry Technol 32:869–876. https://doi.org/10.1080/07373937.2013.873804

Uday KV, Prathyusha NV, Jayanthi SDN, Apte PR (2015) Application of Taguchi method in establishing criticality of parameters that influence cracking characteristics of fine-grained soils. Dry Techn 33(9):1138–1149

Uday KV, Singh DN (2013) Investigations on cracking characteristics of fine-grained soils under varied environmental conditions. Dry Techn 31(11):1255–1266

Venkataramana K, Hanumantha RB, Singh DN, Harendranath CS (2008) Some investigations on cracking characteristics of soils. 12th international association for computer methods and advances in geomechanics conference, 1–6 October 2008. Goa, India, pp 1532–1538

Venkataramana K, Rao BH, Singh DN (2009) A critical review of the methodologies employed for determination of tensile strength of the fine-grained soils. J Test Evaluat 37(2):115–121

Venkatramaiah C (2006) Geotehnical engineering. ISBN: 978–93–86649–70–6. New Age International

Vogel HJ, Hoffmann H, Leopold A, Roth K (2005a) Studies of crack dynamics in clay soil II—a physically based model for crack formation. Geoderma 125:213–223

Vogel HJ, Hoffmann H, Roth K (2005b) Studies of crack dynamics in clay soil I—experimental methods, results, and morphological quantification. Geoderma 125:203–211

Wanare R, Shetty R, Prathyusha J, Singh DN, Kannan KRI (2020) Investigation to quantify suction characteristics of marine soil during drying- and wetting-cycles. Geotech Test J 44(1). DOI: https://doi.org/10.1520/GTJ20190199

Wood C (2006) Recommended guidelines for the maintenance of cricket pitches and outfields at all levels of game. England and Wales Cricket Board in association with the Institute of Groundsmanship

Yesiller N, Miller CJ, Inci GY, Yaldo K (2010) Desiccation and cracking behavior of three compacted landfill liner soils. Eng Geo 57(1–2):105–121

Annexure 1

Pitch preparation Questionnaire/Schedules

Conceptualized and compiled by: Dr. S. B. Singh, Curator, BCCI/JSCA.

This survey is being done in accordance with the Indian Data Protection Act, and the results will only be used to improve the pitches and outfields through analysis. The participant's anonymity is the only thing that matters, so the data will be looked at and published as a whole.

Please enter as much as you wish (not all information will be appropriate for every ground).

Name of the Curator/Groundsman	
Name of the Association	
Educational Qualifications	
Number of matches curated	Ranji-, Test matches- ODI -, T-20 -
Name of the ground/ stadium	
No. of pitches	
Contact No./Email Id	

1. **Pitch Profile**

Number of layers in your wicket profile	2, 3, 4, 5 or multi-layered wicket
Depth of playing surface	
Depth of pitch drainage layer	
Year of laying the new pitch	
whether relayed	Yes/No
If yes, then when	
whether same soil used	Yes/No
Depth of water table below the pitch drainage layer	

(continued)

S. B. Singh et al., *Cricket Pitches*, Springer Transactions in Civil and Environmental Engineering, https://doi.org/10.1007/978-981-99-2913-9

(continued)

Whether subsoil drainage system is installed	Yes/No
If yes, then of what type	

2. Pitch Soil

Pitch soil source	
Whether examined for physical, chemical, or mineralogical properties	Yes/No
If yes, then examined from which institution	
Particle size distributions (PSD)	clay - —, sand - ——, silt - —- in %
Morphological properties	
Physical properties	
Chemical properties	
Predominant minerals	Smectite/ illite/ kaolinite/ mixed
Engineering properties	

3. Pitch grass

Grass variety	
When sowed	
Grass leaf size	broad / narrow
Grass root depth	
4. Irrigation water - Source	
Whether tested	Yes/ No
If Yes, then for what parameters	
5. Weather variables	Mean annual rainfall -
Mean annual temperature -	Maximum temperature-
minimum temperature-	Mean annual evapotranspiration rates-
Mean annual relative humidity-	Elevation- (in Metre, above the Mean Sea Level)
Dew point -	

F: Roller type and rolling schedules: -

What type and weight of rollers do you have-?

Lightweight ——————, Medium weight ——————-

Heavyweight ——

Whether it is manually driven or mechanically driven,

Whether the rollers are tailor-made or supplied by the company-

Kindly provide the roller specifications as-

	Light	Medium	Heavy
Number of drums			
Roller width (cm)			
Roller diameter (cm)			
Roller weight (Kg)			
Mass /Width factor			

At what speed do you generally roll? ——— (Km/hrs).
What time, generally, it takes to complete one rolling session with:
Light roller: ————, Medium roller: ————-, Heavy rollers: —————.

G. **Pitch preparation methods / techniques**

1. What is the optimum grass height that you consider during the initial phases of pitch preparation? ———————————-.
2. Up to what depth do you water the pitch before the start of pitch rolling? ———
3. How much time do you prefer to water the pitch before rolling? —————
4. Do you consider the season, weather conditions, previously available moisture within the pitch profile, pitch soil type, and texture before watering? If yes, please specify.
5. By what method do you measure the depth of water within the pitch profile before rolling? ——————————————————
6. By what method and technique do you determine the time of rolling? ————
7. What is the role of turf grass in drying and desiccating the pitch surface, and how is it different from rolling- ——————————————
8. What is the total rolling and drying time during pitch preparations for different formats of the match? ———, OD I———-, T-20————-.

H. Kindly specify the rolling and watering schedules during the pitch preparation of Test, Ranji, ODI, and T-20 matches.

Days before match	Grass height (mm)	Pitch watering Quantity/ intensity	Type of rolling cross/ diagonal/ Straight, etc	Roller weight (Kg)	No. of passes per Session	Time taken per session	Temp Max./ Min	RH (%)	Dew Point	ET rate
14										
13										
12										
11										
10										

(continued)

(continued)

Days	Grass	Pitch watering	Type of rolling	Roller	No. of	Time taken	Temp	RH	Dew	ET
9										
8										
7										
6										
5										
4										
3										
2										
1										
Total										

Annexure 2

Pitch Quality Index (PQI) / Pitch Quality Standards (PQS)

Conceptualized and compiled by: Dr. S. B. Singh, Curator, BCCI/JSCA.

Properties	Desirable values / properties	Actual values
Morphological Properties, pitch soil structure, pitch construction layers	Pitch soil colour, consistency, cracks number and size, pitch profile depth, number of layers either 2, 3 or multi-layered	
Physical Properties		
Soil Texture		
Clay (< 0.002 mm)	50–70%	
Silt (0.002 – 0.2 mm)	5–20%	
Sand-Fine (0.02–2.5 mm)	10–20%	
Medium (0.25–1.00 mm)	0–10%	
Coarse (1–2.0 mm)	1%	
Crushing Strength	0.8–1.6 MPA	
Cracking Pattern	2–5 per desirable	
	6-10pcs acceptable	
	> 11 Not acceptable	
Linear Shrinkage	8–15%	
Emersion Test		
Dispersion Index	Value 6 is desirable	
FSI- Free Swell Index	< 50	
Expansion 0–20 21–50 51–90 91–130 130	Potential Very low Low Medium High Very High	

(continued)

S. B. Singh et al., *Cricket Pitches*, Springer Transactions in Civil and Environmental Engineering, https://doi.org/10.1007/978-981-99-2913-9

(continued)

Properties	Desirable values / properties		Actual values
Chemical Properties			
Sodium level	< 5 me %		
EC			
H^+, Al^{+3}, Na + , Ca^{+2}, Mg^{+2}, K^+ (exchangeable cations), Ca^{+2}/Mg^{+2} Ratio			
ESP	< 3–5		
pH levels		6–7 in water and 5–6 in $CaCl_2$	
$CaCO_3$ or $MgCO_3$		< 0.15	
		< 1%	
Organic Matter / humus		< 5%	
Engineering Properties			
Atterberg Limits		Liquid limits	
		Plastic limits	
		Shrinkage limits	
		Plasticity Index	
Clegg Hammer Test- Average pitch		(207–424) g	readings
Slow pitch		< 200 g	
Medium Fast pitch		(200–350)	
Fast pitch		> 350 g	
Mineralogical Properties Both direct and indirect methods		Either Kaolinite, Illite, Smectite, or mixed minerals	

Annexure 2a

Pitch Behavioural Analysis Index (PBAI)*

Based on pitch soil Mineralogical, Morphological, physical, chemical, engineering, turf grass conditions, and weather variables. Conceptualized and compiled by: Dr. S. B. Singh, Curator, BCCI/JSCA.

Name of the Ground/ Country-

1	Predominant mineral Type (%)	Smectite
		Illite
		Kaolinite
		Quartz
2	Exchangeable	Calcium
	cations	Magnesium
	cmol (p +) Kg^{-1}	Sodium
		ESP- Exchangeable Sodium %
		EMP- Exchangeable Magnesium %
		Calcium/ Magnesium ratio
		Soil pH
	cmol (p +) Kg^{-1}	CEC- Cation Exchange Capacity
		FSI- Free Swell Index
		COLE
3	Atterberg	LL- Liquid limits
	Limits (%)	PL- Plastic limits
		SL- Shrinkage Limits
		PI- Plasticity Index
4	Particle Size	Soil Texture
	Distribution (%)	Clay- Fine
		Total Clay
		Fine/ Total clay ratio

(continued)

S. B. Singh et al., *Cricket Pitches*, Springer Transactions in Civil and Environmental Engineering, https://doi.org/10.1007/978-981-99-2913-9

(continued)

		Silt
		Sand- Fine
		Medium
		Coarse
		Total Sand
5	Climatic / Weather	Evapotranspiration rate
	conditions	Relative humidity
		Temperature- minimum
		Maximum
		Solar radiation
		Dew Point

Pitch photographs taken during pitch preparation or during different phases of the match showing one complete swell-shrinking cycle, either from full wetting during the initial pitch preparation process to complete drying and desiccation of the pitch surface (until match completion).

Pitch soil structure	Pitch soil colour,	Actual turf grass conditions before mowing
During various stages of pitch preparation (mowing, watering, rolling, brushing, drying, etc.),	Crack initiation and propagation or crumbling (wear and tear) of the pitch surface during different phases of the match	Wear and tear of the match pitch just after completion of the match

Expected Pitch Behaviour:
Gradual wear and tear, cracking or crumbling patterns based on the pitch soil morphological, structural, particle size distribution, chemical, mineralogical, and index properties

*PBAI—Applicable for preliminary pitch behaviour analysis and forecasting only.

Annexure 3

Benchmark Soils for Cricket Pitches of India

S. B. Singh et al., *Cricket Pitches*, Springer Transactions in Civil and Environmental Engineering, https://doi.org/10.1007/978-981-99-2913-9

A Maharastra

S NC	Location	Series-Profile No.	Texture-Mineralogy	Pedon depth	Colour	Particle Size			EC	Ph	BD	WR	WR	CEC	Exchangeable Cations				BS
						Clay	Silt	Sand				33	1500		Ca	Mg	Na	K	
	Konkan Coast																		
1	Bhurapada, Palghar, Thane	Palghar,47A/G17	Fine, Mont.	53+	Dark Greyish Brown	61.7	28.9	9.4	8.5	7.8		34.9	24.8	54.7	9.1	13.03	29.25	2.5	98.5
	These soils are moderately to strongly saline due to inundation of these soils during the high tide.																		
2	Virthan, Palghar, Thane	Virthan, 47A/G16	Fine, Smectite	61-85	Dark Greyish Brown	56.6	27.6	13.8	0.2	8.4	1.94	30.2	18.7	41.8	27.4	4.62	0.52	0.38	79.7
3	Bhurapada, Palghar, Thne	Palghar Series	Fine, Mont. (Calc.)	35-53	Dark Greyish Brown	61.7	28.9	9.4	8.5	7.8		34.9	27.8	47.5	8.6	10.72	21.5	2.25	90.4
	These soils are developed due to alluvial and colluvial deposits near the swampy deposits of creeks.																		
	Western Ghats																		
4	Dahisar, Palghar, Thane	Dahisar,47A/SI/P10	Clayey- skeletal	0-13	Black	48.7	16.4	34.9	0.52	6.5		32.8	16.7	36	20	5.95	1.68	1.14	79.9
5	Haloi, Palghar, Thane	Haloi,47A/SI/P8	Fine, Mont.		Dark Brown	52.7	26.4	20.9	<0.02	7.4		32.5	20.4	48.1	31	14	0.94	0.27	96
6	Chimanipada, Navapur, Dhule	Anjana,46K/SIII/P6	Fine, Mont.	41-55	Very Dark Grey	56.5	28.1	15.4	<0.02	7.1		40	22	50.4	35.3	13	0.38	0.35	97.2
7	Kolada, Navapur, Dhule	Navapur,46G/SII/PI	Clayey- Mixed	14-36	Very Dark Greyish Brown	53.2	20.4	26.4	<0.02	7		36.8	24.2	37.4	23	8	0.3	0.3	84.5
	North Deccan- Maharastra Plateau																		
8	Karkata,Ausa,Latur	Latur,56B/SIV/P10	Clayey,Mont.	43	Dark Brown	59.9	34.5	5.6	<.02	6.9		36.9	25.8	48.2	28	17	0.95	0.54	96.5
9	Jhikri,Jamkhed,Ahmadnagar	Jamkhed,47/N/SII/PI	Clayey,Mont.	25-36	Very Dark Greyish Brown	58.5	29.6	11.9	0.25	7.5	1.9	36.5	24.2	52.3	29.7	12.04	0.42	0.39	81.4
10	Ambulga,Ahmadpur,Latur	Ambulga,56B/SI/P3	Clayey,Mont.	42	Very Dark Greyish Brown	64.2	32.4	3.4	<.20	7.2		37.4	28.1	50	31	15	0.79	0.72	95
11	karlewadi,Ambegaon,Pune	Karlewadi,47F/SIII/P4	Clayey,Mont.	76-90	Dark Brown	54.5	23.5	22	0.3	8.3	1.66	32.2	21.3	48.7	41.4	4.12	0.62	0.61	96
12	Kameri,Valva,Sangli	Valva,47K/SIX/P9	Fine, Mont.	105-150	Very Dark Greyish Brown	58.5	27.3	14.2	0.35	8.1	1.88	37.8	27.8	36.49	25.8	3.13	0.29	0.35	81
13	Bumne, Gargoti, Kolhapur	Bumne, 47L/SIII/P7	Fine, Mont.	115-160	Reddish Brown	62	34	4	<.02	7.5		47.9	39.1	54.4	29.8	19.4	0.4	0.45	91
14	Parpoli, Ajra, Kolhapur	Parpoli, 47L/SIV/P1	Fine,Mixed	80	Dark Reddish Brown	60.7	33.7	5.6	<.02	5.4		32.9	20.3	11.7	5	2	0.48	0.36	67
15	Bamni, Latur	Bamni, 56B/SI/P8	Very Fine, Mont.	110-150	Very Dark Greyish Brown	65.4	31.8	2.8	<.02	7.7		41.6	31.4	50	30.6	13.2	1.19	0.84	95.5
16	Penjalwadi,Bhor,Pune	Penjalwadi,47J/G100	Very Fine,Mont.	55-65	Dark Brown	57.8	30	12.2	0.2	7.7	1.89	43.1	33.6	56.4	41.1	11.67	0.53	0.38	95.2
17	Sirsagaon,Sirur,Pune	Sirsagaon Series	Fine Smectitic	32-55	Dark Brown	46.2	17.4	36.4	0.2	7.9	1.5			49.8	43.1	6.3			
18	Nimone,Sirur,Pune	Nimone Series	Fine, Smectite	76-125	Dark Greyish Brown	59.1	12.7	18.1	0.5	8.5	1.6			56.1	33	8.1			

S NO.	Location	Series-Profile No.	Texture-Mineralogy	Pedon depth	Colour	Particle Size			EC	Ph	BD	WR	WR	CEC	Exchangeable Cations				BS
						Clay	Silt	Sand				33	1500		Ca	Mg	Na	K	
44	Umri, Yeotmal	Umri/55I/SIII/P8	Fine, Mont.	64-82	Light Brownish Grey	53.5	30.8	15.7	0.2	8.4	1.99	29.9	19.6	53	36.6	3.79	0.23	0.78	78.1
45	Loni, Yeotmal	Loni Series	Very Fine Smectitic	65-99	Very Dark Greyish Brown	76.3	18.8	4.9		6.4	1.4	38.3	35.9	63	37.8	27.3	1.3	0.7	106
46	Kinhala, Umrer, Nagpur	Kinhala/55P/SVIII/P9	Fine, Mont.	125-150	Very Dark Greyish Brown	59.5	28	12.5	<.20	8.7	1.99	41.2	30.4	52.3	35.7	10.89	0.3	0.62	90.8
47	Takali, Kuhi, Nagpur	Takali/55O/SIV/P12	Fine, Mont.	125-150	Dark Brown	59.5	26.5	14	<.20	8.2	1.9	38.8	27.6	51.3	38.75	9.16	0.31	0.32	94.6
48	Karla, Katol, Nagpur	Karla Series	Fine Mont.	40-66	Dark Brown	66.1	23.5	10.4	0.3	7.8	1.8	53	26.6	58.4	38.8		16.2	1.3	98
49	Ag. College Farm, Nagpur	Aroli	Fine Mont.	17-42	Very Dark Greyish Brown	60.9	30.3	8.8	0.22	8.4	1.85			58.2	57.4		0.4	0.4	
50	Andhali, Kurkheda, Gadchiroli	Andhali/64D/SIII/PI	Fine, Mont.	125	Dark Greyish Brown	53.5	25.5	21	0.2	8.2	2	36.6	20	42.6	20.7	4.62	0.84	0.39	62.3
51	Palandur, Deori, Bhandara	Palandur/64D/R/2	Fine Loamy Mixed	145	Yellowish Brown	50.4	18.8	30.8	0.06	7		24.3	15.8	20.4	12.55	5.75	0.47	0.43	94.1
52	Bhugaon, Nagpur	Bhugaon/55O/G/104	Fine, Mont.	125	Very Dark Greyish Brown	64.5	26.3	9.2	<.20	8.3	2.01	44.7	34	55.5	37.69	9.9	0.48	0.67	87.3
53	Palaskheda, Chandur, Amravati	Jambha Series	Very Fine, Smec.	20-45	Very Dark Greish Brown	71.5	13.3	15.2	0.15	8.5	1.88	34.5	21.5	49.6	48.3		0.4	0.9	100
				105-145	Very Dark Greish Brown	73	18.3	8.7	0.2	8.6	1.97			49.7	46		2.8	0.9	100
	Clay intersecting slickensides, coarse angular blocks with shinny pressure faces, very hard, very firm, very sticky and plastic, slightly effervescent, strongly alkaline.																		
	Purna Valley	Series	Smectites	Pedon	Colour	Clay	Silt	Sand	EC	Ph	HC	COLE	VSP	CEC	Ca	Mg	Na	K	ESP
54	Asra, NE	Purna Valley	980 gm/kg	27-57	Greish Brown to	72.2	26.3	1.5	0.3	8.3	2.6	0.23	86	61.3	38	14	0.6	1.1	
55	Khairi,NE	Purna Valley	970	85-145	Dark Brown	66.8	28.7	4.5	0.22	8.3	1.8	0.25	95	52.2	22.4	22.8	0.6	1.2	1.1
56	Vihigaon,NE	Purna Valley	960	53-123	"	63.6	31.9	4.5	0.2	8.2	3.1	0.28	100	71.3	42.8	17.6	0.7	1.2	1
57	Ghodasgaon,SW	Purna Valley	980	32-98	"	68.3	30.4	1.3	0.4	8.7	0.3	0.27	100	61.7	28	18.8	6.9	1.2	11.2
58	Thillory,SW	Purna Valley	960	87-134	"	57.2	25.3	17.5	0.33	9.1	0.9	0.21	77	58.3	21.2	16.8	13.6	0.8	23.3
59	Abdulbazar,SW	Purna Valley	970	43-94	"	68.1	28.3	3.6	0.35	8.7	0.8	0.23	86	63.5	27.7	16.2	16.7	1.7	26.2
60	Hingna,SW	Purna Valley	900	57-75	"	69.6	24.8	5.6	0.44	8.4	4.2	0.25	95	53.9	14.6	24.5	1.3	1.2	2.4
61	Paral,SW	Purna Valley	980	69-113	"	66.8	29.4	3.8	0.2	8.8	0.3	0.25	95	61.3	14.8	22.3	12.2	1.1	19.9
62	Kesapur, Hingoli	Kesapur, Hingoli		0-14	Dark Reddish Brown	65.7	32.3	2	0.44	9.08	0.08	0.26		55.65	28.61	23.39	9.13	0.22	16.4
				49-81	Very Dark Grey	71.5	27.4	1	0.22	8.23	2.18	0.31		62.6	49.71	5.91	0.75	0.27	1.2

S NO	Location	Series-Profile No.	Texture-Mineralogy	Pedon depth	Colour	Particle Size Clay	Silt	Sand	EC	Ph	BD	WR 33	WR 1500	CEC	Exchangeable Cations Ca	Mg	Na	K	BS
19	Wagha, Kaij, Beed	Babulgaon, 47N/SIV/P1	Fine, Mont.	88-150	Very Dark Greyish Brown	52.3	29	18.7	<.20	8.1	1.95	30.5	19	51.3	39.4	8.41	0.41	0.61	95.1
20	Targaon, Koregaon, Satara	Targaon, 47K/SIV/P5	Fine, Mont.	165	Brown to Dark Brown	63.7	31.4	4.9	0.8	8.3		52.3	39.2	45	20.2	17.4	6.4	0.6	99
21	Shiware, Shahuwadi, Kolhapur	Shiware, 47L/SVIII/P7	Very Fine, Mont.	165	Dark Brown	66.5	27.4	6.2	0.3	8.1		43.6	32.9	55.5	30.3	17.49	0.85	0.65	88.8
22	Wathar, Koregaon, Satara	Koragaon, 47K/NMK/SIV/P2	Fine,Mont.	150	Very Dark Greyish Brown	65.4	28	6.6	0.45	8	2.05	38.7	28.6	51.2	31.4	14.35	0.98	0.5	92.2
23	Bhoirepathar,Ahmadnagar	Pather,47I/SXII/P3	Fine,Mont.	60-75	Very Dark Greyish Brown	53.5	30	16.5	0.25	8.3	1.69	30.2	19.9	63.9	46.4	4.95	0.37	0.52	81.8
24	Chandhai, Mangrulpur, Akola	Mangrulpir, 55H/SVI/P5	Clayey, Mont.	27-36	Very Dark Greyish Brown	54.6	33	12.4	<.20	7.6	1.94	36.6	23.9	45.8	30.5	9.2	0.35	0.45	90.6
25	Satana,Vajapur,Aurangabad	Vaijapur/47I/SVIII/P9	Fine, Mont.	49-73	Very Dark Greyish Brown	53.5	31.4	15.1	<.20	7.6	1.85	34.4	24.8	53.8	32.5	12.66	0.44	0.45	94.2
26	Ghoshpuri, Ahmadnagar	Ghoshpuri/47J/SVI/P9	Clayey, Mont.	28-45	Very Dark Greyish Brown	43.5	19.3	37.2	0.2	8		37.9	25.9	39	22.6	13.2	0.8	0.54	95
27	Dhulgaon, Mahankal, Sangli	Dhulgaon/47L/SI/P12	Fine, Mont.	35-100	Dark Yellowish Brown	49.5	22.5	28.5	0.3	8.5		31.5	22.9	30.7	11.4	13.9	2.27	0.26	89.3
28	Wadachiwadi, Jintur, Parbhani	Jintur/56A/SII/P4	Fine, Mont.(Cal.)	100-150	Dark Greish Brown	60.5	29.5	10	0.27	8.3	1.96	38.2	27.2	59.66	42.5	7.42	0.43	0.74	85.6
29	Sewagram,Wardha	Wardha/55L/SI/P9	Fine, Mont.(Cal.)116-150		Very Dark Greyish Brown	65.2	26	8.8	0.21	6.6		36.6	27.3	52	35.6	14.35	0.35	0.9	98.5
30	Umrer,Nagpur	Umrer/55P/SX/P14	Fine, Mont.	99-150	Very Dark Greyish Brown	37	8.2	54.8	<.20	8.1	2	33.6	18.3	39.9	26.6	11.33	0.68	0.57	96.7
31	Purandarwardha,Malsiras,Solapur	Purandarwardha/47K/SI/P6	Fine, Mont.(Cal.)80-110		Dark Greyish Brown	59	33.6	7.4	0.35	8.2		43.5	31.5	44	22.8	15.6	3.1	0.36	95
32	Bandhalwadi,Malsiras,Solapur	Bandhalwadi/47O/SIV/P10	Very Fine, Mont.	25-50	Dark Brown	62.4	28.6	9	0.2	8.1		49.1	31	54.7	31.2	12.8	9.4	0.6	91.4
33	Masala, North Solapur, Solapur	Masala/47O/SII/P8	Fine, Mont.(Cal.)	100	Dark Brown	60.8	29.4	9.8	0.25	8.2		42	31.4	63.9	49.4	8.58	0.62	0.75	92.9
34	Kalegaon,Barsi,Solapur	Barsi	Very Fine Mont.	69-114	Very Dark Greyish Brown	76.2	21	2.8	0.5	8.4	1.87			57.5	50.2		6.7	0.6	
35	Thejoda,Satara,Nasik	Thejoda/46L/SIV/P12	Fine, Mont.(Cal.)	50-125	Dark Yellowish Brown	55.5	28.5	16	0.3	8.5	1.79	35.9	22.5	32.9	19.9	7.56	1.13	0.48	88.3
36	Mendki, Katol, Nagpur	Katol/55K/S3/P16	Fine, Mont.(Cal.)	125	Dark Greyish Brown	63.5	30	6.5	<.20	8.4	2	36.1	25.9	54.1	40.77	6.66	0.79	1	90.9
37	Wandli, Katol, Nagpur	Linga Series	Very Fine Smectitic	102-128	Very Dark Greyish Brown	71.2	28.4	0.4		8	1.5	41.2	36.5	61.8	37.8	14.4	0.9	0.8	87
38	Regional Fruit Research, Wandil, Katol, Nagpur	Linga Series	Very Fine Mont.	84-117	Very Dark Greyish Brown	74.6	15.1	10.3	0.2	8.1				69.4	64		4.8	0.6	
39	Sirpur,Deoli,Wardha	Sirpur/55L/SII/P13	Dark Greyish Brown	90-150	Brown	63.1	29	7.9	0.3	8.5		40.6	29.4	43.2	19.6	12.37	0.61	4.26	85.2
40	Jaipur,Kalvan,Nasik	Jaipur/46H/SIII/P13	Fine, Mont.	150	Very Dark Greyish Brown	60.1	32.9	7	<.20	8.2		38	28.6	47.2	26	18	2.64	0.48	99.8
41	Sewagram,Wardha	Sewagram/55L/SI/P5	Loamy Mixed	20-Jan	Yelloiwh Brown	35.5	19.9	44.6	0.3	7.7	2.02	20.3	11.7	34.8	23.8	8.58	0.23	0.28	94.5
42	Talegaon,Karanja,Wardha	Talegaon/55K/SII/P11	Clayey, Mont.	28	Very Dark Greyish Brown	53.7	29.5	16.8	<.20	7.8	1.87	32.4	22.4	52	40.1	10.24	0.31	0.64	98.4
43	Talegaon, Rahuri, Ahmadnagar 1.5 Km northeast of Wambori railway station.	Talegaon Series	Fine, Smectite	2–25	Dark Reddish Brown	48.4	29.3	22.3	0.2	7.8				48.2	30.5	14.7			

S NO.	Location	Series-Profile No.	Texture-Mineralogy	Pedon depth	Colour	Particle Size			EC	Ph	BD	WR	WR	CEC	Exchangeable Cations				BS
						Clay	Silt	Sand				33	1500		Ca	Mg	Na	K	
B	**Madhya Pradesh**																		
1	Pasposi,Lakhanadon,Seoni	Pashposi,55N/SIV/P6	Fine,Mont.	43-70	Dark Brown	57.1	19.3	23.6	<.20	5.9	1.6	37.2	24.3	44.4	24	15.4	0.84	0.37	91
2	Chitukgondi,Sausar,Chhindwara	Chitukgondi,55K/SII/P13	Fine,Mont.(Cal)	35-50	Very Dark Greish Brown	54.6	26.4	19	<.20	7.7	1.7	40.5	28.2	34.2	21.4	10	0.39	0.24	93
3	Amziri,Chhindwara	Amziri,55K,SI/P3	Fine,Mont.	116-150	Dark Greish Brown	65.2	29	5.8	<.20	7.5	1.79	45.5	33.3	40.2	24	14	0.42	0.25	96
4	Makajhir,Lakhanadon,Seoni	Makajhir/55M/P8	Fine,Mont.	73-96	Very Dark Greish Brown	53.7	17.1	19.2	<.20	6.4	1.74	43.4	28.2	46	27.8	14.2	0.88	1	95.6
5	Haidarpur,Burnapur,Khandwa	Haidarpur/55C/SV/P1	Fine,Mont.	90-125	Very Dark Greish Brown	57.3	25.8	18.9	0.2	8.1	1.73	40.2	31.9	52.5	31.6	13.2	2.61	0.43	91
6	Dunova,Multai,Betul	Dunova/55K/SIII/P3	Fine,Mont.	80-100	Very Dark Greish Brown	60.3	25.2	14.5	<.20	7.7	1.73	39.5	28.5	38	24	13	0.67	0.56	101
7	Savli,Sausar,Chhindwara	Salvi/55K/SII/P16	Fine,Mont.	140-150	Very Dark Greish Brown	61.3	21.5	17.2	<.20	8	1.73	44.3	31.5	38	23	12	0.5	0.32	94
8	Hiranchanda,Itarsi,Hosangabad	Hiranchanda/55F/SVII/P5	Fine-Loamy,Kaolinitic	57-91	Dark Brown	62.1	14.2	23.7	<.20	6.6	1.82	38.4	26.5	54	28	20	0.63	0.38	90
9	Loni,Kukshi,Dhar	Loni/46J/SV/P3	Fine,Mont.(Cal)	29-54	Drak Yellowish Brown	53.6	26.3	20.1	1.6	7.6	1.72	32.5	21.6	48.3	31.2	13.6	0.45	0.4	94
10	Jalda,Kasarawad,Khargone	Jalda/46N/SVIII/P7	Fine,Mont.	36-69	Very Dark Greish Brown	54.2	25.3	20.5	<.20	7.6	1.73	33.4	24.6	46.29	25.6	16	1.04	0.52	92
11	Muskari,Sehora,Jabalpur	Muskari/64A/SII/P12	Fine,Mont.	109-145	Very Dark Greish Brown	65.3	23.3	11.4	<.20	7.8	1.82	31.9	22.7	34	19.6	10.8	0.46	0.46	92
12	Gopalpur,Kasarvad,Jabalpur	Gopalpur/46N/SVIII/PI	Fine,Mont.	77-105	Very Dark Grey	57.3	25.3	17.4	<.20	8.3	1.82	37.5	26.9	50	31.2	13.2	3.09	0.56	96
13	Kheri,Jabalpur	Kheri Series	Very Fine Mont.	84-129	Very Dark Greyish Brown	64.4	20.5	15.1	0.1	8	1.94			56.8	55.8		0.2	0.8	
14	Tejgarh,Tendukheda,Damoh	Tejgarh/55M/SIV/PI	Fine,Mont.	57-90	Very Dark Greish Brown	53.4	27.5	19.1	<.20	7.7	1.67	27.5	14.3	30	10	9.5	0.25	0.37	82
15	Baroda Kalam,Hatta,Damoh	Baroda Kalam/54P/G96	Fine,Mont.	100-125	Dark Greish Brown	60	24	16	<.20	7.7	1.82	32.6	19.3	42	32.8	7.8	0.45	0.85	100
16	Mariadar,Hatta,Damoh	Mariadar/54P/SV/P24	Fine,Mont.	102-140	Dark Brown	61.8	21.2	17	<.20	6.8	1.76	26.8	18.5	49	36.2	11.8	0.18	0.42	99
17	Madanpur,Gairatganl,Raisen	Madanpur/55I/7St5/P7	Fine,Mont.	94-130	Dark Yellowish Brown	70	14	10	<.20	7.9	1.83	33.4	23.6	48.6	36.8	12.5	0.46	0.18	91
18	Piparani,Kukshi,Dhar	Piparani/46J/G64	Fine,Mont.	50-75	Very Dark Greish Brown	53	32.2	24.8	0.2	8	1.78	31.6	21.6	54.3	15.2	34.4	0.51	0.44	92
19	Namli,Ratlam	Namli/46M/SV/P2	Fine,Mont.		Very Dark Grey	60.4	23.9	15.7	<.20	8.2	1.82	29.5	19.6	54.3	44	6.9	1.1	0.5	97
20	Gorsonera,Ashoknagar,Guna	Gorsonera/55H/I/P5	Fine,Mont.	97-130	Very Dark Greish Brown	58.4	26	15.6	0.5	7.4	1.78	29.5	19.6	54.4	30.7	17.3	0.43	0.73	90
21	Jindakheri,Sanwer,Indore	Jindakheri/46N/SII/P17	Fine,Mont.	68-95	Dark Greish Brown	57.5	25.5	17	<.20	7.7	1.81	32.5	21.8	49.4	40.4	6.6	0.4	0.55	96
22	Deorikalam,Dubra,Gwalior	Deorikalam/54G/SIV/P14	Fine,Mont.	113-138	Dark Yellowish Brown	53.6	23	23.4	0.4	8.2	1.7	17	10.1	46.8	25.2	19.2	0.5	0.6	97
23	Rapata,Katghora,Bilaspur	Rapata/64M/SII/P3	Fine Loamy, Mixed	32-82	Dark Reddish Brown	39.8	12	48.2	<.20	6.6	1.63	20.4	9.4	21.3	11.4	2.5	0.3	0.3	69
24	Boda,Mungel,Bilaspur	Boda/64J/SVIII/P3	Fine,Mont.	88-115	Dark Greish Brown	60.9	30.7	8.4	<.20	7.6	1.75	29.6	16.2	43.2	32	5	0.81	0.68	88
25	Sirgeri,Mahasamund,Raipur	Sirgeri/64K/VII/P15	Fine, Mixed	125-170	Brown	62.4	23	19.8	<.20	7.2	1.75	32.4	19.6	32.5	22	4	0.6	0.6	83

S NO.	Location	Series-Profile No.	Texture-Mineralogy	Pedon depth	Colour	Particle Size			EC	Ph	BD	WR	WR	CEC	Exchangeable Cations				BS
						Clay	Silt	Sand				33	1500		Ca	Mg	Na	K	
26	Bichanpur,Lormi,Bilaspur	Bichanpur/SVIII/P8	Fine,Mont.(Cal)	109-150	Dark Brown	56	27	17	<.20	7.6	1.8	26.5	14.6	44	28	12	0.7	0.5	93.6
27	Risgaon,Dhamtari,Raipur	Risgaon/64L/SIV/P9	Fine, Mixed	110-150	Dark Red	62.3	26.5	11.2	<.20	6.6	1.77	29.6	14.3	34.6	16	15	0.36	0.43	91
28	Sundra, Panna on Panna to Satna Road	Sundra Series	Fine, Smectitic	70-140	Dark Greish Brown	64.4	20.8	14.8		8.5	1.79			33.2	24	9	2.7	0.5	
C	**Gujrat**																		
1	Bhimdevka, Rapar, Kachch	**Deltaic Plains(mud flats)**	Fine, Smectitic.	117-159	Dark Brown	52.9	18.8	28.3	7.2	8.2	30.6	15.7	35.8	17.2	11.5	4.2	0.6	11.7	
2	Nara, Lalkhapat, Kutch	**Fine silty clay, mud flats**	Fine, Mixed	72-100	Dark Greish Brown	43.8	41.9	14.2	6.2	8.1	34.5	14.9	30.5	15.4	13.6	0.6	0.09	2	
	The Kathiawar Peninsula																		
3	Amrutpur,Dhari,Amreli	Clay to clay loam	Clayey Smectitic(Cal.)	18-36	Very Dark Brown	56.9	22	21.1	0.1	8.2	33	17	38.6	23	18	0.19	0.01		
4	Bhola,Dhoraji,Rajkot	Clay to clay loam	Fine, Smectitic(Cal.)	35-56	Very Dark Greish Brown	50.6	21.9	27.5	3.8	8.6	40	23	44.3	26.8	11.2	3.9	0.5	8.8	96
	Clay, Strong coarse subangular blocky structure, hard ,firm,sticky and plastic,very fine roots,many fine lime nodules,strongly alkaline, clear smooth boundary, violently effervescent.																		
5	Semla,Gondal,Rajkot	Semla Series	Very Fine Smectitic	45-115	Very dark greyish brown	61.9	18.7	19.4	1.1	9	42.2	20.4	56.1	30.8	20	3.8	0.8	6.8	99
6	Gondal, Rajkot	Gondal Series	Clayey Smectitic	27-Jun	Reddish Brown	45.5	17.5	37	0.3	7.5	26.2	13.1	35.4	28.2	15.6	0.5	0.5	1.4	
7	Kalalav,Vallabhipur,Bhavnagar	Fine Smectitic	Clayey Smectitic(Cal.)	42-63	Very Dark Greish Brown	58.5	23.3	18.2	0.7	8	34	17.5	44.3	26.8	11.2	3.9	0.5	8.8	
8	Dholi,Limbda,Surendranagar	Fine Mixed(Cal.), Clay	Fine, Mixed	92-104	Very Dark Brown	76.8	16.5	6.6	1.1	7.7	38.5	21.9	50.6	27	13	0.1	0.3	0.2	
9	Jalia,Bhavnagar,	Clay loam to clay	Fine Mixed(Cal.)	31-40	Very Dark Greish Brown	73.8	3.1	23.1	0.2	8.7	33.3	20.2	49.4	37.6	6	1.27	0.5	2.6	
				31-Dec	Very Dark Greish Brown	60.9	20.3	18.8	0.2	8.8	34	17.9	40.7	34.6	11	1	0.6		
10	Jalsika,Vankaner,Rajkot	Clay to clay loam	Clayey Smectitic(Cal.)	18-47	Very Dark Greish Brown	60.9	20.3	18.8	0.1	8.3	34	17.9	39	35.4	2.6	0.66	0.3		
				47-78	Very Dark Greish Brown	73.8	3.1	23.1	0.1	8.4	33.3	20.2	38.4	36.6	1.2	0.66	0.3		
11	Kagwad,Jetpur,Rajkot	Khagwad Series	Clayey Smectitic(Cal.)	17-35	Dark Brown	61.1	19.7	19.2	0.2	8.1	30.9	17.3	55.2	44.5	8.8	0.4	0.5	0.7	98.2
				35-46	Dark Brown	62.6	12.4	25	0.3	8.1	32	17.3	55.2	44.3	9.2	0.4	0.5	0.7	98.6
12	Khuntwada,Mahuva,Bhavnagar	Khuntwada Series	Fine Smectitic(Cal.)	27-50	Very Dark Brown	51.5	19.2	29.3	0.03	8.7	40.2	15.3	79	68	8.2	0.96	0.5	8.5	107.6
				50-69	Very Dark Brown	73.4	10.6	16	0.02	8.7	45.5	20.6	74.9	64.4	9.4	0.7	0.5	8.5	109.9
13	Lajai,Morvi,Rajkot	Silty Clay to Clay	Fine Smectitic(Cal.)	0-15	Very Dark Greish Brown	58.4	12.8	28.8	0.12	8.2	40.8	16.7	39.4	52	7.4	0.6	0.2		
				15-37	Very Dark Greish Brown	64.2	11.2	24.6	0.1	8.3	42.4	18.2	43.8	52	9	0.7	0.1		
				37-78	Very Dark Greish Brown	63	10.6	26.4	0.11	8.4	41.7	17.8	42.9	54.6	9.4	0.4	0.2		

S NO.	Location	Series-Profile No.	Texture-Mineralogy	Pedon depth	Colour	Particle Size			EC	Ph	BD	WR	WR	CEC	Exchangeable Cations				BS
						Clay	Silt	Sand				33	1500		Ca	Mg	Na	K	
				78-88	Very Dark Greish Brown	62.7	6.3	31.1	0.15	8.4	40.3	17.4	42.8	51.2	11.6	0.4	0.2		
14	Langala,Umrala,Bhavnagar	Clayey	Fine Smectitic(Cal.)	56-115	Very Dark Greish Brown	60.6	22.9	17.5	0.2	8.5	44.7	18.1	41.7	28	11	0.7	0.4		
15	Lilvan,Datha Road,Bhavnagar	Clayey to sady clay loam	Very Fine Smectitic(Cal.)		Dark Brown	78.2	7.5	14.3	0.21	7.7	46.3	21.7	52.8	27.4	10.6	0.06	0.2		
						77.9	7.3	14.9	0.23	7.3	46.1	21.6	52.6	25.2	16.4	0.06	0.4		
						74.4	12	12	0.16	7.3	46.2	21	50.4	28.4	13.2	0.09	0.6		
						63.3	4	4	0.18	7.4	39.8	17.4	43.4	21.4	19.6	0.1	0.5		
16	Meghpur,Jasdan,Rajkot	Clay loam to loam	Fine Smectitic(Cal.)	31-49	Dark Brown	65.2	20	14.8	0.3	8.2	47.2	23.4	61.5	44.2	15.6	0.5	0.5	0.81	
17	Mendarda,Junagarh	Clayey	Fine Smectitic(Cal.)	30-45	Very Dark Greish Brown	86.7	10.7	2.6	0.1	8	60.3	24.2	58.7	42.2	1.8	0.04	0.2		
18	Pipalia,Dhoraji,Rajkot	Clay to sandy clay loam	Fine Smectitic(Cal.)	17-56	Very Dark Greish Brown	68.4	12.1	19.5	0.054	8.2	44.1	19.4	65.2	57.4	6.2	0.35	0.4		
19	Ratanvav,Botad,Bhavnagar	Clay toSilty Clay	Very Fine Smectitic(Cal.)	73-88	Very Dark Greish Brown	86.5	7.9	5.6	0.4	8.4	59.3	23.9	58.8	13	7	0.2	0.7		
				88-102	Dark Yellowish Brown	47	47.8	5.3	0.4	8.5	57.4	16.2	32.8	9	6	0.2	0.8		
20	Sar,Sihor,Bhavnagar	Clay to Clay loam	Very Fine Smectitic(Cal.)	16-29	Dark Yellowish Brown	74	5.9	20.1	0.2	8	54.3	20.4	49.8	26.8	11.2	1.1	0.2	2.2	
21	Bodki,Talaja,Bhavnagar	Clayey(Vavadi Series)	Very Fine Smectitic(Cal.)	40-60	Yellowish Brown	80	6.1	13.9	5.2	7.9	57	22	54.8	38	10	1.9	0.7	3.5	
				60-90	Brownish Yellow	85.2	12.2	2.6	6.2	7.6	60	23.9	58.5	26	26	1.94	0.7	3.3	
22	Charkala,Okhamandal,Jamnagar	Clay	Fine, Mixed, Cal.	114-135	Grey	56.1	38	5.9	41.4	7.9	41.7	17.9	46.3	21.8	10.2	1	2	2.2	
23	Dhareshwar,Bhaxi,Amreli	Clay	Fine Smectitic(Cal.)	15-32	Dark Brown	61.8	13.4	24.7	0.1	8	47.2	17.7	42.3	22	5.8	0.05	0.8		
24	Facharia,Kodinar,Amreli	Clay loam to clay	Clay, Mixed, Cal.	14-25	Very Dark Greish Brown	64.9	18.9	16.2	0.2	8	40	18.9	43.7	32.4	14.6	0.008	0.5		
25	Mithapur,Jafrabad,Amreli	Motimarad Series	Very Fine Smectitic(Cal.)	73-91	Very Dark Greish Brown	72.7	8.7	18.6	0.1	8.3	54.6	20.3	58.8	13	7	0.2	0.7	0.4	
	The Gujrat Plains																		
26	Sanjeli,Zalod,Godhra	Clay loam	Coarse loamy, Smectitic	16-72	Very Dark Brown	62.7	14.3	23	0.6	7.3	52.8	18	42	16	22.8	1.9	0.4		
27	Eru,Navsari	Clay	Fine Smectitic	25-75	Very Dark Greish Brown	63.5	2.5	16	0.11	8.3	45	18.7	47.5	30.8	15.2	1.9	0.5	4	
28	Katkui,Tilakwada,Vadodara	Jalalpur Series	Very Fine Smectitic	22-65	Very Dark Greish Brown	60.9	12.3	26.8	0.2	8.5	51.5	17.4	47.1	22.1	10.4	0.5	0.5	1.06	
29	Karali,Dabhoi,Vadodara	Sandy Clay	Clayey, Mixed	53-75	Very Dark Grey	56.1	13.6	30.3	0.002	8.6	30.2	16.2	39	22.2	15.4	1	0.5	2.6	
30	Suklatirth,Bharuch	Fine Clay, Karmali Series	Very Fine Smectitic	48-92	Very Dark Greish Brown	62.5	20.3	17.2	0.01	8.3	48.5	17.6	51.3	40	10	0.44	0.4	1.1	
31	Tadgam,Navsari	Clay, Maroli Series	Fine Smectitic	132-150	Dark Brown	63.2	26.7	10.1	0.2	8.5	36.8	19	37.4	25.4	9.2	2.5	0.7		
32	Mulad,Surat	Clay	Very Fine Smectitic	120-140	Very Dark Grey	70.5	19.2	10.3	0.25	8.9	57.1	20.4	53.7	21.8	19	6.6	0.4	14.3	

S NO	Location	Series-Profile No.	Texture-Mineralogy	Pedon depth	Colour	Particle Size			EC	Ph	BD	WR	WR	CEC	Exchangeable Cations				BS
						Clay	Silt	Sand				33	1500		Ca	Mg	Na	K	
3[illegible]	Haldhar, Bharuch	Haldhar Series	Fine Smectitic	56-102	Dark Brown	56.5	15.1	28.1	0.08	7.6			46.4	32	12.4	0.4			96.5
				102-130	Dark Yellowish Brown	55.8	18.6	25.4	0.16	7.9			44	31	11.4	0.3			97
	Source : Sohan Lal. Deshpande, S.B and J.Sehgal(1994).Soil Series of India.Soils Bulletin 40. NBSS and LUP, Nagpur, India 684p.																		
D	**Rajasthan**																		
-	Kotri,Bhilwara	Fine Smectitic	40-75	Very Dark Greyish Brown	55.2	25.6	19.2	6.9	8.8	1.1	38.9	19.8	51.6	23.6	15.8	10.1	1.3	98.5	
	Kota, Kota	Fine Smectitic (Cal)	50-65	Very Dark Greyish Brown	58	30	11.1	0.13	7.8	1.3	45.8	17.6	43.7	24.3	16.6	0.7	0.8	97	
3	Govt.Ag.Farm,Digod	Fine, Mont.	27-78	Grey	52	23	25	2.1	8.1	1.58			57						
	Kota. Chambal Series																		
3	Atru, Baran	Fine Smectitic	52-77	Very Dark Grey	58.9	28	9.1	0.23	7.3	1.3	45.8	17.5	42.9	26.8	13.6	0.7	0.5	96.9	
4	Kaushalgarh, Banswara	Fine Smectitic	65-85	Black	55.4	24.1	20.5	0.3	7.3	1.3	40.7	16.6	45.2	22.9	21.4	0.6	0.2	99.7	
5	Amod, Chitorgarh	Fine Smectitic (Cal)	77-100	Very Dark Greyish Brown	52.9	38.5	8.6	0.72	8.1	1.2	36.7	16.4	40.1	26.4	11.5	0.8	0.6	98.1	
6	Sunel, Chhoti, Jhalwar	Fine Smectitic (Cal)	50-70	Very Dark Greyish Brown	55	37.9	7.1	0.54	7.2	1.3	36.7	17	49.4	34.5	12.5	0.8	0.3	97.3	
7	Khando, Banswara	Fine Smectitic	13-30	Very Dark Greyish Brown	59.4	34.5	6.1	0.08	7.1	1.2	49.1	17.7	52.7	32	19	0.6	0.4	98.7	
8	Ramganj Mandi, Kota	Fine Smectitic (Cal)	40-56	Very Dark Greyish Brown	61.2	31.4	7.4	0.18	8	1.5	45.5	16.6	43.9	23.3	18.7	0.7	0.5	96.1	
9	Banswara,	Fine Smectitic	50-70	Very Dark Greyish Brown	55	37.9	7.1	0.54	7.2	1.4	50.3	17	49.4	28.5	18.2	0.7	0.8	96.8	
10	Dag,Jhalawal	Fine Smectitic	13-30	Dark Brown	45.9	35.5	18.6	0.6	7.8	1.6	45.8	14.7	36.7	20.2	14.6	0.8	0.6	98.6	
			30-65	Yellowish Brown	52.6	25.1	22.3	0.3	8.1	1.4	40.5	16.1	42	20.4	18.2	0.7	0.5	94.8	
E	**Andhra Pradesh**																		
1	Itikyala,Koilkuntla,Kurnool	Alur, Deep Cal, Black	Fine ,Smectitic(Cal)	78-108	Very Dark Grey/ Brown	53.1	26.3	20.6		9.3									
2	Billam,Banaganpalli,Kurnool	Kalyandurg, Deep Cal.	Fine, Mixed (Cal).	43-70	Very Dark Greyish Brown	50.7	26.6	22.7		8.8						36			

S NO.	Location	Series-Profile No.	Texture-Mineralogy	Pedon depth	Colour	Particle Size			EC	Ph	BD	WR	WR	CEC	Exchangeable Cations				BS
						Clay	Silt	Sand				33	1500		Ca	Mg	Na	K	
		alluvial clayey soil		70-97	Very Dark Greyish Brown	51.4	28.1	20.5		8.8						1			
3	Moda,Hindpur,Anantpur	Medchal,Deep Cal.	Fine, Mixed (Cal).	62-87	Very Dark Grey	48.8	22.1	29.1		8.2									
		alluvial clayey soil		110-127	Black	60.4	24.9	14.7		8.8									
4	Sollapuram,Oravukonda,Anantpur	Sollapuram Series	Very Fine,Smectitic	102-129	Verydark Grey/Greyish Brown	71.7	22.3	6		8.2	65.2	28.7	14.9	14.1	0.8	18.1	89.7		
5	Patapadu,Banaganpallai,Kurnool	Kurnool,Deep Cal.	Fine ,Smectitic(Cal)	18-50	Dark Brown	54.7	28	54.7		9.8									
		Black soil		79-109	Dark Brown	59.6	28.2	12.2		9.6									
				109-151	Dark Yellowish Brown	62.1	26.3	11.6		9.5									
6	Kanakadripalli,Banaganpalli,Kurnool	Banganapalle,Deep Cal.	Fine ,Smectitic(Cal)	105-127	Very dark grey/greyish brown	67.2	23.4	9.4		9									
		Black soil																	
7	Singarbugda,Narayanhker,Medak	Narayankher, Medium	Very Fine ,Smectitic	0-10	Very dark greish brown	78	17.1	4.9	Tr	6.9	65	41.8	20.3	0.1	0.3		96	0.83	
		deep black soil		13-28	very dark brown	77.3	17.9	4.8	Tr	6.9	69.2	46	18.9	0.2	0.2		94	0.9	
8	Tanplur,Andole-Jogipet,Medak	Amlapuram,Deep Black	Fine ,Smectitic	116-150	Very dark greish brown	65.5	18.6	15.9	0.14	8-Jul	55	44	8.2	2.8	0.2		100	0.79	
9	Elaram,Sangareddy,Medak	Sangareddy,Deep Black	Fine ,Smectitic(Cal)	120-150	Very dark greish brown	62.4	22.6	15	Tr	9-Jan	48.8	nd	nd	7	0.5			0.78	
10	Tornal,Narayankher,Medak	Bhimavaram,Deep Black	Very Fine ,Smectitic	123-150	Very dark greish brown	81.8	17	1.2	0.17	8-Jan	70	47	12.8	0.2	0.6			0.79	
11	Giripalli,Gajwel,Medak	Jammalamadugu,Deep-	Fine ,Smectitic(Cal)	138-150	dark grey	67.1	18.4	14.5	0.13	8-Jan	49.6	36.1	11.9	0.1	0.6	7		0.74	
		Black Cal. soil																	
12	Entapalli,Sangareddy,Medak	Entapalli,	Fine ,Smectitic(Cal)	40-69	Very dark greish brown	53.3	15.3	31.4	0.13	8-Jan	42.6	30.1	6	0.2	0.3	4	86	0.8	
13	Issepet,Andole-Jogipet,Medak	Issepet	Fine loamy, Mixed	31-52	Brown	37.8	11.8	50.4	0	7-Jan	20.8	13	4.7	0.8	0.2		90	0.55	
14	Singarbuguda,Narayankher,Medak	Chinnloni	Very Fine ,Smectitic	28-49	very dark brown	75.4	17.5	7.1		7-Jan	70	47	19.5	0.2	0.2		95	0.93	
15	Dour,Narayankher,Medak	Dour	Very Fine ,Smectitic(Cal).	67-88	Dark Brown	81.5	17.7	0.8		8-Jan	46.8			0.4	0.4	23		0.57	
16	Halgera,Zaheerabad,Medak	Hakgera	Very Fine ,Smectitic(Cal).	53-84	Very Fine ,Smectitic(Cal).	83.8	14	2.2	0.18	8-Jan	69.1			0.1	0.6	12		0.82	
17	Narsingi,Andole-Jogipet,Medak	Nayalkal	Very Fine ,Smectitic(Cal).	39-62	Very Fine ,Smectitic(Cal).	79.5	17.2	3.3	0.6	7-Jan	66.5			0.2	0.8	13		0.84	
18	Pocharam,Andole-Jogipet,Medak	Pocharam	Fragmental, Smectitic	15-25	Dark Reddish Brown	63.4	12.5	24.1		6-Jan	45.4	28.7	11.6	0.2	0.2		90	0.72	
19	Pusalpahad,Narayankher,Medhak	Pusalpahad	Very Fine ,Smectitic	44-68	Very dark greish brown	73.8	16.5	9.7		7-Jan	36.8	20.4	9.7	0.3	0.3		83	0.5	

S NO	Location	Series-Profile No.	Texture-Mineralogy	Pedon depth	Colour	Particle Size			EC	Ph	BD	WR	WR	CEC	Exchangeable Cations				BS
						Clay	Silt	Sand				33	1500		Ca	Mg	Na	K	
2	Walur,Narayankher,Medak	Walur	Clayey,Smectitic	15-34	Dark Reddish Brown	55.4	21.4	23.2		6-Jan	40.9	16.9	10.4	0.3	0.2		68	0.74	
2	Wasar,Narayankher,Medak	Wasar	Very Fine ,Smectitic	90-111	Very dark greish brown	68.6	29	2.4		8-Jan				0.3	0.5	11		0.88	
22	Pattacheru,Ramchandrapuram ,Medak	Kasireddipalli Series	Fine Smectitic	101-130	Very dark greish brown	59.4	33.4	7.2		8-Jan	57.8	35.8	11.5	8.6	0.5		90		
23	Benchmark- Swell Shrink Soils of India		Very Fine Mont.	90-130	Veryy Dark Clay	67.5	21	11.6	0.45	9-Jan	62.9	30.2	18.4	12.8	0.6	6.2		0.93	
24	Chitkul,Narsapur,Medak	Chitkul Series	Fine Smectitic	81-113	Dark Brown	43.1	17.9	39		7-Jan	22.6	17.5	5.5	0.5	0.2	0	105		
	Soils of Laterite Landforms																		
25	Potpalli,Zaheerabad,Medak	Potpalli	Very Fine ,Smectitic	85-112	Dark Brown	73.8	23.7	2.5		7-Jan	49.5	38	10	1.4	0.3		100	0.67	
26	Ranjol, Zaheerabad, Medak	Ranjol	Very Fine ,Smectitic	54-88	Very dark greish brown	70.9	14.8	14.3		8-Jan	27.9			0.4	0.3	7		0.39	
27	Potampalli, Chintur, Khamam	Chintur, Shallow red	Clayey-skeletal, Mixed	Jun-22	Reddish Brown	60.1	3.8	36.1		6.5	20.8	4.5	3.6	0.1	0.7		43	0.35	
		gravellary red soil		22-43	Yellowish Red	63.8	4.9	31.3		6.7	17.4	3.1	3.9	0.2	0.7		45	0.27	
28	Malatippa, Mummadivaram, East-	Deep cal. Black soil	Very Fine ,Smectitic(Cal).	62-84	Dark Greyish Brown	78.7	17.3	4	20	8.2									
	Godavari																		
F	**Chhattisgarh**																		
1	Chandranagar,Bilaigarh,Raipur	Chandranagar	Fine,Mixed	18-48	Very Dark Greyish Brown	54.9	26.2	19.9	0.3	7.2	1.72	26.6	13.4	34	20.4	11.6	0.72	0.28	91.6
2	Kedlu,Bagicha,Raigarh	Kedlu	Fine Loamy, Kaolinitic	55-77	Brown to Strong Brown	50.3	18.3	29.4	<.20	6.6	1.72	21.5	11.5	29.5	13.5	7.5	0.4	0.3	75
3	Junapani,Tahsil,Durg	Junapani	Fine, Mont.	54-87	Very Dark Greyish Brown	56.4	20.2	23.4	<.20	7.2	1.73	22.8	11.5	43.4	26	13	0.79	0.43	92.6
4	Awari,Balod,Durg	Awari	Fine, Mixed	100-125	Olive Brown	61.3	23.5	13.2	<.20	7.4	1.82	30.3	15.3	54	10.3	1.3	0.3	0.2	54
5	Boda,Mungeli,Bilaspur	Boda	Fine, Mont.	88-115	Dark Greyish Brown	60.9	30.7	8.4	<.20	7.6	1.75	29.6	16.2	43.2	32	5	0.81	0.68	88
6	Bichanpur,Lormi,Bilaspur	Bichanpur	Fine, Mont. Cal.	109-150	Dark Brown	56	27	17	<.20	7.6	1.8	26.5	14.6	44	28	12	0.7	0.5	93.6
7	Kohkabor,Khairagarh,Rajnandgaon	Kohkabor	Fine, Mixed	95-120	Dark Brown	60.5	25	14.5	<.20	6.9	1.75	30.3	14.6	36	22.4	10.8	0.65	0.26	94
8	Patharri,Khairagarh,Rajnandgaon	Patharri	Fine, Mont.	0-18	Dark Yellowish Brown	48.4	30.2	21.4	<.20	5.8	1.7	18.5	8.6	34	23	7	0.79	0.69	92
				43-62	Dark Brown	55.9	24.2	19.9	<.20	6.3	1.73	25.6	13.1	33.6	20	8	0.75	0.5	89
9	Khujji,Dongargaon,Rajnandgaon	Khujji	Fine, Mont.	100-125	Dark Brown	54.9	23.2	21.9	>.20	7.8	1.8	38.6	21.5	43.2	27	8	1.2	0.41	84
10	Umariguda,Jagdalpur,Bastar	Umariguda	Fine, Mixed	77-111	Olive Brown	65	20.2	14.8	0.3	7.6	1.79	29.5	14.5	46	30	12	0.8	0.3	93

S NO.	Location	Series-Profile No.	Texture-Mineralogy	Pedon depth	Colour	Particle Size			EC	Ph	BD	WR	WR	CEC	Exchangeable Cations				BS
						Clay	Silt	Sand				33	1500		Ca	Mg	Na	K	
G	**Karnatka**																		
1	Karabadur,Raichur	57E/S3/11		64-94	Dark /very dark-	65.7	23.6	10.7	8.1			60	61.8	15.8	0.9	4.4		9.1	100
	Regional Research Centre, Raichur	Raichur	Fine, Mont.	112-150	Very Dark Grey to Dark Grey	69.3	26.7	4	8	35.7	17.2	70.7	47.4	14.1	3.6	1.7			
2	Nidasanur,Hungund,Bijapur	56D/S3/62		92-117	greyish brown-	71.5	20	8.5	8.8			66.9	54.7	21.9	13.1	5.9		10.3	100
3	Jamakhandi,Bijapur	Jamakhandi	Fine, Mixed	61-89	Dark Reddish Brown	45.5	10.5	44	8.3			38	28.8	6.4	0.5	0.6			
4	Saigon,Bhalki, Bidar	56F/S2/106	Basalt landform	36-58	"	56.3	35.5	8.2	8.5			58.2	64.7	9.3	0.8	5.7		5.8	100
5	Bhatumbra,Bhalki,Bidar	56F/S2/111	Basalt landform	65-100	"	66.1	29.5	4.4	9.1			64.4	55.3	20.5	4	7.5		11	100
	Deep moderately well drained, dark to very dark greish brown calcareous cracking clay to silty clay Soils																		
	Geographically distributed in Bellary, Belgam, Bidar, Bijapur, Dharwar, Gulberga, Raichur, Mysore, Chitradurga.																		
6	Dindwar, Basavanabagewadi,	56D/S1/32	Basalt landform	36-58	Dark brown to very-	56.3	35.5	8.2	8.5			58.2	64.7	9.3	0.8	5.7		5.8	100
	Bijapur				dark greyish brown														
	Moderately deep/ well drained dark brown to very dark greyish brown non calcereous cracking clay to silty clay soils.																		
	Geographically distributed in Bellary, Belgam, Bidar, Bijapur, Dharwar, Gulberga, Hassan, Raichur, Shimoga, Tumkur.																		
	Laterite landforms- Banglore Plateau																		
7	Ramagund, Chincholi,Gulberga	56G/S3/64	Laterite landforms	42-69		68.2	18.4	13.4				9	4.5	1.8	0.08	0.03			71
8	Sirwal,Sahapur,Gulberga	Kagalgomb	Very Fine Mont.	131-172	Dark Yellowish to dark Brown	68.8	22.6	8.6	9	34	17.2	67	39.5	17.5	10.8	0.7	16		
9	Giriappanamane,Somavarpet,	48P/S6/8	Laterite landforms-	122-164		53.6	18.6	27.8	6.1			10.1	4.6	1.8	0.1	0.4			69
	Dist- Kodagu		Maland																
10	Idugundi,Yellapur,Uttara Kannada	48J/S3/12	"	53-90		58.3	20.8	20.9	5.9			12.5	2.9	2.1	0.2	0.3			44
	Metamorphic landforms																		
11	Hulihalli, Ranebennur, Dharwad	48N/S3/20	Metamorphic	97-119	Dark /very dark-	62	26.6	11.4	8.7			63.5	45.1	20.8	1.9	5.4		6.2	100
	geographically distributed in Bellary, Belgam, Bidar, Bijapur, Dharwar, Gulberga, Raichur, Mysore, Chitradurga.																		
	Gobbargumpi,Navalgund,Dharwad	Achmati Series	Very Fine Smectitic	87-152	dark greyish brown	70.1	20.1	9.8	8.9	36	17.5	66.1	36.3	25.2	13.1	0.7		15	114
	North Sahyadris and Western Karnatka Plateau regions.																		

S NO.	Location	Series-Profile No.	Texture-Mineralogy	Pedon depth	Colour	Particle Size			EC	Ph	BD	WR	WR	CEC	Exchangeable Cations				BS
						Clay	Silt	Sand				33	1500		Ca	Mg	Na	K	
12	Ranibennur,Dharwad	Hungund Series	Very Fine Mont.	44-70	Very Dark Greyish Brown	65.1	16.8	18.1	8.9			60	34.3	19	8	0.9		12	
13	Karenhalli,Hiriyur,Chitradurga	57/S7/6	"	91-116	greyish brown-	60.8	28.1	11.1	8.7			65.2	60.8	16.2	3.8	7.1		8.5	100
	Deep moderately well drained, dark to very dark greish brown calcareous cracking clay to silty clay Soils																		
	Geographically distributed in Bellary, Belgam, Bidar, Bijapur, Dharwar, Gulberga, Raichur, Mysore, Chitradurga.																		
	Sedimentary landforms																		
1	Vajramalli, Mudhol, Bijapur	47P/S3/8	Sedimentary landforms	101-150	Dark /very dark-	79.7	11.1	9.2	8.6			63.1	58.4	22.1	12.7	6.1		12.8	100
	Soils of Western Ghats				greyish brown-														
B	Kollur,Coondapur, Dakshina Kannada	48K/S1/10		56-96		66.7	18.9	14.4	6.6			8	3.2	1.4	0.1	0.7			67
4	Kollur,Coondapur, Dakshina Kannada	48K/S1/19		30-61		68.4	16.1	15.5	6			9.7	2.4	0.7	0.2	0.2			
	Soils of Eastern Ghats (South)																		
5	Aralagere, Kollegal, Mysore	58E/S1/5		31-55		32.7	15.8	51.5	6.4			13.2	9.3	4.5	Tr	0.7			100
H	**Tamil Nadu**																		
	The Nilgiris																		
1	Vadavayal,Gudalur,The Nilgiris	58A/S12	Fine Mixed	22-45		58.1	12.2	29.7	0.04	5.7	10.4	4.04	1.1	0.04	0.04	50			
2	Malaperahatti,Gudalur, Nilgiris	58A/S12	Fine Mixed	120-150		63.2	7.5	29.3	0.09	8.1						89			
3	Teppa Kadu, Gudalur,Nilgiris	58A/S12	Clayey Skelatal	50-70		56.5	26.2	17.3	0.05	5.8	9.1	6.04	1.43	0.22	0.44	91			
	South Sahayadri landform																		
4	Killanvayal,Kodaikanal,Anna	58FS5	Clayey, Mixed	29-64		58	2.7	39.3	0.46	5.2	17.52	2.2	2.2	0.53	0.1	16			
	Riverian landform																		
5	Papanasam,Thanjavur	T-273	Fine, Mont.	87-104		58.7	18.7	22.6	0.21	9.2	57.9	14.3	21	25	0.5	100			
6	Putteri,Tittagudi,South Arcot	T-276	Very Fine Mont.	124-154		83.1	6.5	10.4	0.36	8.6	24.4	10.25	7	6.68	0.43	100			
	Lateritic landforms																		
7	Karuvakurchi, Mannargudi, Thanjavur.	T-156	Fine Kaolinitic	154-195		58.3	1.4	40.8	0.95	5.6	10.2	2.52	5.55	0.84	0.28	90			
	Inland Plains																		
8	Kodangipatti, Vilathikulam, Chidambarnar.	T-269/58K	Fine Mont.	38-95		65	15	19	0.39	7.9	65.6	43.5	14.5	5.5	0.6	98			

S NO.	Location	Series-Profile No.	Texture-Mineralogy	Pedon depth	Colour	Particle Size			EC	Ph	BD	WR	WR	CEC	Exchangeable Cations				BS
						Clay	Silt	Sand				33	1500		Ca	Mg	Na	K	
9	Karisalkulam,Vilathikulam, Chidambamar.	T-269/58K/13	Fine Mont.	54-102		61.2	13.8	24	3.2	8.5	57.3	15.5	1	40	0.6	100			
10	TNAU Farm- Coimbatore Variegated brown,very dark greyish brown and white.	Coimbatore	Fine Smectitic	75-124	light Yellowish Brown-	55	21.9	23.1		8.5	46.1	30.5	9	4.5	0.6	97	0.84		
11	Kovilpatti,TNAU Farm,Tooth-hokudi.	Kovilpatti	Very Fine Smectite	104-118	Very dark Grey-Greyish Brown	72.1	17.7	10.2		7.4	65.2	55.5	14	0.9	0.5	109			
12	Annappaanpettai,Papanasam, Thanjavur.	Kalathur	Fine Smectitic	87-104	Dark Grey	58.7	18.7	22.6	0.9	9.2	57.9	14.3	21	25	0.5		0.91		
I	**Kerela**																		
1	Kirukuzhi,Adoor,Pathanamthitta	Anakkatur Series	Clayey Skelatal, Kaolinitic	70-97	Red	58.9	8.2	32.9	5	33.1	20.2	7.3	1.78	0.51	0.03	0.02	32	0.12	
2	Thrikkannamangal,Kottarakkara, Kollam	Thrikkannamangal	Clayey Skelatal, Kaolinitic	67-103	Reddish Brown	63.9	6.1	30	4.8	30.4	20.8	7.5	0.75	0.5	0.05	0.28	22	0.13	
3	Angel Valley,Kanjirapally,Kottayam	Angel Valley Series	Clayey Skelatal, Kaolinitic	35-58	Red	46.3	25	28.7	5	31.1	21.1	7.4	0.44	0.22	0.03	0.12	11	0.16	
4	Tekkumaradi,Muvattupuzha, Ernakulam	Arur Series	Clayey Skelatal, Kaolinitic	86-127	Red	55.5	17.6	26.9	5	33.3	24.2	8.2	0.77	0.53	0.07	0.23	20	0.15	
5	Mekkadembu,muvattupuzha, Kottayam	Mekkadembu Series	Clayey , Kaolinitic	125-151	Strong Brown	62.5	27.8	9.7	6.2	32.5	22.6	8.1	2.55	1.5	0.06	0.21	53	0.13	
6	Churanad,Kunnathur,Kollam	Anayadi Series	Clayey , Kaolinitic	93-125	Yelloish Red	48.5	7.5	44	5.3	25	15.5	5.8	1.31	0.28	0.04	0.38	35		
J	**West Bengal**																		
1	jagganathpur, Daspur,	IGP- Alluvial plains	Very Fine, Mixed	35-58	Gray	78	19.6	2.4		7.3			46.8	17.6	13.3	3.2	1.3		76
	Mednipur	Bengal Basins.	Clay	58-97	Dark Grey	75.9	21.7	2.4		6.2			46.9	16.8	13.6	3	1.2		72
		Jagganathpur Series		97-140	Dark Grey	81.7	15.9	2.4		6.2			48.9	17.2	13.8	3	1.2		72
2	Jagadishpur,Rampurhat,Birbhum	Jagadishpur Series	Fine, Illitic, Silty Clay	94-120	Light Brownish Grey	55.7	39.1	5.2	0.52	7.5	35.3	19	28	19.8	3.8	0.8	0.8		90
	Extensive in Birbhum, Murshidabad and northern Barddhaman.												30.5	10.5	9	1.2	0.7		70
3	Chakprayag,Debra,Mednipur	"- Low lying alluvial -	**Very Fine, Mixed**	16-41	Light Olive Brown	66.1	32.7	1.2		6.3			31.1	11	9	1.8	1.4		75
	"Chakprayag Clay"	Plains	Clay	99-135	Dark Grey	68.1	28.1	3.8		6.7									
	Extensive in Western parts of Hugli districts as well as eastern parts of Mednipur districts.																		

S NO	Location	Series-Profile No.	Texture-Mineralogy	Pedon depth	Colour	Particle Size			EC	Ph	BD	WR	WR	CEC	Exchangeable Cations				BS
						Clay	Silt	Sand				33	1500		Ca	Mg	Na	K	
4	Bansghata,Bangaon,24 Parganas	IGP- Alluvial plains	Fine, Mixed	39-70	Gray	53	40.8	6.2		6.1			27.2	9.6	6.6	1.4	0.6		67
		Bengal Basins.	Silty Clay																
	Sporadic occurance in southern part of Nadia district as well as eastern, central and western part of 24 Prganas.																		
	Sitalpur,Mahishadal,Mednipur	IGP-Dunes	Fine Mixed	39-50	Olive Gray	62.8	34.6	2.6		5.7			36	10	7.6	1.6	0.8		67
[illegible]	Belar,Debra,Mednipur	IGP- old Alluvium	**Very fine, Mixed**	40-70	Grayish Brown	65.6	28.8	5.6	0.4	5.6			33.8	16.4	6.5	0.8	0.4		71.2
7	Burul,Budge-Budge,24 Parganas	IGP-Bengal Basin,	Fine Mixed	29-50	Olive	55.5	44.1	0.4		7.8			28.2	12	10	2	1.4		90
3	Ballartop,Basanti,24 Parganas	Deltaic Coastal	Fine, Illitic	63-90	Dark Gray- Silty Clay	51.9	46.7	1.4	0.87	7	41.2	22.5	17.5	6.4	6.8	2.9	1.2	16.5	98
9	Narayanpur,Kakdwip,24 Parganas	Upper delta plains	Fine,Mixed	90-140	Dark/very dark grey	44.6	48.9	6.5	4.34	7.6	35.4	15.7	17.3	10.5	9.2	6	3.8	17.3	85
		Clay, medium prominent yellowish brown motteles, moderate medium subangular blocky structure, firm, sticky, clear smooth boundary.																	
10	Bankati,Jhargram,West Medinapur	Bankati Series	Fine Mixed	72-108	Brown	59.5	36.1	4.4	0.06	5.8	36	22.9	19.8	7.1	4.6	0.5	0.2		63
				108-137	Gray	66.5	30.5	3	0.02	5.9	41.5	25.5	25.5	9.7	7.5	1	0.4		73
11	Barakadra,Garbeta II,West Mednipur	Barakadra Series	Fine, Illitic	56-83	Pale Brown	46.7	28.9	24.4	0.61	5.9	32.8	17.5	26.2	8.4	6	0.6	0.2		58
K	**Odisha**																		
1	Kodabandh,Borigumma,	Korabandh Series	Fine Mixed	51-72	Dark Gray	54.1	17.8	28.1	0.12	7.3	29.2	15	6.1	2	1.4	84			
	Koraput.Chitika Mati.	**Dark gray Clay, strong coarse angular blocky structure, firm, very sticky and plastic, few very fine pores and roots, pressure faces, gradual smmoth boundry.**																	
		Distributed in Phulbani,Koraput,Gajapati,Nawagarh,Boudh,Kalahandi,Ganjam districts.																	
2	Karada,Angul,	Karada Series	Fine ,Mixed	107-150	Grayish Brown	73.9	15.5	10.6	0.72	8	35.5	23.5	5.9	2.5	0.6	91			
3	Bareguda,Similiguda,Koraput	Bareguda Series	Clayey-skeletal, Mixed	72-104	Dark Red	47.9	20.5	31.6	0.06	7.2	16.6	9	4.5	0.2	0.1	83			
4	Kondraposi,Champua,Kendujhar	Kondraposi Series	Very Fine Mixed	0-16	Brown	83.7	4.9	11.4	0.71	6.2	50	24.2	7	2.6	1.2	70			
				16-35	Dark Grayish Brown	74	14.3	11.7	0.64	6.5	49	25	8	2.6	1.2	75			
5	Pawartala,Daramagarh,Kalahandi	Pawartala Series	Fine ,Mixed	57-75	Dark Gray,Silty clay	46.8	39.3	13.9	0.54	7.1	25.6	14.5	4.3	1.4	1	82			
				75-136	Yellowish Brown,Clay	50.5	36.6	12.9	0.5	7.2	26	14.6	4.3	1.6	1.1	83			
6	Badbanki,Turekela,Balangir	Badbanki Series	Fine ,Mixed	75-120	Dark Brown	52.9	32.9	14.2	0.59	7.9	32	16	10.3	1.7	0.8	90			
7	Badly,Bhawanipatna,Kalahandi	Badly Series	Fine,Mont.	91-120	Olive	63.8	16.7	19.5	0.37	8.2	34.2	13.4	11.5	5.6	0.6	91			
				120-145	Gray	78.9	4.6	16.5	0.31	8.3	35.3	13.3	12.4	6.2	0.6	92			

S NO.	Location	Series-Profile No.	Texture-Mineralogy	Pedon depth	Colour	Particle Size			EC	Ph	BD	WR	WR	CEC	Exchangeable Cations				BS
						Clay	Silt	Sand				33	1500		Ca	Mg	Na	K	
8	Bhudipadar,Balangir	Bhudipadar Series	Fine Mont.	20-53	Very Dark Grayish Brown	50.5	29.9	19.6	0.35	6.9	32.4	20.5	4.4	0.6	0.4	80			
9	Randiapala,Begunia,Khurda	Randiapala Series	Loamy Skeletal	52-66	Dark Red	61.4	6	32.6	0.04	5.5	28	12	3.2	1	0.6	60			
10	Barapada,Jajpur,Cuttack	Barapada Series	Fine, Mixed	0-13	Yellowish Brown,Clay	51.3	36.7	12	0.48	6	24.8	7	5	0.8	0.6	54			
					Gray	59.4	36.9	3.7	0.26	6.5	34.6	14	10	1	0.6	74			
				71-92	Grayish Brown	69.9	27.8	2.3	0.23	6.5	42	16	13	1.4	0.6	74			
				Eastern Coast- Mahanadi upper delta- distributed in Cuttack, Jajpur, Jagatsinghpur,Khurda, Bhadrak districts															
11	Nalibasant,Satyabadi,Puri	Nalibasant Series	Very Fine Mixed	0-15	Brown	55.9	19.6	24.5	0.79	4.5	43.2	10.4	7.8	2	1.1	52			
				38-62	Gray	65.8	21.7	12.5	0.64	5.5	53.3	16	12	2.2	1.2	59			
L	**Bihar**																		
1	Simri,Buxar	Simri Series	Fine , Mixed	0-13	Gray-Clay	49.4	34.6	16	0.53	7	22	12.8	2	1.2	0.8	77			
		Chikni Mitti		64-91	Gray-Silty Clay	49.7	41	9.3	0.38	7.3	30.7	16.1	3.6	3.8	0.9	79			
				91-120	Gray- Clay	66	22.5	12.3	0.54	7.2	35.1	18	5.6	3.7	0.8	80			
		Distributed in Patna,Rohtas,Bhagalpur and Munger.IGP- Alluvial Plains.																	
2	Baruna,Buxar	Baruna Series	Very Fine , Mixed	53-82	Gray-Clay	66	31.8	2.2	1.74	7.4	29.4	13.2	7.8	2.8	1.2	85			
	Barik Chikni Mitti		(Calcareous)	Distributed in Gaya, Aurangabad districts.															
3	Barew,Nawada	Barew Series	Very Fine Mixed	65-96	Very Grayish Brown	67.1	22.3	10.6	0.27	7.5	23.7	15.4	5.3	0.8	0.2	92			
4	Belsar,kahalgaon,	Belsar Series	Fine , Mixed	66-95	Grayish Brown	61.6	35.9	2.5	0.29	8.1	30.7	17.1	8.1	2.4	0.6	92			
	Bhagalpur	Distributed in Lakhisarai,Aurangabad,Jahanabad,Bhagalpur,Patna,Nalanda and Sheikhpura districts. IGP-Recent Alluvial Plains.																	
5	Bhadsai,Arwal,	Bhadsai Series	Very Fine Mont.	Oct-40	Dark Brown	62.1	33.5	4.4	0.58	7.4	32	15.5	8.2	1.3	0.6	80			
	Jahanabad	Barik chikni kalli matti		70-99	Dark Grayish Brown	67.6	28.4	4	0.49	7.8	32	16.7	8	1.6	0.5	84			
IGP- Recent Alluvial Plains- Distributed in Jehanabad,Saran,Bhojpur,Patna,Muzaffarpur,Samastipur,Begusarai,Vaishali,Saharsa,Khagaria,Munger,Buxar and Darbhanga districts.																			
6	Dahiya, Paliganj,	Dahiya Series	Fine , Mixed	84-115	Very Dark Grayish	61.1	30.4	8.5	0.04	7.7	34	15.9	10.8	1.7	0.7	86			
	Patna				Brown	Distributed in Gopalganj,Jahanabad and Patna.													

S NO	Location	Series-Profile No.	Texture-Mineralogy	Pedon depth	Colour	Particle Size			EC	Ph	BD	WR	WR	CEC	Exchangeable Cations				BS
						Clay	Silt	Sand				33	1500		Ca	Mg	Na	K	
7	Merh,Chainpur,	Merh Series	Fine , Mixed	48-95	Light Olive Brown	49.9	40.1	10	0.03	8.4	30	12.8	10	2.6	1	88			
	Kaimur	Pili Chikni Matti.		95-120	Light Olive Brown	51.1	37.3	11.6	0.03	8.4	30.2	13.2	10.2	2.6	1.4	91			
		Vindhyan range (alluvial plain). Extensive in Kaimur,Aurangabad and Rohtas.																	
M	**Assam**																		
-	Kwargaon,Dibrugarh	Khowang	**Very fine mixed**	13-32	Grey	59	39.3	1.7	6.3	18.43	14.98		0.7	0.1					
				32-65	Greyish Brown	74.5	23.8	1.7	6.8	20.48	18.05		0.7	0.15	85.62	0.31			
				65-125	Yellowish Brown	80.5	18.6	0.9	6.8	NA					92.28	0.27			
2	Kalugaon,Sibsagar	Kalugaon	Fine Mixed	15-35	Grey-Clay	60.3	33	6.7	5.6	22.8	11.1		0.5	0.2	52	0.38			
				35-57	Light Grey-Silty Clay	42.7	52	5.3	5.8	18.7	13.3		0.7	0.3	76	0.44			
3	Bangaon,Majuli,Jorhat	Bangaon	Fine Silty	52-125	Grey	29.2	15.7	31.1	7.8	10.2	9.4		0.5	0.2	99	0.65			
4	Thoramukh,Golahat	Golahat	Fine Mixed	35-65	Light Olive Brown	48.5	21.8	29.7	4.9	8.14	3.74		0.41	0.08	52	0.17			
5	Sitalajakhla,Morigaon	Amkata	Fine Mixed	22-54	Dark Reddish Brown	61	12.38	26.62	5	14.2	6.4		0.6	0.4	52	0.23			
6	Dharamtul,Morigaon	Dharamtul	Fine Mixed	0-10	Dark Grey	73	18.8	8.2	4.5	18.6	6.4		0.4	0.3	38	0.25			
			Clay	13-28	Dark Greyish Brown	68.7	14.8	16.7	4.4	14	4		0.4	0.2	33	0.2			
				80-160	Yellowish Brown	62.5	17.9	19.6	4.7	8	3		0.4	0.2	45	0.13			
7	Habilgaon,Boko,Kamrup	Habilgaon	Fine Silty	0-14	Greyish Brown	53.8	30.2	16	6.4	13	10.8		0.3	0.2	87	0.24			
8	Khagrabari,Goalpara	Goalpara	Fine Mixed	76-115	Dark Grey	60	19.5	20.5	5.3	16.13	7.47		0.96	0.21	53.68	0.27			
				115-170	Very Dark Grey	67.2	27.3	5.5	5.5	NA	NA	NA	NA	NA	NA	NA			
9	Bildipan,Karbi	Japrijan	Fine Mixed	110-130	Light Grey	71	27	2	4.7	15.24	6.8		0.74	0.23	50.9	0.21			
10	Badarpur,Hailakandi	Hailakandi	**Very fine mixed**	0-15	Grey	69.5	28	2.5	5.6	20.4	10		0.61	0.25	53.2	0.29			
				15-32	Brown-Dark Brown	65.5	30.9	3.6	5.8	19.86	10.6		0.78	0.16	58.1	0.3			
11	Bhanga,Karimganj	Bhanga	Fine Mixed	70-110	Grayish Brown-Silty Clay	43.5	52	4.5	6.5	15.6	12.6		0.65	0.11	85.6	0.36			
				110-160	Grey-Clay	62	34.9	3.1	6.5	NA	NA	NA	Na	NA	NA	NA			

S NO.	Location	Series-Profile No.	Texture-Mineralogy	Pedon depth	Colour	Particle Size			EC	Ph	BD	WR	WR	CEC	Exchangeable Cations				BS
						Clay	Silt	Sand				33	1500		Ca	Mg	Na	K	
N	**Meghalaya**																		
1	Nongstoin,West Khasi Hills	Nongstoin	Fine ,Mixed	72-130	Dark Red	56.6	32.1	11.3	4.9	5.5	0.6	0.2	0.36	0.12	26				
				177-	Red	54.9	29.1	16	4.9	5.3	0.5	0.1	0.27	0.08	19				
2	Umkrem,West Khasi Hills	Umkrem	Fine ,Mixed	52-100	Yellowish Red	54.9	25.8	19.3	6.8	5.5	2	0.2	0.55	0.26	29				
3	Ramjonggiri,West Karo Hills	Ramjonggiri	Fine ,Mixed	27-50	Grey	42.9	33.7	23.4	6.56	4.6	0.98	0.12	0.18	0.13	21				
4	Myngkre,Jaintia Hills	Myngkre	Fine ,Mixed	75-100	Strong Brown	40.5	18.4	41.1	4.7	4.5	0.8	0.1	0.2	0.09	25				
5	Kherapara,West Garo Hills	Kherapara	Fine ,Mixed	52-85	Grey	57.9	36.9	5.2	22	7.6	7.9	14	0.2	0.22	98				
6	Nongehram,West Garo Hills	Nongenram	Fine ,Mixed	40-60	Yellowish Red	56.1	24.7	19.2	6.9	5.3	0.4	0.4	0.09	0.22	16				
O	**Sikkim**																		
1	Krishagiri ICAR Farm, Longal Hills	Longal -5 Series	Fine Mixed	71-99	Reddish Brown	56.6	41.4	2	12	4.7	0.7	0.3	0.1	0.1	10				
	Imphal	Silty Clay																	
2	Krishagiri ICAR Farm, Longal Hills	Longal -6 Series	Fine Mixed	66-93	Yellowish Red	61	34.2	4.8	20.9	4.3	1.08	0.59	0.27	0.07	10				
	Imphal	Clay																	

SPRINGER NATURE

GPSR Compliance

The European Union's (EU) General Product Safety Regulation (GPSR) is a set of rules that requires consumer products to be safe and our obligations to ensure this.

If you have any concerns about our products, you can contact us on ProductSafety@springernature.com

In case Publisher is established outside the EU, the EU authorized representative is:

Springer Nature Customer Service Center GmbH
Europaplatz 3
69115 Heidelberg, Germany

The manufacturer's authorised representative in the EU is Springer Nature Customer Service Centre GmbH, Europaplatz 3, 69115 Heidelberg, Germany. If you have any concerns regarding our products, please contact ProductSafety@springernature.com

Printed and bound by CPI Group (UK) Ltd, Croydon, CR0 4YY
01/10/2026
02225868-0010